D1284014

$1.09 an hour
and glad to have it...

Conversations with 17 mid-20th Century
Crown Zellerbach millworkers

Sandra Hickson Carter

Willamette Falls Heritage Foundation

Copyright 2011 Willamette Falls Heritage Foundation, Inc.

All rights, including that of the translation into other languages, reserved. No part of this book may be reproduced or transmitted in any form or by any means, electronic or mechanical, including photocopy, recording, or any other information or retrieval system, without permission in writing from the publisher.

Memory is priceless but fallible. Willamette Falls Heritage Foundation assumes no responsibility for the accuracy of statements by contributors to this collection.

Interviews and editing by Sandy Carter

Prior scholarly work:
"Letters from Bob: A GI Re-entering Portland Life in 1945",
Oregon Historical Quarterly, Winter 2005/Volume 106, Number 4, p. 616–641.

Layout and cover design: Anita Jones & Joan Pinkert,
Another Jones Graphics

ISBN: 9781450748438
Library of Congress Control Number: 201094224

Oral history of mill work; Willamette Falls Region-History, local; Stories of papermaking in West Linn, Oregon; manufacturing and mill technology; Mid-20th Century Oregon papermill work; evolution of industrial technology; post-war growth of middle class .

This publication grant-funded by the Cultural Coalition of Clackamas County, Oregon and the Kinsman Foundation.

Printed in the USA

Willamette Falls Heritage Foundation
PO Box 635, West Linn, Oregon 97068
Willamettefalls.org

Dedication

To the upper river loggers, bargemen, brailers,

splicers, rock-pilers, papermakers, plumbers,

millwrights, mechanics, mule drivers, white-hats,

jiggermen, windermen, grindermen, roll buckers,

tug captains, finishers, helpers, bosses,

pipefitters, converters, sawyers, wood-pilers,

welders, horseshoe pitchers and production planners

who lived these stories. But especially to Olaf.

.

Preface

I met Olaf Anderson, my first mill worker, one day in 2004 in the low-ceilinged, noisy basement of the Willamette United Methodist Church in West Linn, Oregon. He and his three sisters, Gudrun, Eleanor, and Ella were sitting at a red-plastic-draped folding table, having a reserved conversation as they waited for the saying of grace and the call to line up for homemade soups and rolls. The monthly buffets were organized for elders of the neighborhood, and although Olaf lived in Gladstone, a few miles away, his sister's caregiver Helena picked up all four of the elderly siblings and a couple of aluminum walkers each month and brought them to Willamette, where Ella lived, to lunch together—the only time they got to see each other. Like the rest of the crowd, they were white-haired, friendly, slow moving, and hard-of-hearing.

It was my first time. My hair is not white, but I'd been encouraged to attend to try to meet some of the people who still remember the stories of early West Linn. When I took the empty chair at their table, Helena leaned over to make introductions. "Olaf was a mill worker," she said. So began my great adventure, capturing the stories of mill work at Willamette Falls.

Contents

WILLAMETTE FALLS HERITAGE FOUNDATION

Introduction

Background

In the winter of 2004-2005, Willamette Falls Heritage Foundation was a three-year-old heritage nonprofit in West Linn, Oregon, focused for the most part on discovering, sharing, and preserving the impressive history of industry and transportation at Willamette Falls. The foundation's board found itself working, by necessity, on public education about the 1873 Willamette Falls Navigation Canal and Locks, which cuts through an operating 1895 paper mill and past an historic hydropower generation plant. The locks were about to lose funding and faced potential closure, after 131 years of connecting the upper and lower Willamette River for boating and shipping.

Around the same time, the board developed a partnership with the McLean House and Park, a few blocks from the industrial area, working with the historic house to host periodic videotaped community conversations about local history. The mill, the locks and the power plant took turns being the focus of those public conversations. Our assignment, as board members, was to try to identify and invite local people who'd had careers at each 'topic site', and when it came the mill's turn, I remembered Olaf Anderson. He was shy and a bit reluctant, but agreed to come and talk about working in our mill.

At the May, 2005 conversation with Olaf and the other panelists, Board member and officer Sherri Burch, who had a passion for oral history and had done deep research on the first long-distance transmission of electricity from Willamette Falls to Portland in 1889, made a point of asking Olaf how we might be able to get in touch with other former mill workers. Olaf thought a moment and then said there was a reunion picnic every summer. I gave him my phone number and the evening ended, people drifted out to their cars, and I thought nothing more about it until my phone rang a month later.

It was Olaf. He said the picnic was in July, if I was interested. I asked him whom I should call to get permission to attend, and he gave me the number of George Droz. As this second contact slipped into place, an idea took shape.

The call to George was daunting, as he stopped me a few words into my introduction and boomed into the phone, "You'll have to speak up, you know—I worked in the mill!" He may have been a bit puzzled at my intentions, but he kindly gave me permission to join the Local 68 reunion in July.

I called Olaf back and asked if he intended to go, and if I could give him a lift, since the reunion is held somewhere east of Beavercreek, practically in the foothills of the Cascades, and his truck mostly goes to the local grocery store these days. He agreed, and the plan perking in my head began to take real shape.

I decided to try to get folks at the picnic to put their names on a signup sheet, agreeing to be interviewed about their careers at Crown Zellerbach in West Linn. If I could sign up 10 or 12 people, I thought, I could put together a grant application and get funding to videotape interviews with them. Then some day, in the board's imaginary Interpretive Center for the West Linn Historic Industrial District (technically just a vision, at this point), visitors could slip into a small, dark theater surrounded by the larger-than-life voices and images of these retired workers, storytelling about what industrial work was really like—the types of mill jobs that built Oregon's middle class in the mid-20th Century.

But first things first.

On reunion day I picked up Olaf and we headed into the countryside, winding our way to the private park where people were gathered under a picnic shelter of massive proportions, the supporting beams and columns made of entire logs. I greeted George for the first time and wondered what I'd gotten myself into, as old-timers milled around, recognizing old friends and spreading out their plastic place settings on the picnic tables.

I'd brought a wonderful old photo of a mill safety committee meeting in the '50s, thinking it would be interesting to these strangers and might draw some to my table if the word got around. Maybe people would be able to identify the men in the picture. Maybe I could strike up a conversation with them and get them to sign up for the project.

In reality, it was Olaf, quietly approaching his old friends and dragging them gently to the clipboard, who enlisted the first twelve recruits to look at the photo and sign up. With their phone numbers I could start setting appointments for interviews. I now felt the project would happen.

Of course life happened, too. One of the men on the list had a stroke before we could talk with him—a stroke that left him angry, confused and unable to find words. Two of the men had serious health issues that lingered and eventually made them decline to participate. Interviews ended up being scheduled and rescheduled around doctor appointments and commitments to grandchildren.

In those early weeks, as I thought about the project more deeply, I realized that there were many jobs in the mill and that most of the men Olaf had helped me sign up were millwrights, as he was. It would not tell the whole story of the mill if we just spoke to millwrights. I would need to get leads to other retirees to fully represent the many processes, operations and social aspects of the mill.

Meanwhile I needed to get the funding in place. I had written a grant to the Kinsman Foundation and was sitting back to wait when I got an unexpected phone call from Roy Paradis, one of the men on my picnic list.

"I have a gentleman I think you should talk to before much longer," said Roy, "He's 96 years old. His name is Rosie Schultze, and he's up in Oregon City in an assisted living center. He has diabetes." I accepted his suggestion immediately, even knowing I had no guarantee the grant would come through or that we'd actually have money to pay for the videography.

I called my friend Melody Ashford, videographer and manager of the Willamette Falls Television cable station in Oregon City, and made her an offer she might have refused, but didn't. "Melody, we really need to save this man's stories—he might not live long enough to wait until we have grant money in hand." Her heart is as big as all outdoors, and she'd been a textile mill worker for Pendleton Woolen Mills in her formative years, so she said yes and I made the appointment for our first interview. Soon after, we had to interview George at Christmas time in 2005 before he left for Arizona for a few months.

The grant came through in early 2006, so over the next few months we continued to add potential subjects to our list, as some panned out and some didn't. We started putting more interviews 'in the can'. Talking with Roy about the need to represent all facets of mill work, I got several leads. The friendship network between different subsets of retirees was my trail of breadcrumbs, and referrals from trusted friends opened many doors to us, in spite of our nosiness, our multiple cases of equipment and cords, our microphone and the very bright lights.

Early on I'd decided to interview in their homes, so people would be more relaxed and we could capture their authentic surroundings. Some weekends we'd make two appointments nearly back to back on Saturday—each of which usually filled two tapes.

Although we were experienced interviewers, we learned on the job to ask people to turn off and unplug their phones. We had little control, however, over wind chimes and wives sneaking through the kitchen or cats jumping on our back. Outside light that was perfect when we started the conversation had often shifted by the second hour of tape. Dogs barked, people got thirsty and needed to go to the restroom in the middle of a fascinating story, and Harold forgot to mention that his daughter, with whom he lived at that time, collected chiming clocks, which were all in the living room in which we set up, and all set for different times.

We asked our subjects to pick the place in the house where they'd be the most comfortable sitting for a while and talking, and most chose their recliners, some of which rocked. For a while I joked with Melody about calling the movie The Lazy Boy Stories.

We interviewed Del Herndon one week before valve bypass surgery, when he was fatigued and having difficulty breathing. We interviewed Chuck Calhoun just after he'd walked back into his house from an unexpected trip to the hospital for the pain in his cancerous hip. Rosie, bless him, died mid-way through the process of capturing the later interviews so never got to see himself on film, which he would have enjoyed immensely.

Midway through the recruitment and interviewing phase, it struck me that I had no women on my list. Thanks to a lucky accident and to Roy we were able to locate and recruit our two Marys, filling that gap. Then toward the end of the interview stage, another happy coincidence: At a chamber mixer a young woman approached me after introductions and said, "You have to interview my grandfather." Although we already had 16 people and the project had grown to be much larger than anticipated, her grandfather was Harold King—a gold mine of memories of union activities at the mill, not a shy man, and a senior with advancing emphysema.

Harold was the last person added. The project had to draw a line and stop adding subjects, as the social and historical ground had been pretty well plowed and the amount of information already accumulated would prove to be very challenging to assimilate and craft into a movie by the end of the year. The oral history collection stage had ended and we'd begun to face the challenge of making both the individual stories and the larger story of the West Linn mill under Crown Zellerbach available and accessible to the public.

I had promised to produce the larger story first, and tried valiantly to squeeze it into 120 minutes. It was not to happen. So after the release of "Grindstones, Boomsticks, Tattletales and Nips—the people and the stories of Crown Zellerbach International, West Linn Division: 1928-1986" I wrote another grant to the Kinsman Foundation, for the editing and production of a second documentary, using the same interviews. We covered the rest of the waterfront in 60 wonderful minutes, and "Friends, fish and $1.09 an hour" joined "Grindstones..." in Clackamas County libraries in 2007.

The two documentaries sort information into many common categories of recollection, from strikes, to the 1964 flood that inundated the mill, to papermaking and river barging. During the off-line editing of our 30-plus hours of raw interview footage, I was repeatedly struck by the amount of valuable and inspiring information that had to be left out of the movies, which were focused on describing the overall world of papermill work. Acting on a desire to make the original interviews in their entirety available for research and genealogical exploration, we applied for a grant from the Oregon Cultural Trust for the purpose of having the audio pulled from the video and transcribed for eventual publication in print. That part of oral history is the piece least likely to happen, but in the next year, that was accomplished and we'd gotten a Juan Young Trust grant to produce local history lesson plans for area junior highs, trying to get the stories into the hearts and minds of young people.

In 2009 we wrote another grant, this time to the Clackamas County Cultural Coalition, to publish the transcribed interviews as a book. When the book draft grew to 500 pages we returned to Kinsman, asking to contribute some additional support for the printing of these amazing memories.

With the publication of this book, six years of effort has finally resulted in the heritage resource that we imagined. But I still dream of another project; a collection of the women's stories of Crown Zellerbach. Memories from the wives, mothers and daughters of mill workers, as well as stories from some of the dwindling group of women who worked there, before and after—as Harold would say—"you couldn't HAVE a women's rate…you had to have a RATE!"

Methods

The interviews were conducted informally, working generally from a stock list of topics I wanted to ask about, but allowing the subject to go where it pleased. With reticent men, I needed to prime the pump by recounting stories we'd heard in earlier interviews. It is for that reason that the chapters are arranged by the dates of the conversations.

Sometimes we captured information before the 'official' start of the interview—sometimes after it ended. George started bringing out photos and samples of old papers after we'd already torn down, so we set back up again to save those contributions. Others wanted to talk more about their retirement years, and we kept that later history in, for a richer picture.

We tried to take our cues from the men and women. Some men's wives left the house, some stayed hidden and some wanted to actively participate. It was difficult to exclude their input, so, inasmuch as it contributed to the conversation, if a wife spoke, her pertinent comments are shown. I was able, by formatting and by a deep editing of my own comments and questions, to whittle 150 pages from the original transcriptions.

Individual speech patterns, mannerisms and moments of reflection have been retained. I edited out only incomplete thoughts that did not contribute content, excess doubling back on a point, or the third, fourth and fifth "No" or "Yeah" in a long string.

In all, it may be an imperfect effort, but it has been the most inspiring and important work I've ever done, and if you listen closely you may hear their voices coming through in each chapter.

Above, looking north, West Linn mill on the upper left, 1922 Bridge at upper right. Circa 1950.

Looking west from Highway 99E across the Willamette. West Linn mill and TW Sullivan
Powerhouse. Nearest, Hawley Pulp and Paper Company Powerhouse, Oregon City. Circa 1940

Aerial photos courtesy Aaron Schell and Scott Bouck

Above, looking northeast, at Willamette Falls Navigation Canal and Locks, lower Willamette River, Oregon City Bridge and Oregon City Elevator, upper right. 2005

Above, looking southeast, at upper canal basin, Willamette River and falls. 2005

Ownership timeline for the paper mill on the west side of Willamette Falls

4800 Mill St., West Linn, OR, 97068

1889	**Willamette Pulp & Paper Company** Located on Moore's Island, former site of Linn City Merged with Crown-Columbia in 1914.
1914	**Crown-Willamette Pulp and Paper Company** Early products included newsprint, creped toweling and gun cotton. Merged with Zellerbach Paper Company, San Francisco in 1928.
1928	**Crown-Zellerbach International, West Linn Division** In 1947, Crown-Zellerbach pioneered the coated paper process—influencing paper manufacturing across the west. Employment peaked at approximately 2,000 workers after World War II.
1986	**James River Corporation**
1990	**Simpson Paper Company**
1996	**Closed** Laid off remaining approximately 350 workers and went on the market
1997	**West Linn Paper Company** Bought by Belgravia Investments Limited, Vancouver, B.C.
2010	WLPC operates three paper machines, manufacturing and shipping 700 tons of coated paper daily, with a workforce of approximately 240 people.

William 'Rosie' Schultze

Interview October 27, 2005

Died June 9, 2006

…And they had a lot of trouble on a machine over there [in Holland], so out of the organization they picked me to go. And the mill manager called me into the office. You know, they don't call you into the office unless they're going to give you hell.

Fire you.

Yeah. But he told me what they wanted, and I says, "What are you doing, getting down to the bottom of the barrel?" "No, we figure you can do some good." So I was over there for about a year. That was really interesting. You know, Holland and England, and I went to Hamburg, Germany, and I went all over.

Did you get to take your family?

Yeah, 1975. Took the wife. I've been to Spain, too. I went from Spain over to Africa. Morocco, yeah. It was interesting.

[Roy Paradis sat in on the interview]

Are you a storyteller?

Oh, I've got a few that I could tell, I guess.

I'm going to ask, for the record, your name and your age and what year you retired.

Okay. Well, my name is William Sherman Schultze. I got the nickname of Rosie when I was in high school. I went through all my life, working years, by the name of Rosie. I had a fellow come in from the watchman shack. I was working with a crew of about six men. Asked for Bill Schultze, and they didn't know who it was. So I went by the name of Rosie all my years. And after I retired, I told them that my name was Bill.

And how old are you?

I'm 96 years old. I went to work in the middle of May 1927.

What was your first job?

My first job was a roll bucker on No. 7 paper machine.

Do you remember what you did on your first day at work?

Well, I was the lowest part of the paper machine. That's the roll bucker. I was to wrap rolls and transport them out to the finishing room. Put the paper, roll your roll, put them on a dolly, push them through the machine room on to the other room, the finishing room.

How old were you?

I was 17 years old.

Wow, it sounds like it was hard physical work.

Well, the job was easy, but I was going to tell you a little story now. Okay. The machine tender on No. 7 paper machine was a fellow by the name of Al Fredericks. Al Fredericks…told me when I left that job that night, to be sure to bring in a can of Copenhagen and a packet of Beechnut. And so I did. I took him seriously. So the next day when I came in, I brought that in and handed it to him. He said, "No, kid, that's for you." I says, "You mean I've got to chew that?" "Yes, you're going to chew it if you're going to be a papermaker." So to please him, I took a chew. And in five minutes I was white all over, and I says, that was enough of that. I never chewed any more after that. And I never smoked all my life.

You started in '27?

Yeah, 1927. I retired in 1972. That would be 46 years. I only worked for one company, Crown Zellerbach Corporation. The whole time. I had a number of different jobs. That's important?

Mm-hmm.

Well, I worked on No. 7 maybe three or four days, and they transferred me to No. 5 paper machine. And I was on No. 5 paper machine, and they transferred me over to No. 6 paper machine. And while I was on No. 6 paper machine a short time, an opening come up on No. 9 paper machine for a second helper. And the scale was only about two or three cents more, so nobody would take it, but I took it. And I stayed on No. 9 from then on.

So what was the pay, do you remember?

The pay? The base pay, yes. It was 43 cents an hour. And the Depression came along in 1930, and they cut that back five cents, to 38 cents an hour. But after well, maybe this is a little too early but the unions came in in 1934…And first I didn't join, but I joined the union right away. There was two paper mills, one, Publisher's Paper Company in Oregon City and Crown Zellerbach in West Linn. I was elected president of the papermakers' union in 1939. I held that for two years. Both mills was the same union.

What was it back then?

    Roy Paradis: *International Pulp & Paper.*

Yeah.

When was the big strike?

Well, my dad worked there; I don't know what year that was.

Your dad did work there before you did?

He worked in the wood mill, yeah.

And so that's when the big strike was, back in his time?

That was a big strike. They had quite a fight, yeah.

Did you like that job, being the union president?

It was pretty good. I got to go to a wage conference and help get the wages up a little bit. And the year I went to the wage conference—see, they have paper machines on the coast in Stockton and Port Angeles—well, there was several paper companies and they brought them all up to scale. That was the first time. And then things got better. When I got to No. 9 paper machine, I worked there as a second helper, not too long, and I got to—there was three helpers: first helper, second helper and the roll bucker. I got advanced to first helper not long after that. And I wasn't long as the first helper too long, years go by, but I got promoted to third hand on the paper machine. That was the winderman. See, there's a back tender, machine tender and a third hand. And they had three helpers on No. 9.

Is No. 9 one of the new machines or one of the old machines?

It's an old machine. It's been there a long time. But I guess it's still running sometimes.

So each machine had like five or six men on it per shift.

Well, No. 5 and 6 each had five men. And No. 1, 2 and 3 had four men. No. 4 had five men.

But that's per shift, right…

Yeah, right, shift.

…so there would be like 15 men who knew how to work that machine all together?

I got to be on the winder. I ran the winder there, the third hand, for 16 years. And they finally got mad at me. I always always in the back of my mind was to learn the next job. "What's coming up?" So I always paid attention and tried to do that. When it come to being promoted, I was ready for it. So I run that winder for 16 years, and machines, and over the period of that time, there was a lot of wires, felts, dryer felts and stuff been put on, so I was pulled off there and put on the bull gang. ...I went back and worked day shifts instead of shift work. But then I was on the bull gang for a while.

Now, you have to tell me what that is, because I don't understand. A bull gang?

A bull gang is a group of men that comes in and helps put the felts and wires and stuff on. That's always—we called it a bull gang.

So did the machines ever catch fire?

Well, No. 9 had caught fire on the back side. It run out of oil, and the bearings got too hot or something. But most of the fires were from the dust and stuff up in the rafters or someplace. They'd get a fire once in a while. But paper machines were pretty good … but I can recall No. 9 having a fire on the back side.

I've been through there and seen how big these things are. They must have been so noisy.

Yeah, they were noisy...we wear earplugs now. When I first started, nobody wore earplugs or anything like that, you know, and those couch rolls and the machines make a lot of noise.

So you never worked down in the pulp-making part of the mill?

No. Well, when I was a kid and I first started, I think I piled wood for a couple of days. And I told the boss, I said, "That's not for me." The blocks were about this long and about this square, and it was waterlogged. So they put me on the machine.

You went from machine 4 to 5 to 9?

Yeah.

And you liked that?

Yeah.

What kind of paper were you making?

On one and two we run telephone directory, yellow paper and white paper mostly. No. 3 you could run telephone directory and butcher paper and, oh, several other grades of paper. No. 4 run toweling and newsprint. No. 9 was only on newsprint. No. 5 and 6 are the first—when I first started there were both running newsprint.

I read in some booklet that the mill used to make guncotton. Were you involved in that?

We made a trial run on No. 3 for paper to wrap dynamite in. We just made a trial run on it one time on No. 3. And No. 3 used to run butcher paper, a lot of butcher paper. And No. 9, 5 and 6 were all running newsprint at the time I started.

Donna Dunn, daughter of Rosie Schultze, posing for magazine photo with a roll of telephone directory paper.

So you weren't there when they were making any guncotton?
I never heard of that.

So doesn't butcher paper have a coating on it?
Butcher paper? No.

There was no shiny side?
No. It was shiny. It was meat wrap and stuff. It's already in the stock, and it came out of the dry end of the paper machine ready to wrap stuff in.

Can you talk about making a coated paper, what machine, and how that worked?
Well, they started making coated paper I don't know when it was. But I was assistant to the head guy on 5 and 6 when we went to coating paper. So I was interested in that. And when I was transferred to Holland, when this one machine I was supposed to work on was trying to make coated paper…they were having a heck of a time, so they thought I could straighten them out.

Why did it take a year?
Why, it didn't take a year; they just left me there that long. Then after I was there two or three weeks, I had them running. They weren't making enough paper to pay their wages, and Crown didn't go for that. So I was sent over there by the

mill manager in West Linn to get this machine running, which I got it running right away. But they had different grades of paper to make, too. They had to make coated paper and the other kinds of paper. I don't know what all kinds of paper they made.

So what was your last job at the mill when you retired?

When I retired, I was paper machine superintendent. When I went into the mill, I was a roll bucker.

That's a career!

Yeah, just like I said before, I was always trying to learn what was ahead of me. And I was pulled off the bull gang and made boss machine tender when one of the boss machine tenders retired. I don't know what year that was. Fred Yoder retired, and then I had a shift. I had to travel with a shift to all the machines everywhere. I worked eight to four and four to 12, and 12 to eight.

How did you like it when you got to be boss?

Well, I figured this way: if we knew what we were doing, and doing the right thing, I didn't have to worry too much. It's like Roy and I—we had all the experience. Well, I had all the experience from day one up there. And when I was on the winder running No. 9 for 16 years, I was pulled to every job that opened up. I mean, like putting the wire on—I was always available. So I got so I was running the crews and those wires, and the machine tender should have been doing it instead of that. So that's why. I kept myself going.

You say your dad worked there. Did any other men in your family work at the mill?

No. Well, my daughter worked for the telephone company, and she came down to the mill to inspect the paper one time.

So you must have been working there when there were some big floods?

Yes, and the big flood came along.

I know about '64, but was there a big flood before that?

Oh, we had several floods, I guess, but the biggest flood was I think the '64. I had a log come in through the window in Mill C, I think it was, where the paper machines were. That's how high the water was. And when they started up after cleanup I was put on it; I was to follow around the electricians to get their motors running the right way. I don't know; there was a lot of motors, and they got one that went the wrong way. I was probably boss machine tender by '64.

Do you remember that day it crested? Do you remember the flood?

Yeah, the machines were all down, you know. Nobody was running.

They stopped everything?

Oh, yeah.

How high did the water come?

Well, I guess it got around the turbines and somehow into Mill C room, I think. I was down in the other end of the mill at that time, and they had these high ceilings, and they jacked the motors up to the ceilings—cable jacked them up so they didn't get wet. Yeah, they had big hooks up there, and they'd put a chain block on them and just jack them up.

Are these the motors that actually ran the machines?

Oh, nothing like that. Them motors would take up most of this room, some of them.

So when the flood came, and the log came into the window, into the welding shop...

Yeah, I knew that, but I wasn't up there. I was down at the other end of the mill. In fact, I was coming in for inspection every day. I never missed a day.

So can you talk about the different crews of men working the different shifts and rotating?

Well, take No. 1 and 3. No. 1 and 3, they run alongside of each other. And both were running directory, let's say. One might have been running white, and the other could be running yellow. And each machine has a machine tender. And each machine has a back tender. And each machine has a winderman. And each machine has a roll bucker.

Okay. And then they work five days?

The crew, yeah, they'd work five days. Well, they'd be on a shift. They may have been on day shift; then they'd work five days. If they were on swing shift, they'd work another five days. It's a rotation around.

So there were four crews, so that somebody was always working.

Yeah. Each machine had a different crew of men.

What was the worst thing about working at the mill? I mean, it sounds really hard.

Well, it was really what you call strenuous work, but I never really had any bad things about it. Probably did at the time, but I can't think of a thing like that. Might have been mad at some guy or somebody or other. I can remember one time, on the No. 9 winder I was on the winder and was putting a splice up. And the sheet was about 150 inches wide, I guess. So you stretch the tape about three-quarters of an inch, all the way across there. And another fellow comes along and taps that with a hot iron. Well, he got down to the end, and I was holding that, but he didn't stop. I turned around and almost slugged him, but I didn't. It hurt.

Now, didn't men get killed on the job, or injured, or things fall on them, or...?

I never really got injured on the job. We always looked out for each other. In the paper mill, after you get down through the dryers, you've got calender stacks, and there are probably eight or nine rolls piled on top of each other. And the paper goes through the top, and goes around this way and this way, and it puts a finish on. See, this paper comes off it wet, all wet, just a bunch of soft mush. And it comes off on a wire. And this wire runs...it has a forming board, table rolls and a couch roll, suction boxes to draw the water out of that stock when it comes out of the head box, before it goes into the felts. And it goes into the felts. They've got weights on the presses, and they're all suction, and it pulls out the water. And it goes probably two felts, then they've probably got a smoothing press to smooth it out, and then it goes into the dryers and goes around and around and around. You get down to the dry end, and it's dried out enough to put it through the calender stack, where No. 9 had run seven nips most of the time. It started up here, and they'd go down and around and around. Around onto a reel.

And they got it down to the bottom, and automatically you had to put it on the reel by hand. Somebody, a helper, was there with a spear and cut the sheet, and you'd just grab that and throw it in there... And he went across there where it would wind up. And about 20 minutes or 30 minutes or 40 minutes, whatever it was the length of time to make the size of reel they'd turn it up, and the winderman gets it. The winderman sets the slitters to cut the sheet. See, this is one wide sheet, and maybe you cut it into 62-inch rolls, 30-inch rolls, 34, run it off on the winder. Then they have to transfer them off of the winder onto the drums, two drums running together to turn this reel over slow so they could wrap it. They have to be wrapped. Rolls have to be wrapped. Then they're wrapped, then the roll bucker weighs them and takes them to the finishing room.

Now, is this the way it was being done in '75, when you were working there, or is this earlier technology?

That process mostly is all the same. Yeah, they might have improved on it a little bit, but you still had to get them through and they still had to go through the winder, and they had to be cut to ship. See, that one big wide sheet on 9 was probably 150 inches wide, and you couldn't ship that anyplace.

Do you know where they shipped, what kind of cities they shipped to in those days?

In those days they had a steamboat out there they used to load, and they had barges. The barges had tugboats driving them, and trucks come down there and hauled paper away.

A lot of people think a calender [correct spelling] is something with months on it, and they don't know calender is in papermaking. Is there a simple way to explain?

These are calender stacks. They're stacks, straight up and down. You go through the top, and that puts the finish on the paper. I don't know where they got the name, but they had that before I was born, I guess.

Now, what job did your dad do in the mill?

Oh, he worked in the wood mill; I don't know what. Worked for a guy he used to call Guy Reddick, they called him. Guy Reddick for the boss. No, I didn't want anything to do with the wood mill.

Why?

Why? Well, first thing, they bring the logs in up there on a conveyor, out of this [zoom] out of the water upper ramp, and put them into—it goes into a deal that's covered. And then they turn the high-pressure hoses on them to peel the bark off. After the bark is peeled off, the logs would go on through into what they call the sawers that would saw them into length. But they had these conveyors where the log come in, and it tipped up and go down, and they peeled it off. They had it over at Camas, I think it was, a fellow got caught in it one time, and it made hamburger out of him.

Boy. So it's more dangerous work down at the wood…

Oh, yeah, it was dangerous in the wood mill as far as I could see. Of course, we had fellows on the paper machines that got killed. There were several. While I was there, there was a couple of them killed. Remember Stanley Reddaway? You've got a lot of ingoing nips. You got felts coming down on a roll. They had a roll down on No. 5 in the basement, where they run the felt. And you had to get up on a platform, and he was going across there, picking the crumbs off that roll. And he got up to the part of the nip—pfffft!—away he went.

Oh, my goodness, that's awful.

Yeah.

So when you worked there, what was the size of the workforce at the mill?

At the time I worked there [means time of retirement], Crown employed just under 700 people.

That's a lot. And how many were over at Publishers'? Do you know?

Oh, I don't know. Publishers' had a big machine that trimmed 234, I think. A sheet of paper coming over, 234 inches wide. And they had three small machines, two or three small machines. And the big machine probably had six men on it, and I don't know much about the other ones.

Did any of the machines from Oregon City ever come over to the West Linn mill?

No. You don't transfer them around like that. There's too much stuff to handle. My God, woman, you've got—in fact, No. 9 had some 50 dryers. How are you going to send it over to the other mill?

Thank you for clarifying that. Now I understand why they just keep these machines and keep working on them…because they're huge, and they're there.

They're permanently set there.

Was there a social side to it, with picnics and stuff?

Yes. A lot of us at the mill belonged to the Horseshoe Club. The supervisors all had a Horseshoe Cub. And they had a picnic every year. And at Christmas time we had a big party. Other times, I don't know.

Where did you have the parties?

At the hotel. There used to be a hotel there.

Tour group in front of the West Linn Inn

The West Linn Inn?

The West Linn Inn, yeah. Yeah, it was a big place.

Did it have a bowling alley in the basement?

They had a pool hall there; I know that. But I don't think they had a bowling alley. But they did have a poolroom for playing pool. You couldn't play pool unless you was 18, and we used to sneak in there, round the back door, and get in once in a while and play a game of pool.

So they had the big parties at the West Linn Inn?

Yeah. Yeah, well, at the Christmas party, of course you took your wife. The Horseshoe Club was all men. I almost won the championship horseshoe playing one year. I didn't quite make it. I couldn't beat out…I can't say his name either… in supercalenders.

You want to tell me about the Horseshoe Club?

Yeah. I'll tell you about it. But don't get embarrassed. Well, they had an initiation every year, when all the new recruits would have to line up in a circle. And each recruit got a picture and little saying of a part of a horse for the Horseshoe Club. And they started…shall I tell her, Roy?

Roy Paradis: Go ahead and tell.

They started at the rear end, passed each man [a slip of paper], and asked them what they said. They got down to the last man, and you know what they said? "Shove it up your ass."

And then you were a member?

Then you was a member of the Horseshoe Club. Isn't that right, Roy? [laughing]

Now, okay, tell me about the Horseshoe Club. Where did you practice?

Oh, I had the courtyard in my place sometimes. We didn't really get together and practice. We just played horseshoes.

You had a horseshoe set in your yard?

I think the horseshoe pegs are 40 feet apart, anyway, something like that. Women's is shortened up 10 feet, but the men played about 40 feet. And we didn't have any women in our Horseshoe Club.

Roy Paradis: I want to kind of straighten something out about the Horseshoe Club. The club wasn't mainly a pitching Horseshoe Club. It was a supervisors' organization.

Oh, okay. So it was like a social club.

Roy Paradis: It was, yeah. But it didn't have any hourly people, see. I just wanted to straighten that out because you were getting off on the wrong trail there.

Yeah. So let's see. Do you have any old pictures from when you worked at the mill, of the old days?

Right here. I have pictures from Holland.

So this is 1965?

Sixty-eight. Now, this is my introduction to the paper mill.

Oh, in Dutch. That's wonderful. Do you read it?

I can't read it. This is a crew of men. I had one shift I worked with.

Rosie Schultze in Zellerbach-owned paper mill in Holland, making the machine run—1965

So what was it like working with people in Holland? Did they speak English?

Well, they were all friendly. All this little town was called Apeldoorn, and I'd go up and down the street, sometimes I'd be out on the side of the street. Somebody would keep waving at you. And they had Russians to work in the basement. In the beater room they used a lot of Russian people to put the paper back in to beat it up, when they used unused paper, waste paper. No, I got along really good with them. I made a lot of changes on this machine. And the millwrights come to me…one day I come into work at 8 o'clock in the morning, and they couldn't get the sheet through the dry end, and wasted about two hours trying to get it through. I stood there and watched it…I told the millwrights to take this roll out there. The rolls come in with this tail, and then it went down like this and over. Well, that's what we call a reversed bend in paper. And that's where it was breaking off every time they'd bring the tail through it.

I watched them there, and I dampened the sheet up to make it a little tougher, and it still wouldn't work. So when they were going to shut down in the afternoon to do something, I told the millwrights to take that roll out so you don't have that bend down there. Oh, no, it won't work. I says, well, leave the brackets in there. If it don't work, you can put the roll back in in about 10 minutes. It's a small roll, about so big around. When they washed up in the afternoon and come down, and I made a special point to be down at that point when the tail got there. Pssst! It went right through like that. And they didn't have any trouble. And the millwrights are standing over there in the corner, and I looked at them over there, and I rubbed my stomach like this. Told them everything's all right. See, ahead of the nip they have what they call a Mount Hope roll, where you can adjust it on the end to make the curve so no wrinkles go through. And they didn't even have to change that.

That's great.

They took that roll out, and they never had it off the reel all the rest of the night.

You know, when you talk about the machines it sounds like you were an engineer. But you must have just gone from high school into the mill.

No, I took cooking in high school.

You took cooking in high school?

I did. No, but I always try to be ready for the next step-up. There are a lot of step-ups in the paper business. And I was always ready. Roy has the same experience. One time Roy went down to California, and quit the mill in West Linn. And he came to me one day when I'd been in Holland already, and he asked me about it, and I says, well, if you think it's going to help you and create an advantage, take it. That's what I always did. I was always prepared. When I was on No. 9 all those years, a backtending job came up, and I thought I should have got it. I was a bit mad about it. But another guy that had more seniority than I did got it, a fellow by the name of Ellis Jones. I can remember him very clearly. He got the— that job. But they decided that I'd been a winder long enough, so that's when they put me on the floor, and I come and worked day shift.

And after I was on the floor for a while, Yoder, one of the boss machine tenders, retired, and I got his job. And I had his job for a while, and that's working with all the machines on eight hours and 12, eight hours and 12, with no other help.

You were the big shot. Now let's see, where was I? Oh, I was on there for a while, and then on the coating end. See, there's two ends in the mill here in West Linn: the uncoated paper machines and the coated paper machines. So you had a different superintendent for the coated paper makers, and I was working with the uncoated papers. So I was transferred to get experience with the coated papers, when I was assistant to a fellow by the name of Ed Haas in Mill D. And he called me in to go over to Holland.

Did you have good bosses and bad bosses? I know that's kind of a sensitive question.

All right. Yeah, I'll answer that question. These guys are all passed away. A fellow by the name of Clarence Schearer, well, he was the boss machine tender, and I used to relieve him. And invariably I'd have a crack in the wire someplace. And he should have had it before I got there, so I didn't think too much of him.

And who was your favorite boss?

Me. No, the bosses were all pretty good. What was the guy's name from Port Angeles?

Roy Paradis: Norm Tracy?

Yeah, Norm Tracy. He was a pretty good guy, but it seems like each one of those guys are looking out for themselves. And if their shift made more paper than the other shift, they couldn't see…

So there was competition?

Yeah, there was competition. If you got a crack in the wire, a little crack in the wire, you can feel it where the wire is running and fast, so you'd have to feel it, you should shut down and do something about it. They wouldn't want to shut down, so they could run it over to the next guy.

Oh. So there was competition between the shifts.

Oh, yeah. Yeah. And another guy was Joe Kozick. He was a boss, and I relieved him on No. 1 paper machine one time. And they had a crack. And he hadn't gone yet. And I jumped him about it and said, why didn't you shut down on your shift? "Oh," he said, "I couldn't; I didn't know what to do." He was pretty new.

Can you explain—when you talk about a crack in the wire…

Yeah, on the edge.

…there's a big wire screen that the pulp sits on…

A wire screen, and if you got a little hole in the wire you've got to do something, in the base of the wire. We had to patch it. And I did quite a bit of patching. You put a little patch of the same kind of wire over that, and cut it in a shape so the air will go over the paper. And you had to be very careful with your iron, when you put your iron down. If you didn't take your finger off the thing at the right time, you'd burn a hole in it, see? So I was down in—I was in Mill D, down on the uncoated paper machine, and a fellow by the name of Carter was on 5 and 6. And he got a hole in the wire. And he had a new guy down there helping him and breaking in, I guess, and they got a hole in the wire. So he put this new guy up there to patch this wire, on the base of the wire. And I was standing down there quite a ways away. You could see he was lifting that needle off the wire, and before he took his finger off or whatever, it made a spark. If it made a spark, it burned a hole. So they started up, saw the hole, and they had to shut down and put a wire on.

What job do you think is the most important job in the mill?

What's the most important job in the mill? That would have to be a machine tender, I guess. I never did get to be a machine tender.

What do they do?

Well, it's hard to explain. They had wire crews, crews that come down with wires or felts or dryer felts, and they needed extra men all the time, and I was always available, so I knew what was going on. And a lot of people didn't like to do that. You got extra pay for it, see. I think when they put a wire on, I think they got—what was it? Two hours? You got two hours more for each wire you put on. You couldn't do it yourself; you had to have a whole crew.

Do you remember your last day at work? What happened? Did they have a party for you?

Yeah, they had a party. And let's see, this Bob Carter was one of the other guys. They had a party for me. When I retired, the next day I was in Honolulu. And when I got to Honolulu, at the hotel I was registered at I was a little late getting there. And they had a big bowl oh, it was a big bowl like this full of fruit. And a bottle of good whiskey sitting right in the middle of it. It was for me, for one of the gifts. This is one of the gifts I got right here when I retired.

So the mill sent you over that bowl with the whiskey in it to Hawaii?

The mill didn't; the employees did. Yeah. Not the mill. I don't think the mill give me anything.

Did you miss your friends after you retired?

Oh, I had so many things going that I didn't miss the friends. Right away I joined a square dance club and took square dance lessons. After I got the square dance lessons so I knew what to do, I got four other couples and my wife and we formed a little club. And we entertained in nursing homes and places like this. In fact, I had one dance club here. The first time I had the square dancers here to entertain. I'll tell you a little story. I had this little club, five couples of us, and we'd travel around. I was a big shot and handled the—we went to quite a few places in Portland and in Oregon City. And I went way out to Troutdale one time.

I had a call to go out there. Got up to this place; it was up on a kind of a mound, a hillside, all high wire fence all around it. And I didn't know what we was getting into. So we went in, and it was an all men's place, and there were some pretty tough-looking guys. But anyhow, the girls wore fancy dresses, you know. And all the women—I don't know whether you know—they wear pettipants that comes right around the knees. Well, some of these guys would lay down on the floor and look up to watch and see the girls. After Tip and I seen what they were doing, I told the big shot over there, you'll have to get those men off the floor or we're leaving. So we left. There used to be a lot of square dancing. Portland State used to have a square dance. There was one gal, every time I went, she never wore pettipants. And it was a bum sight, I tell you. Yeah, it was plain ridiculous.

Mm-hmm. So do you remember what your pay was when you started at the mill?

Well, the pay, the base pay, when I started was 43 cents an hour. And the Depression came on in 1930, and they cut it back five cents an hour, to 38 cents. And then I was working on No. 6, I guess, or 9 anyhow, so this job came up on No. 9 that paid two or three cents more an hour. Nobody wanted it, but I took it, and it paid off for me. And in 1934, when the oldest daughter was born, practically the same week I got promoted to the winder on No. 9, which paid 90 cents an hour.

Ooh!

The union was in by that time. I joined the union, and like I said, in '39 I was president of the union.

—So what kind of effect did the war have on the...

Well, their wages didn't change, but the help was different. We went on six-hour shifts there for a while. I don't know why we did that, but we did. They run six-hour shifts. At that time I was living back of The Castle [a restaurant off River Road in Gladstone, demolished in early 2000s]. And I had two acres of peaches. And my wife had an aunt that married a German guy that was in Hitler's army. And he's coming over to the States. He didn't have any place to go, and he was a good carpenter, so I had him build me a chicken house, 20 by 80. I got 500 chickens, raised a bunch of chicks during those six-hour times. And I had chickens, and the most eggs I got in a day was 364; that's a case of eggs. I had them for a couple years, and I sold them off and went out of the chicken business. The people from—it used to be Skid Row in Oregon City, down on Abernethy, where the bums used to come and camp out overnight. And if they got the wind to come down to my place for a handout for eggs or something, so one time I gave them some eggs, and then that brought them back every other day for free eggs.

Did you sell peaches too?

I did sell peaches, yeah.

Did the men get drafted, or did they get to keep their jobs during the war?

Oh, yeah, they kept their jobs. The mill went on six hours. It didn't stay there very long. I don't know the reason for it. That would mean you had to put another shift on. But during the war they were shorthanded.

Did some men get drafted?

Well, yeah, some of the got drafted. And my boss got drafted, a fellow by the name of Ken Thompson. He was drafted. And I was married and had two kids, so I was 4A or whatever it was. I never did get to war. My brother joined. I took my brother to Ft. Lewis in 1941, and he went down to California for training. Then they shipped him to Iceland. And he was quite a skier. And I had all his skiing equipment at my house. And after he got out of training, they shipped him to Iceland, so he wrote a letter to me and said, I want all my clothes. When my brother come back out of the service, he used to go to Ft. Lewis all the time.

Don't you think you have to be really handy to be able to work with the machines?

Well, that's true.

Is there's anything else about your time at the mill, about making paper during the time that you were down there?

I can't think of any stories, really.

Did you work with people at the Sullivan Plant and the Corps of Engineers at the locks?

Well, PGE had a station down there at the falls. I don't know what they called it. I had a pretty good friend that worked over there. I can't think of his name..

So I have one more question. How many men were in the Horseshoe Club?

Oh, about 25, 30 or more than that.

And did you go out and have a beer or…?

No I never went out. Never did socialize with them. Just at the mill thing. Like a horse—like I say, we had a circle, and each little slip had a part of a horse on it.

Yeah, got that.

You got that? [grinning]

Got that. Okay, when you were working at the mill, you packed your lunch. What would you take?

Well, I'd take a sandwich and I used to the last 30 years or so I lived in West Linn. I could walk and come home for lunch. When I was on day shift. See, I got off of shift work pretty early and went well, it wasn't that early off on shift work either.

When I went to Holland, I was assistant manager of the coated paper crew and that was shift work. I used to have to take the night shifts, and Ed wasn't there at nights.

So you had to bring food if it was at night, right?

Well, I didn't eat too much. I had one fellow in our crew by the name of Bert Elledge. His wife—we used to carry our lunch buckets, you know and she was a good pie maker. And he'd always have a piece of pie or so in his lunch. And when he'd get in sometimes and we'd see him go to the restroom, we'd go over to his lunch bucket and get the pie out and eat it and he'd come back and he'd open his lunch and he wouldn't have his pie. He used to take us out to lunch down at Tebo's, and they had good pie. And this one time I went to lunch I guess Bert and I together, and a couple other salesmen and went down to Tebo's for lunch. And he wanted a piece of pie, and they said, no, you don't get any pie; you've got some in your lunch bucket. He got back to his lunch bucket, and the pie was gone.

Poor guy.

Well, you had to enjoy your job or it wasn't worthwhile being there, right? I enjoyed my job. All the time I was there, I enjoyed it. And I put in a lot of hours. I remember one time No. 9 had been down for repairs, and they started up, and this other boss and I, we worked 24 hours around the clock before we got that thing running that time. When you got to be a supervisor, it don't mean how many hours you put in.

So you worked yourself out of the union when you were a supervisor?

Well, yeah, I dropped out of the union, yeah.

But were you still friends with the union, or did you have to get adversarial?

Well, the guys are all dead.

Right. But I mean when you were working with them…you were their boss.

Oh, you had stay friendly with them, yeah. It was friends….

George Droz

Interview December 20, 2005

You kind of have to speak up for me. I can't hear too well.

Okay. Can you tell me your name, your age, where you were born?

Yeah, my name's George, I'm 78 years old, and I was born in Wathena, Kansas.

Tell about coming to Oregon.

Now that was a trip. We came, and I drove an old '32 Chevy. And the family came, and my stepdad had a '34 Plymouth, pulled a four-wheel trailer. We came 1,985 miles, and it took us 10 days to make it, because it was during the war—World War II, that is—and I spent most of my time repairing tires. Sometimes I'd go a half a mile; sometimes I'd go 10 miles before I had a flat tire, and I'd have to—and they had inner tubes in them then, and you'd have to take it off, patch the tube, put it back in and pump it up with a hand pump.

It sounds like you got to be an expert at that after 1,900 miles.

Well, we did our share of them, yeah.

Why did you come to Oregon?

To work in the shipyards. Things were tough in the Midwest at that time. I had one of the top-paying jobs in the packinghouse back there, at Armour Packing in St. Joe, Missouri, paying 85 cents an hour. And I came out here, and I started as a helper in the shipyards for $1.27 an hour.

That's pretty good.

Yes, it was. And besides, there was work everywhere out here. We picked beans. We picked prunes. I worked a 4-to-12 shift 4 to 12:30, and in the days we could go out and pick beans and stuff like that. I mean, we never saw such…I mean, we were dirt-poor back there, I'll tell you.

How did you move from the shipyard to the mill work?

Well, I went to the service, and when I come back, of course, you know the shipyards were closed down. And I worked around, worked in the woods for a while, then I worked at Foundation Worsted, which makes cloth, for a while. And that didn't work out, so I went down and I got a job at the mill.

Where were you living then?

I was living out here, out towards the butte; it's a little farther out. [Four Corners, Clackamas County]

What was the first job that you applied for, or did you just take whatever they had?

At the mill? I went to work in the grinder room on July the 10th, 1947, piling wood. Yeah, I was piling wood there, and I piled wood for No. 25 grinder line; that was upstairs. I can talk about that a little, too.

What did you do with the wood?

Well, we piled the wood, and they had grinders. They ground the wood up to make pulp, and No. 25 line was an electric line, run by electric motors, big electric motors. There was five stones up there, and you had to pile wood for those five stones.

By 'five stones' you mean grinding stones?

Five grinders. Yeah, each grinder stone had three pockets, and you put them blocks of wood in each one of those pockets, and it would grind it up into pulp. And then it went down and they made paper out of it. But they had the 25 line upstairs, and then they had the No. 2 grinder room, and then they had No. 3 grinder room. They had No. 9, 10, 11, 12, 13 and 14, I think, in No. 2 grinder room, and then they had 15, 16, 17, 18 in No. 3.

Tell me what you mean by all those numbers.

Well, each one of those is a line, they call it. It has three grinder stones machines in each line. And a grinder man ran two stones, so you had two stones on one line and another two stones on another line, and then they had what they called a jackrabbit. He went from one stone on one line to one stone on the other line, back and forth, you see, so that took care of the three stones. So three grinder men ran six stones.

What happened when the stones got dull?

They had what they called a jiggerman that came along, and they'd run a test on these pulps. They had a pit right where the pulp come out of the stone, they had a pit there. And they'd come along at the first of the shift, and they'd take a dip of that, and they'd take it to the lab and they'd test it. And if the fibers weren't—if they were too long or too short they just weren't doing what they were supposed to do, they had what they called a jiggerman. He come down, and he had a burr that he'd lift up the back end of the stone there, and they had a guide. They would put a little burr on there and run it across the stone and rough the stone up, so that sharpened it up.

Eventually the stones would wear out, though, right?

Eventually. Then that's when the millwrights would have to change them. It was a little bit of a chore because the stones weighed several ton.

Do you know what they did with them?

They had a core, and they would send them in to Norton. It was a Norton stone, and I suppose they sent them to Norton someplace. And they would refurnish them, put new the stones were put in there like brickwork, and they were cemented in some way. So there were just several sections of stone in there, and it was a big round stone; it was about five feet in diameter.

I heard that in the old days they would take the worn-out ones and use them for riprap along the river.

Well, I don't remember doing that, because they were expensive stones. They would refill them as much as they could.

Maybe that was before, when they actually used solid stone.

It might have been.

Okay. So you were hired in '47?

Yes, July the 10th, 1947 was when I went to work for the mill. I retired from the mill January the 1st, 1983.

So that's 36 years. Wow.

No. 25 grinder line, I started work up there. It was right up by the sawmill. And they would pile wood on top of the floor over there for the weekends, because the sawmill shut down on the weekends. Saturday they repaired the sawmill, and it was down Sunday, and they started back up again Monday.

Every week?

Yes. So they had a pile of wood in there I suppose it probably wouldn't all fit in this room. It was a big pile of wood. I got transferred down into No. 2 grinder room and piled wood, and then eventually got on as a grinder man down there. I was on the 4-to-12 shift that day—I came to work, and I didn't have any work in No. 2 grinder room. That was in, I believe, the spring of '48. I got down there, and all that upstairs, with the wood on it and everything, was in No. 2 grinder room. The floor had caved in. And one of the grinder men got his ankle sprained, I guess. That's all. They heard it coming, and they ran and got away from it. And it was all down there.

So No. 2 grinder room was down, so I went to No. 3 grinder room as a grinder man down there. And they had the four lines down there, and I can't remember now which one it was, what the number is but anyway, they had a flume, they called it, for these two stones, and the wood would float down in this water; it was about two feet deep. It was just a flume that come down through there, and they would pick the wood out of there and pile it. But the other two lines, they had a pile from the sawmill they'd dump back in there, and then they had to haul it on carts. The carts held a third of a cord of wood, and you'd push it down there. But in No. 2 grinder room they had tracks, little tracks. And these cars had wheels on them like a railroad car. And you would put the wood on a cart like I say, a third of a cord and push it down the track. And when you got to the line they had a turntable. You'd get on that turntable, and you'd turn it, and then you'd run it along the tracks along side the stone so the grinder men could get the wood off the cart.

You had to be awfully strong.

It was a job. I mean, it was a job. We had one fellow came in there one time I can remember. He was bragging about all of the work that he'd done in an aircraft factory. And he wasn't afraid of work at all. Of course, No. 2 grinder room was so foggy you couldn't see from one end to the next. I mean, just the steam from everything, you know; there's so much humidity there. And he worked down there for about an hour, and he came along, and he didn't—I heard him ask one of the guys, "Where's those stairs I came down?" So they told him, and he shook his head and he said, "A man would sure have to love a woman to work down here." And I never saw him again.

Photo of grinding room, courtesy of Ken Cameron.

When I was talking to Rosie, he was very clear he never worked in ground wood and he never wanted to because it was such a terrible place to work.

It wasn't a very nice place, I mean, it was a lot of work, but I worked in the grinder rooms until December of 1948. Then I kept bidding on jobs day jobs, because I hated the shift work and I finally got onto the general maintenance crew in December of 1948.

How do you bid on a job? How does that work?

Well, they posted them on the bulletin board down there, when they needed somebody—just like hiring out of another job someplace. They just put the bid up there, and then they would interview you, and when I got on the gang there was 10 of us that got on at the same time because see, it was shortly after the war, and they needed a lot of help. See, there was—when I went to work at the mill there, there was over 2,000 people working there.

Wow!

Well, see, they had the grinder rooms, they had the sawmill, and they had the converting department, they had the toweling department, they had the acid plant, they had the digesters, they had the bleach plant—did I say they had 10 machines? Ten paper machines. And they had the supercalender department, they had the mixer department, and the coating plant, and the supercalenders. That took a lot of people.

And it was seven days a week.

It was seven days a week, 24 hours. You know, they had four shifts.

Oh, it was four groups of men.

There was four shifts, so one shift was off at all times, see, the way they worked. They would work five days, then you'd go to work at midnight, you worked five days. You'd get off at eight o'clock this morning, you went back to work at eight o'clock tomorrow morning, at a day shift. So you'd work days for five days, then you would have two days off and go to work at four o'clock. Then between four o'clock shift and the midnight shift, I think you had three days.

Boy, that's hard.

It was kind of hard to get used to, yeah.

You must have been happy you moved to days. Well, you planned it that way.

Yeah, I did, because well, I was building the house, and they always worked on Sundays. Sundays was the day they shut the paper machines down for repairs. So I had Wednesday and Thursday off for several years.

That's a good day of the week to have off. So you can go shopping…

Well, yeah, but it's not very good if you've got somebody to help you build a house like my father-in-law was helping me build a house, and he had a grocery store, and he had Sunday off, so…it was several years before I got Sunday off.

So you came out with your family, just your immediate family, or…

Yes. Just the immediate family came. Later on, after my brother got out of the service, he came out here, and there was my brother, myself and a half-brother that started to work at the mill. But I was the only one that stayed.

Why did they leave?

Well, of course my older brother, he was disgusted about working shift work. He didn't like to work the shift work. And I don't know; I kept trying to get him to bid on a job, because I bid on it and finally got onto the day job. But he didn't seem inclined to do that, so he finally quit and he went logging. He was a faller—they had a lot of little gypo outfits around after the war; they had a lot of little sawmills around and stuff, and a lot of timber, so they needed a lot of people to furnish wood for them.

When you bid on a job, does that mean you offered to work it for X number of cents an hour?

No, no. No. The mill had base pay, and so when you bid on a job, you started in as a starting helper at base pay on these jobs. Your time went on at the mill—I mean it wasn't like starting a new job in that sense. Your seniority just kept on. It didn't interrupt that at all. It was just a matter of changing jobs.

And were you in the union?

I was at that time, yeah. When we first started, why, we didn't belong to the union. They didn't have a union there. And you didn't have to join a union if you didn't want to, because there wasn't that many people belonged yet. And they eventually sweetened the pot, so to speak, so that it would go. They offered us your initiation fee, which I think was $30 at that time, which was quite a bit of money, and they would waive that if you would join. So I finally joined. And then eventually everybody got to be union; it got to be a closed shop. But at the beginning it wasn't.

Rosie was the head of the paper workers' union there for a couple years. I don't know what years those were…

Yeah, it was before he got on salary, because when you got on salary you didn't belong to the union. Because, see, I had to get a withdrawal from the union when I went on salary.

So do you remember the very first day and what that was like? What your impressions of the mill were?

I went to work July the 10th at midnight, piling wood. And it was a job, piling wood for that No. 25 line up there. And the first three days I worked there I packed my lunch home; I never had time to eat it. And then the boss came up on the fourth day—the fourth night—and he could see what was going on, and he kind of changed things a little bit, because they had five stones on that No. 25 line up there, so one of the grinder men was supposed to run three stones, and one was supposed to run two stones and help me pile wood. And of course, they weren't doing that. So they changed that.

You were new, and they were taking advantage of you.

Well, new boy on the job, you know. But the fourth night I got to eat my lunch.

That's good. So have you counted how many jobs you did at the mill?

Well, I worked in general maintenance. Then I got into the departments. See, each department—I mean the paper machines and the supercalenders—all those different departments had a group of millwrights that did the work in there. But when they had a problem or when a machine went down or something like that, then they had general maintenance guys come in and fill in the crews enough to do all the jobs. So after several years I got into one of the departments. And I worked in one of the departments, and I finally worked up to number two man in Mill C, which was No. 10, 1, 2, 3 and 4 paper machines. And I was number two man there. And then when one of the fellows retired down in Mill K, which was the No. 9 paper machine and the mixer department and the basement there, with all the pumps in it and whatnot, why, they offered me the job of being the lead millwright in that department, and I took it.

So you worked in the basement?

Well, no, I worked as a millwright on that machine. I worked in the basement, I worked on the paper machines, and I worked in the mixer department, you know, keeping the machinery going. Then when the machines went down, why, you had to have a list of things that needed doing; you went over it with the paper makers. Rosie was one of them and I went over with, you know, decide the things that they wanted done, and the things that we knew needed doing, we got them all together and made a list. And then you had to figure out how many men you needed, how many hours it was going to take to do the job and so forth, and then they scheduled it and shut the machine down, and you did it.

And you had to have it done when you said you'd have it done, because they were going to be there to start back up again.

Well, the paper makers were there to start the machines. Yeah, you had to know about how long it's going to take to do the job.

I got a tour down in one of the lower-level places in the basement. And there was like a vat built into the floor with a mixer in it.

Slush maker.

And there were stories about people falling in...

We never had anybody fall in, but I mean, that's one of the old horror stories that they tell. Nobody that we ever knew of fell into a slush maker. Those slush makers were something new that they put in after years, because when I was in general maintenance they had what they called beaters. They had a beater underneath each paper machine. The paper would come down, when they had a break or something, it would come into the basement or from the rewinders, all the tails— all the scrap paper came into the basement, and the beater men put that into the beater and they beat it all up into new pulp again. And they pumped it back out, and they used it over again, reclaiming the paper. And then eventually they got rid of those and put in the slush makers because that eliminated one man—and a lot of maintenance, because those beaters were made out of wood, and we had to make them. I helped Roy Allen build several beaters down there, yeah, when I was in general maintenance. And that's what the general maintenance did. They did work like that around on the docks and whatnot.

So can you talk about the sulfur room and these big acid things. Can you talk about the chemicals and whether it was safe or not?

Well, yes—the acid plant, they called it, had sulfur burners, two big sulfur burners in there. And they burned sulfur and made sulfuric acid. And the sulfuric acid was put into the digesters with steam, and the digesters held they had five of them—and one of them held eight ton, and the other held 10 ton of chips. And they would put sulfuric, sulfur acid, and steam, and they would cook those things for eight hours. And they were under such terrific pressure that when they opened the valve it would blow all the pulp out. The chips that was in there was just pulp, because they just cooked them, and they would blow them into a blow pit, and then they would use them from there to make pulp out of. The chippers was down below there, and the blocks of wood come from the sawmill down to the chippers, and they were just a big blade. They had knives on them that run around, and the guys fed wood into those, and it chipped those chips up, and then they took them up and they put up in the loft. The loft ran the full length of the building up above the digesters, and when they needed to fill the digesters they'd go along and they'd pull a big lever, and it slid a chute back, and the chips all fell down into the digester to fill it. Then they would pump a certain amount of acid in there, you know. And after so long a time those digesters had brick in them, like firebrick, and they were put in there with litharge, they called it, which was a mixture of glycerin and red lead. And you would take and put those in there. The brick masons put them in, but we had to mix the hod for them.

But the first digester we chipped out, all we had was glass goggles. Didn't have a mask; didn't have earplugs. And it had a round opening about 18 inches in diameter at the top of this big digester, and so you'd have to take a scaffold that you made—it was out of 4 x 4s—you folded it up and you let it down in there, and then a guy would have to do down what they called a Jacob's Ladder, which is just a cable, two cables, with rungs bolted on it. You'd go down there and you'd straighten those 4 x 4s out, and then they would hand planks down to you, and you'd build your staging inside. And they'd hang that staging from four ropes from the top coming down. And they would put five or six people down in there with chipping guns to chip those bricks out. The first ones we chipped out, like I say, we had goggles, no earplugs, and you know what being in a big metal would be with those chipping guns going. And as I say, after the first one that they chipped out, I spit up blood for two weeks afterwards because of that acid in there, you know, because that dust was just—you couldn't hardly see in there for the dust when you're chipping the bricks out. So eventually, though, they got masks and earplugs, but it was a while before they did that.

What year was that?

I can't remember the year. Well, when they got the earplugs and stuff it must have been about 1949, somewhere along in there.

So did safety committees really start working when the union came?

Well, we really didn't have too much problem with safety. And I can't remember when they started their safety meetings. It must have been about oh, in the mid'50s they started having safety meetings. And they tried having the safety meetings right after lunch, and it didn't work out very well, because the safety meetings were supposed to be an hour or less, but there would be some guys— two, maybe, or three at the most would always have something that they had seen, and they'd save it up. Instead of telling about it ahead of time, they'd save it up, and then all they wanted to do was talk, and then pretty soon the safety meeting was running over an hour and 15 minutes, an hour and a half. And so finally they got smart and started having them at four o'clock. Well, the guys got off work at quarter to five, and so the safety meetings ended at quarter to five. Well, they didn't save up so much stuff, you know. When they found something that was haywire, they would report it, and it would get taken care of. They found a board that was loose, for instance, or broken, and they wouldn't say anything about it until at the meeting. Or something that wasn't safe. But when they started having them in the evening like that, why, then they would bring that to their attention, and it got taken care of.

Now on the docks—there was something about the docks I wanted to talk about. When I first started working down there, they had everything was done with hand trucks. They didn't have any forklifts. They hand trucked all the paper. And these docks was wooden, and they were wood blocks. They would take 4 x 6s and they would cut a 4-inch block off of them, and they would put those down, and they were put in with tar, like making a brick wall or something, you know, put down. And when you get a leak in the roof, those things would swell up, and you'd come to work in the morning, and you'd have a four-foot-high hump there someplace because it got wet. So you'd have to take those blocks all out and fit new blocks in there and pour the tar, and tar them down again. Then along about oh, when they started getting forklifts it was around 1952, I think, somewhere along in there, they started getting forklifts and they could quit trucking the paper, because before they always had to take the trucked paper and truck them up into the barges. Wherever they print the paper, they had to truck it by hand trucks, push it by hand. Well, when they got the forklifts, they were too hard on this wooden floor, so they started putting quarter-inch plating, steel plates, down. And that really helped.

It would be slippery, though, wouldn't it?

No. There usually wasn't much water in there because it was all covered.

Were the docks on the locks or out on the river or both?

Oh, they were alongside of the locks. They were on the inside. And it was all covered. And then they would run the paper up into the barges. The barges sat in the locks.

How much did these rolls of paper weigh? How big were they?

Well, they were the regular-sized rolls. They weighed over a ton. They were big rolls of paper. You'd get them on a hand truck and get them balanced just right. And then at first they had a ramp that they could raise and lower so that when the water level changed, why then they'd have to change the ramp. So they tried to keep the water level pretty low so there wasn't too much of an incline to push them, roll them up. But in the converting department, they would hand truck the rolls up there, and what they did up there was, they took the rolls of paper and they had cutters, they made flat sheet paper out of it. Big flat sheets. But they would roll the hand trucks up there with the paper on them. It had to be laying down in order to put it on the machine to roll it off to cut the paper the right way—they had big truck tires up there, and they would just push them over and let them bounce on them truck tires to keep from breaking the rolls.

You say one person moved a big roll of paper, that's 2,000 pounds, alone?

Yeah, one person.

That's a very strong hand truck.

Well, yes. They were big, but it was work. I mean, it was hard work in a paper mill. Eventually they got hand trucks, and they got them so that they would turn, they would turn the rolls and stuff. But…Wymore Trucks come down the docks and loaded the paper on. You probably heard about that. And they would run the paper over into the warehouse across the locks, and then they had an incline; they'd run them up. And this deck up there was level with the back of the trucks. And they would run the rolls up there and flop them over, and then they would roll them in the trucks. That's the way they got them into all those Wymore trucks; they rolled them in. It was a loading dock, yeah. But later on they got forklifts, you know. That was after they got the semis in. I'll have to show you something about the trucks that they had there when they went down there before. I've got a picture of them.

Can you talk about this loading on the barges?

Oh, yes. They loaded until they got a barge loaded, and then Western Transportation took them down to Portland to a warehouse, and then they were shipped from there. See, you had—another branch of Crown was Western Transportation. They had tugs, and they would bring the barges up through the locks, and then they would also—they brought chlorine up in the lower river and parked it out down there, and also the clay barges. The clay is just Georgia dirt, and they brought that in by railroad cars, and it was unloaded up the river at Pulp Siding, which is up above Canemah. There's a railroad siding up there, and they would bring the clay up there, and they had blowers that they would blow it out of the railroad cars into barges. They would bring it down to the lower river, and then they had what they called airvators, and they would suck the clay from there up into the clay plant to store it up there. It was quite an operation. It took a lot of people.

I heard that the mill had its own tugs because they would take the wood from log rafts coming down the river.

Well, that's a different batch. That's where Art Dorrance worked on the river. They had what they called upper river loggers that built the rafts up the river, and they rafted them down to the sawmill. And that's what those tugs did. I know the Crown Z was one of them; that was the biggest tug. And he had two or three smaller tugs, yeah. And that's all they did. They didn't move the clay barges. The clay barges was actually moved by water power.

You understand how the locks work down there? Okay. Well, all they'd have to do, see, is they lower water below, and that creates a current, and they would float down, and they would put the tugs on them down lower, and they would take them down the river then. But they could float them right down—there wasn't enough room in the locks for the tug and the barge both.

So they just guided from above Canemah, and then picked them up with the tug at the bottom.

Yeah, when they brought them in—they had the Corps of Engineers down there running the locks. And they would run a cable down to it when they bring them up, and when they opened up the locks, they had a winch, and they'd just start the winch, and it would just pull the barges right up into place, to get them up there.

Okay. Did you ever interact with the fish ladder people and PGE?

No, we never—we didn't do anything with those. No. That wasn't PGE there, anyway. That was the PEP Company, Portland Electric Power. They have a little generating plant there, you know.

Well, actually it's PGE now.

Well, that may be, but it was PEP Company at that time, Portland Electric Power. And see, when the water in the lower river got too high, the water pressure coming through the tunnels, down through the turbines, wouldn't turn the grinders fast enough to grind wood. The ones that weren't on electric motors would shut down. But the ones that were on electric motors would keep grinding. Then when the water pressure was low in the lower river and they had a lot of flow through there, and they didn't need quite so much pulp, they would shut one or two of the lines down, and those motors would generate electricity that they would sell back to the power plant.

That's pretty slick.

Mm-hmm. So they worked both ways.

Okay. So tell me about after you were promoted into management.

When I got onto management from hourly, it was pretty tough, because I'd been used to doing all the work, you know, myself, with the other fellows. And it was completely different. I had to learn to keep my hands off things, because I got into trouble with the union lots of times just doing a little something that—I mean, it didn't amount to anything to me—like move the crane when they was working. The guys were up on the machine working, and they needed the overhead crane moved a couple of feet or so. I'd move the crane, and I'd have a grievance. You know, it just didn't take much. And it was hard to do. It took a long time to kind of get from doing that. But of course, I knew all the fellows that I worked with, that was on my crew and everything, and a lot of them, they understood, and they were glad to have some help at times.

So how many years were you in management before you retired?

I was promoted to maintenance foreman in 1969. I don't remember what month. I was maintenance foreman from then until I retired in '83. And I started out as a foreman in what they called the Flying Squad. And I had been on the Flying

Squad as number two man before I got into the departments. But the Flying Squad, it was about five men and a foreman, and when they had any emergency on any machines, no matter where you were or what you were doing, that's the group that went and tried to get it taken care of immediately, see, because when those machines are down they weren't making any money. Now, like No. 6 paper machine there, I remember that at one time there it was six dollars a minute came off the back end of that thing. So they didn't like to have it down very long. So then after I got out of that, then I got into Mill D, which was the big machines, you know, the coated paper machines. And I was foreman over that. I had a crew of five men, and then I had what they called the shift millwrights; there was four of them. And there was one man worked as a shift millwright. He worked the four shifts. And anything would happen during the night or something that he couldn't take care of like when they lost a back-tender rope or something like that, he could take care of that but if they had a breakdown or something like that, why, he'd call me. And I'd have to call a crew, and we'd go in and see if we couldn't get it taken care of.

Did you ever see a machine fire?

Well, I can only remember one fire. We had a fire on No. 9 paper machine, yeah, and it burnt the clothing off of it. But it really wasn't all that bad; we got it down in a hurry. We had our own fire department down there, you know. Guys that belonged to the fire department, they did their jobs just like a volunteer fire department—whenever they had a fire, which is not very often,they had small things that happened, you know. A lot of times the fires were caused from the friction from the bearings. A lot of those machines down there had Babbitt bearings. We used to have to pour their Babbitt bearings for them. When they would run out of lubricant or something, it would get hot. And that would start a little small fire. It didn't amount to much of anything. But the one fire that I remember that amounted to anything was the one on the No. 9 paper machine, and it burnt the clothing off of it, so we were down for a couple of days to get that straightened out.

Tell me what you mean by 'the clothing'.

Well, it's the clothing. They have felts; they had a dryer felt that run over all of the dryers, and that was the clothing, see? And then on the wet end they had wet felts that run on the wet end there, see? See, on a paper machine they started out on the front end of it. They had the fourdrinier; the flow came out of the head box onto the fourdrinier, and that kind of formed the sheet. And it went over a couch, they called it, which sucked some of the water out. And then it went over the first press, which took some more water out, and that had a wet felt, a wool felt, on it. And that squoze some more of the water out. And then it went over to the second press, and it got some more of it out. And it was a real soft sheet then;

it was really fragile. And it would go into the dryers. The dryers had a dryer felt on it, and the sheet would go through the dryers and dry it, and then it would go through the calender stack and then onto the reel, and roll them up into a roll of paper. Because the dryers all dried it. They were steam.

It's just the same today?

It's a little more sophisticated than it used to be, but it's the same process. I don't know; did you want to know some of the papers that they made on the paper machines?

Yeah, that would be fine, too.

All right. No. 10 paper machine, it was fairly slow. It was an old-time machine, and they used to make a lot of lettuce crate pad, they called it. These boxes that they put lettuce in, they would pad them with that. And the soft paper, it was sort of like these egg cartons, and it wouldn't crush the lettuce. So that was one of the things that they made on No. 10. It was heavy paper that they can make there. And let's see one, two and three they made mostly newsprint. No. 4 paper machine, they made a lot of toweling, which they converted up in the converting department. You know, they had interfold toweling machines up there, and they had women that ran those machines, and you know, it folded them up, and they banded it, and it was in those little packages, those interfold towels. And then they had winders up there that they made the roll towels out of. And they made that paper on No. 4 paper machines, and a lot of that there. And of course they did some others there. And No. 3 paper machine they were out of sequence, you know; four was there, and then three was on the other side of it. See, they had one, two, four, and then three. On No. 3 they made a lot of telephone paper—the yellow pages and the white pages—a lot of telephone paper. Specialty papers. No. 7 paper machine, they made a lot of it—it was a real slow machine and they made pineapple mulch paper and Crezon and body wrap and that kind of stuff on No. 7. This is some of the pineapple mulch wrap paper.

What was it used for?

Sent over to Hawaii, and they have—see these marks? This is a reject roll, actually; that's the reason I got it, because this would have a mark in here. This mark didn't show up, so this is a reject. And they would take this over there to Hawaii, and they would take a knife and they would cut here and here, and they'd stick a pineapple plant in there. And this kept the weeds and everything down, see? And then this would all go back to the soil afterwards. It was before Visqueen, you know, that black stuff.

So it's a ground cloth just like they use out in the nurseries.

And you can have this if you want it.

Wow. Do they still make it?

No, no, no. No, when Visqueen came in, why, this went out the door. And this is a piece of Crezon paper.

What was it for?

They put that on each side of plywood, for building. For building materials. They attached it. We didn't attach it, but some other place attached this to plywood, and it made a good slick finish—I guess you could paint it or anything you want to.

How do you spell Crezon?

C-r-e-z-o-n, I think it was.

Very interesting. Thank you for that.

Then they made body wrap, which they wrapped the paper with to protect it, to keep it from getting damaged. And then No. 8 paper machine, they made just practically rugs on it, because it was an old machine that—well, No. 7 also that came around the Horn. They were real old machines, so they made real heavy paper on them, anything they needed that was real heavy. And they took No. 8 out, and they sent it down to one of the developing countries, because it just got too expensive to operate and everything. Because there was a fellow named Stone that lived up here on Upper Highland Road, was a machine tender on that machine, and when he retired they shut the machine down, and eventually took it out and sent it down to one of the developing countries. Then No. 9 paper machine made newsprint mostly. And five and six paper machines made coated paper, all coated paper.

Coated papers is what they do today…

That's all they make, is coated paper there now. They've got three machines. The machines they've got down there now, they're different numbers, but the ones that they're actually running was No. 9 paper machine, five and six. No. 9 paper machine was built in 1927. The reason I remember that, that's the year I was born. And No. 6 paper machine was built in 1947. Five is a little older than that, but I don't remember when it was built.

So they used to do a lot of recycling and making phone books and newsprint, which is what the mill on the other side of the river does now, I think.

Hmm. I don't know. I have never been into that. When I started to work down there, that mill was Hawley's, Hawley's Mill. It's Blue Heron now, I believe.

Right. It's changed a couple of times.

Yes. "Times Mirror" had it at one time.

"Times Mirror"? Okay. What was your best job? Or what you liked best about it.

Well, one thing was the pay, and it was steady. I mean, you didn't have to worry about having a job. They used to have what they called a pin dinner for everybody when they had their five-year service in. They had the pin dinner in the spring, and they had one in the fall. And if you had a five-year anniversary, why, you fell into one of those dinners. And I can remember my first pin dinner. J. D. Zellerbach was CEO at that time, and he used to make all of the pin dinners; that was one of his functions. He would come and he would talk to the people. And he told us right then and there, he said, "Anybody that works for me for five years has a job for life." I can remember that, and I thought that was pretty nice. And so anyway, I enjoyed working there. And it got a little bit hectic the last few years, but it was a good job. It was steady, and I mean, you're raising a family and building a house and such; you need a steady job, something you can count on.

What was the worst job?

The worst part was working in the acid plant, the bleach plant, and Mill B. Because you had to do work in the acid plant, which was sulfur, and they made sulfuric acid. And if you got a break in any of the lines down there—all the lines were lead lines, to keep the acid from eating them up. Of course, years later they went to stainless steel. But at that time they had lead; everything was lead that had anything to do with the acid. And that stuff would hit you in the nose, and I'm telling you, you couldn't hardly breathe from it. And we didn't have gas masks. I can remember one time we had to go down into the bleach plant. They had a three-quarter-inch pipe broke, and that chlorine was just boiling out of it. And we had to fix that pipe while stuff was still running out of it; we had to take the broken pieces out and put new ones in. And did you ever breathe chlorine? You can't hardly breathe. I mean, it makes you sick; I mean really sick. So that and chipping out the digesters was one of the worst jobs that I had.

We could talk about that for a long time. But I was going to ask you if any flooding happened when you were there?

Yes, 1964 they had a flood that flooded everything. I mean, it was a major flood. Every year we had floods. I mean, we had to pull motors that was in the basement. You'd pull those motors to keep them from getting wet. But in 1964, that flood, it came from the basement clear up to the upper floor in the mill. Everything was shut down except the generator up there at the power plant. And I had to go in; I remember Christmas Day; that was the day that the flood broke and it started easing off. And I went in at four o'clock, and I had to keep pumps going in there, because we'd keep pumping the water out to keep it from getting into the generators, because if that power panel went down, why, everything because everything else was down in the mill. And I can remember being in there, and about 30 feet up above us there was a hole about that big around, and there was water shooting about 10 feet out of that hole, because the water was that far above the generator room, and we were down there with pumps, keeping all the water pumped out so that it would keep those machines going. That was a big deal, and it took us months to get that thing going again afterwards, because everything everything electrical in each one of the departments had to be—because a lot of it was downstairs, and it was all ruined. They had to all be replaced. And the motors had to be rewound. When we first got the notice about it, we went in, and they said, "Raise all the motors two feet; get them up two feet high." So we did; we went through and we got them all up. And we barely got them raised up, and they said, well, we better raise them up four feet. So we raised them up four feet. Then they said, "Well, better get them up higher." So we pulled them clear up to the ceiling, left them hanging on chain blocks at the ceiling and wrapped plastic around them, and they still got wet. And there was logs that came through the mill down there. Fred Bietschek and his partner was upper river loggers at that time, they came down there, and I can remember when the flood went down, they sawed those logs up out of the basement. Some of them were, you know, 30-inch logs. Well, regular trees is what they were; I mean trees that flooded out and come down. And I don't know how they ever got in that mill, but they were in the basement of that mill. Big trees.

[After a break…]

Are you ready? Okay. On the dryers they have a manhole; it's not very big. And they have a siphon inside of the dryer that siphons the water back out from the steam that goes in there. It condenses and goes to water, and it siphoned it back out, and it would go back to the boilers and be recycled. But when the siphon would break, it would fill up with water, and it would break journals, because the head and the journal was all cast in one piece, and they were cast, and they would break. So when that would break, then they'd have to shut the machine down, and they'd have to go in and take the gear off and everything. And one fellow would have to go in this manhole with a big ram, they called it. It was a big hunk of steel rod, you know, see, about three inches in diameter and about four to five feet long. And he'd pound on that head to get the head out—because there's no way of getting hold of it to take it out. And that would get the thing out. But what I was wanting to tell about was, Harold Bancke was one of the smaller guys you'd have to get in there. Well, he had claustrophobia, and we didn't know it. And of course, all the millwrights wore overalls, bib overalls, because, you know, you couldn't wear a belt because with all the tools you carried and everything, it would pull your pants off. So we carried big overalls. Well, he started to get in there, and he got in about halfway and he swelled up like a toad. We couldn't get him either way. So we finally got his pants unbuckled, pulled them down, and we greased him. Finally got around and got enough grease around there, and we got him out. And I'll tell you, that boy was scared! He was claustrophobic.

So he just held his breath?

Yeah, yeah, and we couldn't get him out; couldn't go either way.

You didn't make him do that again?

Well, no, he didn't get to do that again, no. Well, I only had to go in there a couple of times because I got a little wider in the shoulders, you know. As you get older, you get a little more weight, so you can't get in those small places. And also we got a little different methods of doing it too, so, you know.

Did you want to mention about the silver paint on there? Ooh, this is so brittle [mulch paper].

It is oil. It was oil, it was kind of a crepe paper type, and then it went through this oil bath, and then it went through this mechanism we had that had these little rubber pads on it that were spaced just so, so whenever they came around they made the ink marks on it just right. And then this stuff was wrapped up in body wrap, and it was shipped to Hawaii. And those people over there couldn't read who it went to, so we had to put a different color tape on each batch of rolls. And so when the people went to pick it up, they just got all those rolls with that color tape on it, and they took it to that plantation.

Very interesting. And if they used the same machine for this and the Crezon, you'd have to clean it?

Oh, yes. When you went off of pineapple mulch paper, you had to steam-clean everything. Everything had to be cleaned up solid. And the first few rolls of any other paper you ran, whether it was body wrap, Crezon or any other kind of paper, had to be recycled because it invariably had a little bit of this residue on it from this other run that they had. So they would get enough orders to run that, and they might run it for two weeks, because they didn't want to run it today and then clean up and then run it again tomorrow. I mean, it was an all-day job to clean the machine up.

And what does body paper do?

Body wrap was what they wrapped the regular rolls of paper in to protect them, because if you just shipped a roll of paper just like it came off the machine, there's no protection for it, and it would get all skinned up, and people didn't want to buy skinned-up paper. When they bought paper they wanted to use every bit of it, so…

So is it like butcher paper?

It's almost as heavy as that Crezon. It was just a real heavy paper. And that's what they made they made that on No. 7, and I think No. 10 they made some of it also. But the No. 10 was mostly that really heavy stuff like lettuce crate pad and that kind of stuff.

Okay. The next question. Strikes. Did you see labor problems down there?

Yes, yes. I was—before I got on salary they had a strike, the first one that they'd ever had down there. I was lead millwright in Mill D at that time, before I got on salary. And I was the last person out, because we had to go up, and we covered the vents up on the roof. When the machines are running, they've got the steam comes out of those; you probably have seen those vents. Well, when the machines are down and it's raining, water will come down in that, and it will spoil the clothing on the machines and stuff. So we had to cover those, and I was the last one out on that strike. And it lasted I can't remember; I think it was 10 days. And they called me up and wanted me to come to work. And I said, well, I can't come to work; we're on strike. No, they settled it. Well, I didn't know anything about it. I mean, they didn't ask us to vote on it or anything. So I said okay, so I went back to work. I was the first one back in. And sure enough, it was settled, and we had to vote on it afterwards, and of course they accepted it. So that was the first one. The second strike I was in on, I was on salary. And that lasted from September through to March. I don't remember what year that was. That must have been about '77, I think. Somewhere along in '77, '78, in there. I can't remember the exact year.

There's a story about the West Linn Inn that it was built during the big strike in the early part of the century, and it went up in like 30 days or something…

That's just another one of those horror stories, I think, because well, they didn't have a union; how could they have a strike?

I wondered if there was an earlier union, and they broke it…

I'm trying to remember the name. It wasn't the Association of Western Pulp & Paper Workers; I can't seem to remember it, it was a national union to begin with. And they had a big problem with them some way or another, and then they broke away and they started Western Pulp & Paper Workers. But that other union was the first one there, because well, when I started work there, you didn't have to belong to the union. So I think that's just something somebody got started. I don't really believe that, because they built that inn to house workers for the mill. That's for sure. But they didn't have scab workers there. It was a hotel.

Was there a bowling alley in the basement?

Now that I can't tell you; I never was in the basement. They had a restaurant. They had a good restaurant there. Seavey was the chef there. I mean, he was good. Yeah. He used to come out when they had a picnic. You know, they had a picnic every year for the salaried people. Well, the lead men had a picnic, and then the salaried people had it, but the regular workers didn't have a picnic. And it was just a little extra for them. And Seavey, he always did all the cooking for the picnics. He would furnish all the food.

Okay. How did the reunion picnic get started? Did you organize that?

Well, the way the picnics got started that I recall is, the lead men in the paper departments and the maintenance part had their crew and they had a little more responsibility; they got paid a little bit more. They got paid anywhere from a nickel to 10 cents an hour more than the regular workers, because the workers were scheduled as D Helpers, Senior Helpers, Class C, Class B, and Class A. And then later they changed them to journeyman and everything, so the Class A was the journeyman. And then they had what their lead man was a plus, A plus. And they got paid five to 10 cents an hour more. And all of these fellows got together and decided, you know, that they'd like to have a picnic. And they presented it to management, and as a perk, I guess you might call it, or something, they agreed to it. So once a year you would have a picnic. And a lot of times they had it out at [Hatton] Park, which worked out pretty good. The reason for that was, the salaried people had already had a picnic. And so they thought, well, you know, they were like straw bosses, and so it went over.

So the retired people started having their own picnic? Was it the same men?

No. This was everybody. Well, first thing is, we started having the retiree picnic after I retired, and another fellow that I worked with—we'd been good friends. Boyd Ringo was the fellow—and he had a pond, and he had a place up there. And we got to talking one day, and we hadn't seen all these fellows, you know. Once you retire, you don't get to see them. Not necessarily the salaried people, but the men that you worked with. You know, you grew up with them. So we got to talking and decided, well, let's see if we could have a picnic. And so I got a bunch of names from the mill and ones that I could remember, and I got their addresses and whatnot, and so we started calling them on the phone. And there must have been about 20 or 25 of us to begin with that I got, which wasn't too big a job, see? And so Boyd furnished the place, and I did the legwork, more or less, to get the people started coming, and then it just grew from there. And so it got to be such a good thing; everybody liked it so well, so we decided well, we'd keep it going. And we decided whichever one of us was left, we'd keep it going, you know. If I went before Boyd, why, he'd keep it going, or vice versa. And so that's the way it's happened. He died five years ago next month, I guess. And so I've kept it going since. But now I mail out letters, you know. Of course, we have been doing that for years. We mail out a little brochure to all the names that I have on the list. And everybody that comes up there, we check them off to make sure that they're there, and the ones that we don't have the names for, we have them put their name down and their address, and then I can send them an invitation for the next year. So with the wife's help, we've kept that going.

So what was the first year you did it?

I believe it was '84, wasn't it? I don't know. This is our 20th year, I think, last year.

Twenty years? You've seen it grow, and now it's starting to get smaller

Yeah, we're losing a lot of the fellows. At first it started out as the pulp and paper workers, not the paper machines. See, they were paper makers. Theirs is a different union. That was Local 166. We were Local 68. So it started out as Local 68 employees. And then some of the paper makers got wind of it, and they wanted to know if they could come out. Well, yeah, sure, go ahead, you know, because then it just kept—we're losing a lot of the old folks, so there's not near as many. We only had, I think, 60-some this last year.

And that was including the families?

Yeah, that's the wives and everybody.

Good potluck, though.

Oh yeah, they've got a lot of good cooks. This is the 21st year, mm-hmm.

…[looking at photo]

Well, now, I don't know about the year. I'll have to see. These pictures here, this is what it looked like after the rockslide. Now, when they first started this you see this walkway right here? This walkway was where the trucks used to come down.

Collapsed hillside during blasting for new truck access to warehouse on the locks.
Circa 1950s.

Boy, those are old trucks. This is after the cleanup.

Yeah. That goes up past the office. That's where the trucks used to come down. All of these trucks used to come down through here. This was just a big rock wall here, just like this. Now, that was quite a trick when you come down on the day shift, because you walked down there, and you'd have to get up on the timber alongside so that the trucks could come down by there. These are the trucks that they used at that time. See, they weren't semis. They were making this ramp here, and they had a problem. They put a little too much powder in, and they broke the ramp down here. That was where that was from. That whole thing came down. I'm trying to think of what year. I'm not sure. It must have been it was in the mid-'50s, I suppose. That was really a hairy little walk down that walkway when they had those trucks coming down, because they came to work at the same time the day workers did.

Do you remember how long it took to get this cleaned up?

No. It must have been a month or something like that. I would say that they made that ramp, and they must have been there six, eight months getting that all blasted out. It was just a big rock wall they had to get out of there, and then they had to pave that. And then make an approach to it up above, see?

How did they get down into the mill parking lot?

They didn't have that parking lot at that time

That's where wood was stacked in the pictures that I saw...

Yeah, well, I don't remember that wood being down there. You know where that ramp comes down, then it kind of jogs over, and it goes down into the lower parking lot? Well, that ramp was where the trucks came down onto this other ramp, the little ramp, and went down by the office and came down to the lower warehouse. That's where they hauled the paper out of. And that was one of the maintenance jobs, was keeping that ramp all rebuilt, because they tore them up quite a bit.

So did a lot of people live in West Linn and just walk to work?

Yeah, some of them did. I mean, they just lived up on top of the hill. But that lower parking lot was made—gad, I can't remember when—it must have been along towards the late '50s and stuff, but that was for the supervisors. I even got to park down there after I got on salary. They had your stall, and it was numbered, and that was assigned to you, and that was your parking stall. But the hourly guys parked up on top of the hill in the other parking lots.

So when they were blasting for I-205, what effect did that have on the mill?

It didn't bother us at all.

Somebody told me that they would shut down the mill for an hour every day at noon because the blasting was making the paper machines get off.

They never shut those mills. It's a job to shut a paper machine down and start it back up again. They didn't do that. See, now, the paper machines used to shut down when I started working in the maintenance gang, they'd shut down on Sunday for repairs. That's why everybody had to work Sunday. That was the only day. And they would try and do all the clothing repairs at that time. Well, it got so that then they got better clothing, and the machines would run longer. And it cost so much to shut a machine down and start it back up again that they would run it—see, those fourdrinier wires that they got plastic wires afterwards—would run 30 days unless they had a problem with them some way or another. If they didn't have a breakdown they would schedule all their work, and then they would shut the machine down then. They didn't shut those machines down unless they absolutely had to. They wouldn't shut them down for an hour, because once you shut a machine down, sometimes it took six to seven hours to get it going again. It upset the whole machine.

So that's another myth—

Yes. Those machines were hard to get going. I can remember one time they called me. Just got home, we'd had supper. Must have been about eight o'clock, and they was having trouble with No. 6 paper machine. Couldn't make paper. So they said the fan pump was haywire, so I had to come down and change the unit. So I called the crew and went in, and that's a four-hour job with three guys. We got that changed, and they tried to start up, and it wouldn't make paper. Well, they said, it's the secondary fan pump. Well, so we had another unit up above. I went and got it, and they started changing it, and they run out of time for the guys. Had to call another crew. So I called the other crew in, and we got that changed. And they still couldn't make paper, and said, well, it's the third one. Well, No. 9 paper machine had a fan pump the same thing, only it was a left-hand. So I had to call in a couple of machinists, and they had to take the impeller off and the bearings off and change it around so that it would work for No. 6 paper machine. We changed that, and we got done it was about five o'clock in the morning and I went upstairs, and there was nobody around; everybody was gone. Of course, before, they had all the mill management people there and the engineers, everybody was there, before. But when I come up, nobody there, so we just went home.

I come back at eight o'clock in the morning, it was making paper. So Gib Hanson was the shift foreman that came on, and I went down and I said, "What happened, Gib? You're making paper." "Well," he said, "I came in, there was nobody around or anything, so," he said, "I just went ahead and started her up." But what had happened is, he got the valves all set right—everything had to be set right for it to work. The night before, at five o'clock, one of the boilers had went down up there, and they had lost the steam from that one boiler, so they had to start conserving. So they took the steam out of the wire pit on No. 6. Well, when they took the steam out, they didn't compensate for it, and the machine was just out of balance, and it wouldn't run. There was nothing wrong with the pumps at all. So I mean, you know, it's that touchy. So they weren't about to shut those machines down for an hour. No, no way.

Scene in the sawmill end of the operation, of spruce log salvaged from Tillamook Burn forest fire.

Okay, I'm going to hold up these pictures to jog your memory:

This is a spruce log in the sawmill. Now, see, the Tillamook Burn, they logged the Tillamook Burn off. Remember well, you've heard about the Tillamook Burn. Well, when they logged that off, we got most of the logs in the paper mill for paper. Well, because they were burned and whatnot. And this was one of the big spruce logs that came into the mill, and they made paper out of.

Okay. There's also a picture of logs in the river.

This picture here is the upper river. This is the pond, they called it. The upper river loggers rafted logs into these rafts. And there's one of the Crown tugs that bring them down, and they brought them into this pond. This is the millpond. And they had a conveyor that brought that spruce log up into the sawmill to be sawed into blocks of wood. And they went from there into the drum barkers. And they were just big cylinders with big long strips of steel in them that was shaped, had a contour on them. And they would roll, and that wood would roll around and around in those drum barkers and knock the bark off of them.

Can you imagine how loud it was in there?

Oh yeah, they banged around.

How did you use up every bit of a tree?

Well, that hog fuel went down to the boilers, and they burned it in the boilers to make steam. All of the bark and the sawdust and stuff that came from the wood went down to the boiler room. And they had oil; it was real thick oil—Bunker fuel, they called it—the same stuff that they used in boats years ago. And it's real—it's just like tar. And that would burn. And with this wet hog fuel and stuff, it would make that burn and generate steam. Then later on they went to gas. But I mean you saw probably down by that lower parking lot, there's two big tanks down there. Well, that's those tanks is where they had that bunker fuel. And they had heated lines running from that up so they could pump it. Otherwise it would just set up like cement. So they had to heat it in order to pump it up. So they had coils in there, which is similar to what they had up there for the sulfur burners. They had a couple of tanks up there that had steam coils in it. They'd dump the sulfur in it, and it melted the sulfur, and then it flowed into the sulfur burners, and they burned it and made sulfuric acid. But anyway, they had to melt that first before they could burn it.

Do you have any stories or any memories about the white house?

We had to go up there and do a little repairs now and then, but I didn't spend any time up there. About the only time that I was up at the main office is when I went on salary. Had to go up there and get oriented.

The pictures of the slide show buildings where the locks park is now. Do you remember those buildings?

Well, they were there when I left. That was No. 6 Warehouse, they called it. That's where they loaded the rolls of paper onto the trucks, to truck them out of there. Wymore Trucking was the one that trucked them out.

But in what is now the park by the locks, there were Corps duplexes where the people that operated the locks lived.

They didn't live there. They used that as a weather station. I mean, they got in out of the weather. They were just small little buildings. And they had their shop there. They had to do their work and stuff. They didn't live there. I knew some of those guys that worked on the locks, and they came back and forth to work. But they did have a shop in there where they did their work, and that's where they stayed in there, out of the weather, until they had to get out there. And like I say, that's where they were when they used those cables to pull the barges up the locks.

Did it make any difference what kind of wood you turned into pulp?

No, no. We used hemlock, fir and spruce. That was the type of wood that we got. When I was piling wood I always liked to get spruce. It had a lot of pitch on it, but when you piled that on the cart and took it over, it took forever for it to grind for some reason. Now, fir, it just melted through there. And hemlock was so heavy you couldn't hardly pick the blocks up. That stuff was I mean, it was like lead.

And these came out of the river? They're wet?

Oh yeah. Yeah, it was all wet wood. It come right up out of the millpond, and they sawed it, and it went down conveyors. They had conveyors that shifted it over to one area or to the other, and that's where it went.

Did it make any difference what different woods were in the pulp?

No, no. Now, see, a lot of the pulp that they had for specialty paper, such as like in the Mill D, where they made the coated paper and stuff they put the paper to the bleach. You know, from the bleach plant they had chlorine, and they bleached it, see? And that changed it, and it was all about the same then; it just all made white, fluffy pulp.

Any characters at the mill that you worked with that really stand out?

Oh yeah. They had one fellow there; they called him Donald Duck when I worked in the grinder rooms. And he piled wood on the flume over in No. 3 grinder room. And when he'd get behind and get excited—because normally they took a pike pole, you know, and they'd spear the wood and pick it up and put it on a cart for them to take it off. But when he would get excited and a little bit behind, he'd throw the pike pole down and he'd just get right down inside the flume, stand right there and throw the wood out by hand. He was quite a character. They called him Donald Duck because he walked like Donald Duck, and his feet was out like that. And he waddled. Of course, he had some little problem; I don't know what his problem was, but he had a little problem. We didn't really pick on him or anything, because, you know, he did his job.

Do you remember Rosie very well?

Oh yeah, I worked with Rosie for years. He was in Mill C, and see, I was number two man in Mill C for yeah, we were good friends for years.

He said, "I just always look for opportunities, and the mill would give me opportunities."

Yeah, well, along toward the end of my employment down there Huntington Rubber Mills up in Washington every year would take some of the salaried people on a fishing trip over to the coast. And my boss, Don Lindstrom, asked me if I wanted to go one year. And so him and I went from Crown and went with them. And then they took a couple of guys from over across the river. And we went to Warrenton. And the guy there—they called him Ribeye—he had a place over there where we bunked, spent the night, and he took us out on his boat the next morning, and we went fishing. So I did get one trip, but that's all. Oh, a lot of the fellows down there, some of the engineers and that, took trips back to Beloit, Wisconsin, where they built the paper machinery. And some of them went down to a mill at St. Francisville, Louisiana, and they had one at Bogalusa. And they made some coated paper too, I guess, down there. And so they'd go down there for one thing and another. And during the strike, when I was on salary, why, they had salaried people from St. Francisville and Bogalusa both come up, and they worked 10-day shifts up there, and then they'd fly them back home again for five days, and then they'd fly them back again.

How long did that strike last?

That was the one that lasted from September through March. And they settled for less than they offered them in October. In October they presented a package to the union, and they wouldn't accept it because they wanted another pair of shoes. They gave them one pair of shoes a year, but they wanted two pair—a $30 pair of shoes. And they were out until March. They didn't get their second pair of shoes. They got less benefits and money than they would have taken in October. But it just starved out. Mm-hmm, that's what they wanted: two pair of shoes.

One of the guys at the mill who's been working there 20-some years said on his first day he thought everybody at the mill was so nice—they took him around and gave him a tour and they even gave him his own pair of rubber boots.

Yeah, well, when I went to work on the midnight shift, I just wore regular shoes that I'd been wearing, until the day shift, when I could buy my boots out of the storeroom. But you had to buy them. They didn't give them to you; you bought them. I never had a pair of shoes given to me down there. When I was on hourly, I had to buy my shoes. Now, I did get a pair of boots down there one time; it was hunting boots, you know. And the reason I got that is because I was in the maintenance department, and they wanted a paper. They were going to have a little paper for the department, and they wanted a name for it. And so they had a contest for names, and I came up with the name of Toolbox Chatter. For the newsletter. And the winner got a pair of boots out of it, so I got a pair of them. Fact is, I've still got them.

Well, I wonder if, in Rosie's stuff, you know, there's that newsletter.

Well, this was just for the maintenance department. Rosie Schultze was a paper maker. They were two different departments, you know.

They didn't socialize, apparently.

Oh yes, we knew all all the paper makers and all that, we were all friends and all that. But the paper makers made paper; the maintenance department took care of all the repairs and everything. Now, the only thing that the paper makers did was change clothing; they changed the wires. They actually wouldn't trust to have a millwright to change a wire because they were pretty fragile. And then they had a crew that did the felts; they put the dryer felts on. Oscar Chandler and a guy by the name of Brown—I can't remember his first name —was the two guys that did a lot of the dryer felt work. And paper makers put the wet felts on and the wires on the fourdrinier.

So everybody was a specialist...

Yeah, more or less, yeah, because they had a welding department, and those guys did the welding and the burning. They had a machine shop; they did the machine shop work. And the millwrights didn't do their welding and the burning, but they did later on. After I retired, why, they had to they combined the crafts. But yeah, they had different crafts when I was working there.

So you haven't been down to the mill for a while?

I went back one time since I retired, and that was for a retirement for some people, but that was years ago. Yeah, they presented us a little plaque, brass plaque, supposed to be something to let you into the mill if you ever wanted to go in again, but I never did use it.

Oh, really?

But like I mentioned before, they had pin dinners. Every five years you had a pin dinner. And what that amount to was, you either could get a belt buckle or a tie clasp. At 25 years you had your choice of a wristwatch, a silver tea set, or a clock. I got a clock, and it's still setting there running now. It's an atmosphere clock. You never wind it. Don't touch it.

Oh, you can't mess with it?

You can't move it. If you move it, you've got to stop it. Yeah, 25 years, mm-hmm. And a lot of people took those. And this friend of mine, Boyd Ringo up there, he took a silver tea set for his wife. But I was selfish; I took a clock. Some of the people wanted wristwatches, but I had a wristwatch. [Grinning...]

Olaf Anderson

Interview February 23, 2006

Could you tell me your name and age and when you were hired at the mill?

My name is Olaf Anderson, and I was hired in July 1952. It's hard to remember. And I started out in the yard department, which is the lowest part of the mill.

Do you remember your first day in the yard department?

Yes. We were what they call piling laps. It's just about like blankets, only they're wet pulp, just folded over and over and made into laps. And we would make a big pile. And what they were doing is saving pulp. They would take—then later on they would take it out and run it through their equipment and make paper out of it.

What happened on your first day?

No, I don't remember too much. I remember I had a hard time finding my way around down there. And the only thing I can remember when I went down there was there was a big pile of rocks. It's what they call limestone; they made their acid with limestone. And so I found the yard department. And they have a supervisor comes out there and sends each person out to do something. And he said, you go out on the lap pile. And so I had to follow the others that had been there a while. And we went out; they use a forklift, lift up a pallet board, and you climb up on the lap pile, and you put the laps on the forklift. And then they haul it away and bring back another deal, and that's what you spend the day in, piling laps, which is common labor, you know.

So you got really strong arms.

Yeah. And before that I had been an appliance repairman, which is not too much physical. But I built up.

Why did you decide to apply at the mill?

The highest pay where I worked was $1.75, and the lowest pay in the mill was $1.72. So I lost three cents by going to the mill, but I was starting at the bottom going up. And that was at the top at appliance repair. That's the reason.

What year was that?

Nineteen fifty-two.

You weren't from Portland, though. Where did you come from?

Oh, I lived here [Gladstone] since 1946, but I used to work in Portland, W. L. May Company in Portland, as appliance repair, which is now just parts; they don't do any repair any more. I got out of the service, and my brother was working there, and that's why I went up there.

To May or to the mill?

May. The mill—well, I knew some people there, because there was a lot of people there, but I just went there because before I went in the service I worked for Hawley Pulp and Paper Company, which is in Oregon City. And when I came back, they didn't really want me back. I could have got a job, but I mean they had too many people came back. So I went to work at the appliance repair.

Did anybody else in your family go to work there?

I had a brother-in-law, Walter Peterson, who lives in Willamette. He worked down there as a welder.

Ella's husband.

Yes. No other relatives. I had another brother-in-law that worked there later on. Louie Hagedorn, which is Eleanor's husband. He worked up in the logs, up the river from the mill.

Your sons didn't follow in your footsteps?

No. I didn't really encourage them to work down there, because it was going downhill. And as you can see now, it is downhill. I think when I came to work there, there was 1,400 people there, and when I left there was about 400, so—

What year did you retire?

Nineteen eighty-five.

So how many years is that?

Thirty-three. More or less.

And there were 1,400 people at the mill when you went there?

That's what they told me; I never counted them. Actually, I didn't see all of them.

Can you tell me how your promotions went after that first job?

Well, in the yard department they do a lot of different things, so I worked various deals. They'd lend them out to the chippers, where they chipped up the wood to make chips for making pulp. And I did a few things like that. And then they have a bidding system when a job comes up, and I bid on millwrights. And fortunately

I got it. It's way up. The yard department was one level, and you didn't go up in that. So by going into the millwrights, you could go steps. Started out as a helper, and D, C, B and A, go up through the—you have to take tests, and you have to be there so long.

You have to be real mechanical, don't you?

Not everyone was really mechanical, but it is a job that requires a certain amount of skill. Yeah. It's an interesting job. It was not exactly what I had planned on doing with my life. I enjoy mechanical work, but this was heavy; a lot of it was heavy work. The last seven years down there was totally different. They put me on as a winder—I don't know what you call it, winder expert or something like that—the person that's kept all the rewinders in the mill, which was probably about 15 winders, had to keep them going. And it was a job I enjoyed. Only job I ever had in the mill that I enjoyed.

Millwright grouping, with George Droz, sitting, far left; Olaf Anderson, front row, third from left; and Ed Witherspoon, standing, third from right. Circa 1960s.

Out of 33 years, you just liked the last seven?

The last seven years. In the morning, the supervisor would say, "Well, Ole, what are you going to do today?" And I'd tell him, and they'd say okay. He didn't tell me what to do. He asked me so he could write it down.

That would be nice.

It was nice. I was totally my own boss. And it was something I fell into, and I guess I turned out good at it. When I retired they said they used to call in winder experts, but they didn't have to after I was down there. So it was interesting.

Well, it's good to be respected and to have that degree of freedom.

I did a few things that weren't exactly kosher. According to the union, you're supposed to do your job, and that's it. You're not supposed to really do anything on your own. And they had some slitters down there made by Jagenberg Corporation in Germany, and they were horrible.

Tell me what a slitter does.

Paper is made in rolls that are about up to 15 feet long and about 5 feet in diameter when they're laying down. And then they put them on a rewinder, and it cuts them into the length that they sell. Sometimes they're only 20 inches, and sometimes they're almost the full length. It depends on who buys them. And the slitters are on the rewinders, and they put them in there, and they set the slitters so they cut the paper in the right length as it goes through. It does through mighty fast, by the way. And I made one slitter using parts off of the Jagenberg and parts I scrounged up. And I asked them to try it. And they tried it, and right away they wanted more. So I think I made about 50 of them down there for them. And I'm not supposed to be building stuff. But it was interesting.

Well, that probably saved them a lot of money.

It probably did; I don't doubt it.

You were on the Flying Squad?

Yeah. That was before I was on the winders. You probably know what a Flying Squad is—it's a squad that when the paper machine goes down, they send them to the paper machine to repair it. Or if something else comes up, there's a group of, oh, varies from five to 10 people, and they send them wherever they wanted to, and they'd get the job done. If they had to install a pump, which is quite a job, really, sometime pouring cement and making a base and so on, they'd just send the group in there, and they'd do it. And it was an interesting job. It was something different every day. It wasn't the same thing.

So those people all worked at their regular jobs until a problem happened?

No, they were in the Flying Squad all the time. There was always something had to be done. No, if they weren't working on a breakdown or something like that, then they were installing equipment or something like that. There was always something to do. There wasn't any standing around.

So they literally did anything and everything?

Yeah, we did everything: carpenter work, cement work. We didn't do plumbing and we didn't do welding or electrical work, but we did everything else. They had an electric department and a plumbing department, so they took care of that, but we did everything else.

Well, during the '64 flood was everybody turned into a Flying Squad?

They called everybody in, and we all went in, and just before the flood we raised motors up. From the basement we took all the motors out, then on the upper floors—sometimes we'd just raise them but we tried to get them up high enough so the water wouldn't get them. And then after the flood, you come back in, you put everything back in again. Everything was a mess down there.

So where did you work in the mill?

I worked on all the paper machines. I also worked up in the wood mill, or the sawmill, and I worked in the grinder room, just short times at each place. And I worked in all the different paper machines, so I covered just about the entire mill at one time or the other. I worked in the acid plant, which makes acid, and I worked in the bleach plant, which bleached the paper to make white paper. And so I covered almost every maintenance job there was down there.

I bet they were sad to see you go.

I don't know as to that. We were on strike. I never went back after the strike, so I really didn't see much about anything.

So there was a big strike in '85?

Yeah. Eighty-four or '85. It was a long strike. I think it was about it probably started in the first of '85. I think it was six months, something like that.

Was it already down to 400 people, like you said?

Probably at that time it was about 400 people, yeah. And I could have gone back, but I was 62 or 63; I don't remember now. Somewhere in there. And before I left I had a damaged shoulder and a rotary cuff operation, and the doctor said you can't ever go back and handle any heavy equipment any more. So I just left.

When you were in the mill, did you get on different shifts?

Once in a great while we would have to come in at night and do something, but I really was never on shift work. They had a few maintenance people, that were shift workers, but normally it was day work. Sometimes you had to work all night also because you had a job that had to be done, but no, I didn't work shift work, fortunately.

Well, how about safety when you first started working there, you know, in the '50s.

Well, there was a lot of bad things when I first started working there, and it was slowly getting better. It takes a while because for years the object was to get the paper out; it was the number one priority. And then I don't know what caused it, but somebody started thinking they had to look out for the people, too. And it did get pretty good at the last. We had a problem with hearing. That's why I wear a hearing aid, because of the noise down there. But I never got involved in anything real serious. My back has been hurt, but I still seem to survive, so...

Did they have a lot of injuries when you first started working there?

No, I can't say they had an awful lot of them. They had a certain number, but it was not a great deal. I mean, considering all the work that was done over there, I would say it was not very bad. I never thought about it, but thinking back, I can't think of too many people getting hurt. There were a couple of people that got killed down there, but with that many people down there, there was a small amount.

So did you guys socialize at all? Or did people just work and take their lunch pail home...

Yeah, that's mostly what they did. They worked and went home. No, I can't remember any socializing. I think some of them down there went and gave out loans at the credit union and then got different positions on that. That was on their off time. But as far as socializing otherwise, no. I suppose some of them went hunting together and so on.

I imagine there were a lot of people who fished back then.

There was a lot of people that fished. But mostly, people that fished, fished alone or with one other person or something like that. There wasn't really a group of them that went out or anything. I didn't do any fishing, except I had to do a certain amount with my son because he was an avid fisherman. He was a guide also, so he was quite busy.

Do you remember the Native Americans down at the falls fishing?

Yeah, they were eeling. They would come down in a boat. At a certain time of the year they'd come down. I think it was in the fall the eels would go up the river, and the Indians would be down there catching them a variety of ways. I didn't have too much time to watch them too much, but once in a while you'd see them out there. In the last probably five years I didn't see any. But then I wasn't in an area where I could see the falls, either. But there were Indians down there. I don't know what they did with eel.

George was saying that he watched a gull grab an eel and just gulp it down all the way whole.

Yeah, I saw one down there, and it stayed there just about all day on that rock out in the middle of the river because it couldn't fly. It got a little too heavy. Those [eels] are awful.

He said they would swim up through the drainpipes in the pocket grinder room.

They would. So would the salmon. We had salmon come right up into the mill. Some of them went out of the mill, too, in the lunch buckets.

How did they get into the mill?

Well, a paper mill is a wet place, and they have drains in the floor. And the drain might be about 18 inches square, and they go on for a long ways, and it goes out in the river. Well, the fish would see that, and they'd swim up the drain instead of—they didn't have ladders to begin with, so they'd go up the drain, and they'd end up in the mill. And there was always somebody there waiting for them. I never got in on that. That was mostly on night shift, though.

Were you involved with what was going on down in the locks?

No, we didn't have too much to do with the locks. We had that bridge that raised and lowered, and once in a while we had to do some maintenance work on it. But as far as the locks are concerned, that was taken care of by the State, I believe. They had a lockmaster, and I never got too involved in that.

Were logs coming through the locks to the mill?

Yeah. I was just trying to think. They did have some logs go through, but I can't really think of them going through the mill, because most of the logs we had were in the upper river, and they would just float them down to the mill and run them right into the sawmill. They wouldn't have to go through the locks. But if they did have logs to go through the locks, I didn't really get involved.

So when you left, the pulp part of the operation had shut down?

Yeah, the grinder room was shut down. When I left they were still making ground wood with a peanut butter machine. I know that sounds odd. It was two giant motors and two disks about four feet in diameter with cutters on them, and they'd come together. And one of them had openings going in it, and it dropped chips from the chippers into there, and these motors was running opposite directions, and they ground that into a pulp. And they did still use that, that type of ground wood. But I think they used one or two of the grinders too, but the grinders were just about all gone. And one thing that did quit was cooked pulp from the digesters, if you know what a digester is. You put chips in it. It's probably at least 50 feet high and about 20, 30 feet across. And it looks like a big thermos bottle, with a lid on it that's about that thick out of iron and big bolts.

They'd fill it full and dump it in, then they dump acid in with it, and then steam in there, and they boil it. And then they blow it out. It blows out itself because it's got steam pressure in there.

And then it goes to where?

It goes into big bins, more or less, and from there on it goes over the screens. The screens, they were probably about four to six feet in diameter and probably about 10 feet long, just a rotor deal with a mesh on it. The pulp would come in there, and there would be water running on it, and they had a vacuum on it that sucked the water out of the middle part of the screen, and so it would be washed, more or less. And then it would come out of there more or less dry. In Mill B, I think they had four or five screens.

Does that wash the acid out of it?

Yeah, there would still be some acid because it had been blown off from the digesters, and it would take the acid out and anything else that was in it. And from there some of it went through a Jordan, if you know what a Jordan is. It's a cone-shaped device that fits into a cone-shaped hole, and it spins. And it cuts up anything that wasn't cooked up; it cut it up so it would come through there. And then they went through other screens where the pulp would go through, but the sticks and stones would not. So they had quite an array of equipment down there to take care of it, and it's hard to remember all they did.

And you had to work on all of those things?

Yeah. Yeah, sometimes you didn't know what you were working on to begin with, but you did learn.

When the water was washed off, it was river water—what happened to the water?

At that time it went in the river. Later on it was pumped up to Willamette [neighborhood]. They had a big slough up there. I don't know what they call that area. But a giant slough, or a pond, they had about four aerators that threw the water up in the air and down, and then after so long, it would settle out. And I'm not too sure what happened. I'm thinking it went through the whole process and then ended up in the river again.

Do you recall any things that the Flying Squad had to do?

Not that the Flying Squad had to do. Of course, I was only on the Flying Squad maybe six months or so. We did have some fires down there, and there was a fire under one of the paper machines. And I don't know if you know those soda-acid fire extinguishers—you have to turn them upside down. Well, I know the guy grabbed it and turned it upside down, and then they flipped over and no water comes out. Of course, it wouldn't. He was supposed to leave it upside down. So I run over and turned it up again.

But the fire is leaping up while you're doing this?

Yeah, it got out. Fortunately it didn't get very big, but it could be real serious. There's a lot of paper down there. Other than that, no, we didn't really have any fires. Mostly it was floods and things like that.

George told me a story about pulling the motors up in 1964. Do you remember that?

I remember they called us in, and we went down and we raised all the motors about six feet instead of hauling them out, because they were hauling them out from lower places, and we were raising those up. And they called us in later on to raise them up higher again. Finally, they got wet anyway. Nobody knew how much water was coming. Well, the one person that used to figure this out had retired. And he was very good at estimating how much water was going to come. And of course, when he was gone, then they no longer had that. And so they were guessing, and they missed it by quite a ways. So they lost a lot of stuff, equipment, by not doing everything at once.

It must have taken a long time just to clean stuff out.

They were hauling stuff out of there for months afterwards. Electric shop, they had to get rid of almost everything they had down there. All the switches and everything were all soaked, you know. And then the welding shop: you can't have a welding rod that gets wet, so you dumped out tons of it. Yeah, it was a mess. I think they were ready to close the place down. But somehow or other they kept going. It was the closest they came to shutting the mill down. I thought it was going to have a flood like that this year.

I went to the mill the other day and I asked if they got their feet wet the last few weeks with all this rain. And he goes, "This? No, not a problem unless it looks like 1964."

Sixty-four was a bad year, although we had a lot of bad years. But '64 was the worst; they weren't prepared for it. A flood is not so awfully bad if you're prepared for it, if you expect it. But they didn't expect it in '64. All the other years they could think ahead, and they were quite good at it. But somehow or other they missed in '64.

How do you take the motor that might weigh tons, and clean it up after a flood?

You mean after it's got wet? You take it to the electric shop and let them worry about it. Because it's electric, and you can take the motors out; you can do whatever you want to, but you're not supposed to do any repairs on electrical work.

So that was out of the millwright's scope? That was electrician stuff?

Right. When you raised a motor, you had to call an electrician to disconnect it. We didn't do any electrical work. I don't know; I think it was for safety reasons, I suppose.

MAKING PAPER

WEST LINN DIVISION

CROWN ZELLERBACH CORPORATION

No. 372   March, 1965

Story of Christmas Flood Recorded by Cameras

FLOOD FACTS:

. . . The mill began closing down on December 22 as the flooding Willamette River continued to rise.

. . . Flood crest came on Christmas Day. High water mark at the mill (tailrace) was 52.2 feet. (The previous high was 44.5 feet in 1923.) At the flood's crest there were 16-1/2 feet of water in the grinder rooms, the grinders being completely submerged. Water covered the parking lot alongside the office building. It stood 7-1/2 feet above the floor level in the basements in the paper machine rooms.

. . . Some 200 electric motors were removed from machines. After the flood 438 motors had to be cleaned and dried. CZ electricians from as far away as Bogalusa, La., were brought in along with others from Northwest mills to aid in repairing the electrical system.

. . . The mill resumed regular operation during the week of January 11.

. . . Cost of the flood damage was estimated at $1-1/2 million.

SECOND FLOOD —

On January 28 it was again necessary to close the mill due to flooding. On this occasion the water reached a depth of 7-1/2 feet in areas where it had been 16-1/2 feet deep in December. Production resumed on February 2, 3 and 4.

BOATS BECAME the only method of transportation in some parts of the mill as the water rose relentlessly. From left are: Densal LaClef, Louis Schaber, Ervin Heilman and Bob McCleary.

WITH GRINDER ROOMS already flooded, groundwood employees turned to filling sand-bags on December 23. From left Bernard Santo (holding bag), Tom Chapman, James Murphy and Lyle Hooper.

HASTILY REMOVED motors and supplies cover floor in Mill C.

1964–65 flood stories in CZ newsletter.

Probably some union rules, too.

Union rules, probably, and safety rules and so on. Actually, they were probably good rules, because some people shouldn't mess with electricity. I have two rentals, and I have to do some electrical work every once in a while. And I find out people do some awful stupid things.

You told me about what comes out of the mill into the air and how it's changed—about pollution from the mill.

At that time, I had one acre here; now I have two. And it was all fenced in with stock fence or cattle fence. And it seemed like about every three years I had to replace the fence, because the galvanizing would just go off of it. And sometimes you couldn't hardly stand to be around because of the sulfur smell. And slowly but surely it started to get better. And now I think I've had the fence there over 10 years, well over 10 years, and it isn't bad yet; it's good. So it was an awful lot of sulfur dioxide in the air, and I don't know how people lived, but they did. Now you go down there, and every once in a while someone will say, look at all that smoke coming out of those mills. There is no smoke in the air of the mill. Steam is the only thing that comes out of that mill any more. I drove out with my son and his friend, and this friend said, "Boy, they ought to blow that mill up. Look at all that smoke coming out of there!" And I said, "If you look down there, you'll find that there isn't a single bit of smoke. It's all steam." And I said, "They have to have steam come out because they have to dry the paper. It's got to go somewhere. There's nothing wrong with steam." But in his mind he had the idea that because it looked like smoke, it was smoke. But right now there's nothing coming out of that mill. They have a boiler house that burns sometimes a certain amount of oil; mostly it's trash wood and so on, bark and so on. And the smokestack, the smoke goes up, and it has showers in it that wash down everything so what comes out of there is nothing.

So they wash the smoke from the boiler house?

They wash the smoke. And then they pump it up on the hill, and I don't know what happens to it after it goes up on the hill. That's not my department. I worked on the pump down there and a few things like that. But yeah, the smoke is all washed, which is quite a job. They have sprinklers up above that and it does a good job too.

They really use the river, don't they?

Yeah. Well, river water is used for a certain amount, but they send it up to the filter plant. Or do you know about the filter plant? Right up on the hill, above the mill on the West Linn side, there's some big ponds up there. They pump the river water in and let it settle out. And it moves from one pond to the other, and it finally comes back to the mill, and it's called filtered water. And once a year or so you've got to wash out because it gets a big layer of mud and so on. And they used to wash it into the river, and now they say they can't, which I can't quite understand. That's where it came from.

But anyway, the filter plant, they add stuff to the water to make the stuff get together and sink. But they say the water is almost pure enough to drink out of there, and they say they could make it pure enough to drink. See, they can't use river water to make paper. There's too much dirt in it. Paper is quite sanitary. It's—what do you think about that piece of paper you have there? That's mostly sulfite, I believe. Sulfite is cooked paper, cooked chips. Ground wood, it would be like newspaper, sort of brownish a little bit. But white like that is usually cooked pulp, and it's got longer fibers in it, and it's real sanitary paper.

This is actually recycled paper, too.

Yeah, well, even if it's recycled, it's still the same pulp in that it's just rechewed. We recycled a certain amount of paper down there. You dump down in a vat, and violent things would take place down there. And you ended up with pulp. It's a place I never wanted to fall into. But there were something like pumps down there, quite violent, and they really stirred stuff up. Big paper and everything like that just ended up as mash.

People say that there's a ghost down in the basement in the old mill.

Oh, a ghost? No, I don't I never saw a ghost. Let's see, old paper under No. 7 or 8 paper machine was probably the oldest part of the mill, where they had rock walls and so on. But I worked down there, and nobody ever bothered me. And I can't think of any other place where any ghosts would be.

So when you went to work, you took your lunch?

I took my lunch.

Was there a noon whistle?

Yeah, there was for years, but I'm not too sure that they used that in the last year or so. You know, you get so used to things that you don't hear them. But yeah, we used to hear it here. And there was a lot of times there were two whistles, one to stop for lunch and one to go back to work. And they had a five o'clock whistle. I don't remember whether they had an eight o'clock whistle or not. But they did have whistles, and they were quite loud, because you could hear them here.

This [Gladstone] is downwind, right?

Well, depends on which way the wind is blowing. But you could usually hear them. They were very loud whistles. They were just as loud here as they were in the mill, as far as that's concerned. They carried very good.

So when you were in the sawmill part of the mill, isn't that where they had that big tumbler?

Oh, you're talking about the barker, yeah, drum barker.

Were you ever connected with that, or did you work on it, or how did it work?

Yeah. They had the bars riveted or welded across with openings on them. And sometimes they'd break loose and we'd have to go either rivet them or do some welding on them, or something like that. But we didn't have too much to do. Sometimes on the drive, we had to work on the drive. You could figure it had several tons of wood blocks in there, and it was quite rugged. And all the equipment that run it was quite rugged.

It would have to be really noisy.

It was noisy. It was very noisy. And it done a good job. It's something like a dryer in here, only if you put a bunch of rocks in the dryer here, only this cylinder was probably at least 10 feet in diameter and probably about 30 feet long. I'm just guessing. And so it was quite massive. And the blocks would go in one end and come out the other.

On a conveyor belt?

Yeah.

And the bark came off?

The bark would go through the openings in the drum barker. They had gaps about this wide, and the bark would come out. And there was always water running in there, and the bark would come out. There was a conveyor underneath that hauled that away, hauled it to the boiler house, and they'd burn it.

So that was what they call hog fuel? What's hog fuel?

That's hog fuel. Hog fuel is anything that is waste. Mostly it's bark and sticks and chips and things like that that weren't usable. I don't know why they called it hog fuel, but when they drag the logs in to be sawed up, of course there was a certain amount of bark come off of them. That had to be taken away, and when they sawed all the sawdust, saw chips would have to be sent away. And then they sent it to the barkers, and of course all of that had to be cleaned up. So there was a tremendous amount of hog fuel. I don't know why it was called hog fuel, but it was a word they used. When you work down there you learn different words. You don't think too much about them. That's called hog fuel, so that's hog fuel. There's so many different things they come up with. It has to have a name. So who named it that, I don't know.

So these guys talk about supercalenders—to me a calendar is something on the wall with dates on it.

A supercalender is—you know what a wringer on a washing machine is? Only it's tremendous; it's about 16 feet wide, and the rolls are probably about 18, 24 inches in diameter. And there's a steel roll with a cloth roll in between them. Now, when you're talking about cloth, they have a shaft, and they cut out sections of cloth with a big hole in the middle. And they drop them down on this shaft until they get it so high, and then they put a tremendous pressure on it to squeeze it down. And then they screw on the end cap, and then they put it in a lathe, and it looks like—you know what white Teflon looks like? That's what it looks like. It looks a lot like plastic. And so they have a cloth roll and a metal roll and a cloth roll. And the paper goes back and forth in between them, and it irons it, makes it shiny.

Coated paper.

Yeah. When they run through the paper machine, they have a coating section, and they put clay on, a layer of clay on both sides of the paper. And then they come through, and the paper is still kind of dull like that. But if you run it through a supercalender, it turns out real shiny. It's not very good for blowing your nose on. It doesn't crinkle very good.

Not very absorbent.

No, they're not. Yeah, it prints a lot better and it looks a lot better. It's not waterproof or anything, but it's messy. Some of that paper is quite messy because it's got quite a bit of clay on it. When I worked down there, we had a clay barge come up to Pulp Siding. It's about two miles south of Oregon City, there's an area they call Pulp Siding, and that's where they—we used to bring in railroad cars of clay. And we would empty those with a big vacuum. It was a vacuum, six-inch, I think, vacuum hose, and we'd load it on the barge. And then they'd haul the barge down to the mill, and they would vacuum it off the barge into the silos. It was a dust. It looked like talcum powder. And then later on I heard it come in in a slurry; in other words, it had water mixed with it. So I never got involved in that, but we handled dry clay. And what they did, back in the Midwest they had vacuum deals to pick up grain. And it picks it up into sort of bags, and then a deal turns, and it shakes the grain out and goes to a different place. Well, then they found out they could pick up clay with it, so it's the same thing. So we were using grain vacuum deals. But we're talking about a vacuum there. It's a big vacuum with six-inch pipe.

I think George talked about this. I think he said that the clay barges didn't have any power; they would use gravity down the river, and tugs would guide them.

Probably, yeah, because they were loaded upriver. Of course, they had to be hauled upriver with tugboats, but coming down they probably came down without any, because the river does go downstream and takes everything with it. Although I think they must have had barges on them to steer them, because we were on the east side of the river. That's where Pulp Siding is. And the locks are on the west side of the river. So it had to get across the river, and in between is the falls. And I wouldn't want a barge going over the falls. So they had to have some steerage in there somewhere along. I don't remember. I never really had much to do with it. But I think they had to have a tug to get them across the river.

George gave me a little sample of plastic mesh. He said in the old days on the supercalenders they had a copper-wire mesh, but they stopped using copper because it got too expensive or it broke, and they went to this plastic.

That was on the paper machines. See, it seems impossible, but they've got a big vat they call a head box, where they run the pulp and the water in. And then they've got a slit across the full length of it that they can open various amounts, and they can also bend it a little bit. And they run out on this wire; it's called a Fourdrinier wire, and it used to be copper. But I think they've changed a lot of them to plastic. And it goes on this wire, and of course a certain amount of water dribbles through, and then they have suction boxes across, underneath the screen, that they have a vacuum on. And the screen slides across them with their slots, and that sucks some more water out.

So does it still work the same way now?

It does. That's the way they do it now, yeah, as far as I know. And when they get to the end of the screen—I don't know how long a Fourdrinier wire is, but it's 20, 30 feet, something like that, just as a wild guess—then it's picked up off of there onto the rolls and goes through the dryers, which are enormous drums with steam in them. And it's got a canvas that goes back and forth.

They cram a lot of linear mileage into a very short little space, don't they?

They do, yeah. One dryer won't do the job, so they have to have maybe 20 or 30 dryers, and they just put them one right next to the other, a roll on the bottom and a roll on top, with canvas in between.

So it must be hot in there.

It's hot. It's hot. When the paper machine shuts down, sometimes you have to repair a bearing on one of the dryers, and you have to crawl in there. And I'll tell you, it's hot. You have to be careful how you stand, because I don't know how hot it is. They used to be able to take readings off of them, but they were probably hotter than 90 degrees or something like that. It's a bad job in the summer. It's a bad job in the winter too, if you're dressed for the outside and you go in there. You're probably better dressed for it in the summer than you are in the wintertime, although you have to—you can't have bare skin out. If you touch the dryers, you burn. That's all there is to it. So it's interesting; you learn a lot; you learn what not to touch.

Did you ever see pineapple mulch paper?

No, never saw that.

There was this other paper they used in old construction called Crezon a very slick, thick paper they used to nail on walls.

Brown paper? I have some of that.

Do you? What kinds of papers did you see made while you were down there for 33 years?

What kind of paper? Let's see. Of course, newsprint and coated paper for all the magazines and so on. That's mostly what I saw, but the paper you were talking about was made on No. 8 paper machine. It was coated with Melmar, I believe it's called, or something like that. It made it very waterproof. And they put it on the outside of plywood and things like that. And they made almost cardboard down there in Nos. 7 and 8. They're very slow paper machines, and they could make very thick paper on it. But that was one of the main ones. Crezon I think was the name of it. Yeah, I have some out there. Some of it didn't turn out good, and they'd sell it down there. And it's very good to put between floors when you're building a house or something like that.

So did you ever see any guncotton or any kinds of strange fibers come out of there?

No. 10 paper machine tried to make the paper that they make coffee filters out of. I don't know how good it turned out. They tried and tried, and I know they worked on it. And then they made toweling, paper toweling. No. 4 paper machine used to make paper toweling. And that was interesting. You know how toweling is sort of crinkled a little bit? The paper goes over a roll, and then they've got a square-edged blade that lays there. When the paper hits it, it kind of crinkles a little bit, and then it goes on up over it. So that's what makes that crinkly deal in towels.

Would you be interested in seeing the mill now?

Probably could get a tour. A couple years after I retired, I went back to the mill. I just walked down there and asked if I could go through. And I went through; there wasn't any problem. I don't know if you could do that any more because it's changed hands about a dozen times. So I don't know.

So you mentioned one thing I'm curious about, and that's the converting department. What was the converting department?

This converting department is something that I don't know too much about, but the converting department was either making toweling or toilet tissue. They didn't make the paper. They got the rolls up there, and they put them on their machines, and they made either the toweling or the toilet tissue. They had that; they also had the finishing room, which you may not have heard about; they'd send the rolls up to the finishing room, and they'd put them on a machine, and this blade would come around and cut them, and they'd pile them up, big pile of sheets of paper. And then they'd move that over to a different place and do a certain amount of inspection, and then they'd wrap it and send it up in square sheets instead of in a roll. A lot of the paper was sold in rolls, but this was sold in sheets. And that's what the finishing department did. It evidently was necessary. I did repair some of the stuff up there, but you could take a pile of paper about a foot high and shove them on the cutter and shove them through there. And then a big deal would come down and clamp them, and then you pushed the button and the knife would cut. And it was just a perfect cut-down. A pile of paper that thick, you know, you couldn't do that with an ax. They cut it with no effort whatsoever.

Who kept that sharp? Was that one of your jobs?

I think the blades were sent someplace. I don't know where they were sent, but you replaced the blades every once in a while. The blades didn't come straight; it was a long blade—just say it's eight or 10 feet long. And the blade when it came down, it wouldn't come straight down; it would come down a little bit like that. And when it let up, the edge of that pile of paper, you know, it was just smooth as glass. And then they'd turn it—the table had little ball bearings stuck all over, but the balls were loose. And they had pressure underneath it. And when you turned the paper it would slide over that, and haul the paper so it would just slide around. It weighed probably several hundred pounds, and it would just slide like nothing.

They were so clever.

Yeah. They did a lot of things that it's real hard to figure out who come up with the idea, but somebody come up with the idea. They would move big rolls that way. They'd lay down something, and they'd get air underneath it. Air is a very good deal. The air you know, it may only have 30 pounds' pressure, but by the time you get it under a big 30 pounds per square inch you can lift a lot of weight with it.

You must have had the chance to be a supervisor. How come you decided not?

Well, I don't know if I ever had a chance to be. I was a part-time foreman a time or two, but it was just for a short time when somebody was on vacation, but no, I didn't—I can't say that I had. At Hawley Pulp & Paper Company I was a floor man, which is the same as foreman. I had an opportunity to be a lead man, but that was not a foreman. A lead man would have a crew of a few people, but the foreman was a company man. Let's put it that way. It's the difference between being a union man and a company man. And I never was a company man. I never tried. It's not something I was too interested in. And maybe they weren't too interested in me having it.

Sometimes when you have a person in the right job, you don't want them to be promoted.

It's hard to say. I never thought too much about it. One man down the mill went in the office and told them that he'd been working in the mill for years. He says, "If I don't get this next foreman's job, I'm quitting." And they said, "Go ahead." And he did.

So you didn't want to press your luck?

No. I wasn't that interested in being a foreman. That's a kind of thankless job in a way. It's a totally different deal. Foremen got chewed out a lot more than the workers did because there was always somebody over them. If something didn't turn out, they were the ones that got it.

Did you ever meet Mr. Zellerbach at one of these recognition things?

No. I think I've seen him. They had people come there maybe once a year or once in a while. But they'd go through in a tour that would have so many of the office people take them through, and they were kind of protected, I guess maybe trying to save them from a punch in the nose or something. I don't know. I never had anything against them. I know we went through a tremendous cleanup job before they went through. That I know.

I bet you did.

Yeah. Even had to go down the basement and clean up. Because they would know when they were coming, and so they cleaned up.

I found pictures of hands, online, missing something from accidents in paper mills.

We didn't have too much of that down there. I mean, see, I've got all my hands here. And I worked on all those machines. But no, I don't know of anybody who had anything cut off. They could have. I mean, it was so simple to have done. But mostly the people got hurt either in their back or something like that.

So when you retired, did you miss it?

Uh no. Let's say when I retired, I almost ran away. I started building things, and I was so busy building that I didn't have any time. I have a garage that isn't a garage; it's a mess. But at one time I had an awful lot of things I was building out there. They have a door press, a panel press. I had to build one. It was 20 feet long and about six feet high. They'd put in the bottom part of a door panel, and then they'd put each section in, then they'd put the top. Then they'd step on the deal, and cylinders would come down and squeeze it. Well, they'd glue them as they did it, and then squeeze them together. And then they'd staple them. Then they'd go on to the next one. And I built one of those. And I only built one of those, but multiple drill presses, I built two, three, I think four of those.

Yeah, I enjoyed that. I really did, because nobody bothered you. You'd go ahead and do what you have to do, and if it turns out, fine. If it doesn't, then you change what you have to. And I have like a drill press and a multi-cutter, there's one in Seattle, there's one in Portland, there's one in San Francisco, and one in Los Angeles.

That you made?

Mm-hmm. They'd come up here, see Harris Garage Doors, it was a big place. And they'd see them there, so they wanted them. So I built them.

So that was your second career.

That was my second career.

Or your third or however many. Tell us what happened when you got drafted.

I got drafted, sent up to Ft. Lewis, Washington. All of a sudden you're not your own boss any more. You're in a big room with a whole bunch of bunks. And I went through there, and then right away they put me on a train; the first time I'd ever been on a train. They went down to Fresno, California, and the first thing they said there, "Don't take any oranges." There was orange fields all around, orange groves, and they said, "Don't take any oranges." So, all right. They went through basic training, which is a horrible thing. Then they sent me to Amarillo, Texas, and I went to airplane and engine mechanic school. Then from there I went up to Milwaukee, Wisconsin, was there about a week, and they sent me down to Kansas City, Missouri, where they had airplane engine mechanic school again. And then they sent me to St. Joe, where they had airplane inspection school. And then they sent me to Nashville, Tennessee, and they had some sort of a school there; I don't remember what it was. From there I went down to Miami Beach and waited for the boat. No, I didn't wait for the boat; I waited for the airplane. And then I went through Bermuda to the Azores and England and went through North Africa. I was in Iran for a few hours. And I ended up in Karachi, which was India at the time; it isn't any more. And went down to Kermatola, India, and I stayed there for a year and half, I guess. And we flew trips from India over to China, hauling gasoline, bombs, whatever they wanted over there. It took about five hours to get there. Then you'd refuel, you'd dump off all the gasoline you could, and then you'd put some in our tanks and come back again. And then you'd work on the plane. Then somebody would go back again. It was just a constant.

So you were a mechanic on the plane?

Yeah, I started out as an aerial engineer, which is a mechanic on a plane. And at the last I was called a crew chief, which was one step above aerial mechanic, aerial engineer. And it was interesting. There again, you had control of everything mechanical on the plane. In other words, the pilot had control of the controls, and that was it. If anything went wrong, it was you. I enjoyed it. It was good. It was interesting. First of all, you've got to remember I was born and raised on a farm. We had 380 acres in the middle of North Dakota, and the nearest neighbor was half a mile away, and you maybe saw him once a month, or her. And so I wasn't too familiar with people or associating with people. Then all of a sudden I'm stuck on this. So it was different. I learned a lot.

You must have come out here in car— did your family bring you out…?

Yeah, all my sisters are here. The only ones left was my mother and two younger brothers. I was 15. My mother decided to come out here. She never told me anything, but it dawned on me later on that she didn't raise turkeys that year. Normally she raised at least 100 turkeys. We didn't plant potatoes. Normally we'd plant a big patch of potatoes. A lot of things we didn't do, but at 15, you know, you've got a lot of other things on your mind. And then all of a sudden she said she was going, and we had a sale and sold most everything… We had a ton-and-a-half Model A truck with a grain bed in it, a grain box on it. And 1928, I think, or '29 and I assumed we were going out there. And then one day she says, would you go up and see if Howard will drive out with us? And I said, well, why is that? I was kind of stupid; I was 15 years old. And so I said okay. And I walked up. They was about two miles away, so I walked up there. And he wasn't interested. So I walked back and I said, no, I said, he's not going. And she says okay. So the next day we got in the truck, we loaded everything we could in the truck, put a canvas over it. Two kids sat on the chest in the back, and my mother and I sat in the front. I drove out here. I had never seen a roadmap or anything like that. I learned a lot in a short time.

I think a lot of people didn't think we were going to make it. I had my doubts, and I—I think we took a wrong turn someplace. When we were coming back we ended up way up on top of the mountains. And there was a road just wide enough for the truck. And I thought, now what? Meet somebody, what am I going to do? You could look down over the side there, and way down there was a little toy train chugging along. Of course, then I was from the flatlands. And we finally got down. And then I remember we come down from Spokane, and the roads were gravel there. And it was driving out of Spokane, and the truck, of course, the springs didn't give much on it. You know what a washboard road is? Going down, and all of a sudden you're sliding sideways. That was one of the times that kind of scared me a little bit, because you can't do anything. Driving slow would shake you all to pieces; drive fast, you start sliding. So you don't know what to do anyway.

So did you come to West Linn?

No, I went through Oregon City. Gudrun lived in Oregon City. Went through and we went down to Salem, where Amanda was. And when I came to start out here, I asked somebody on the way, I said, how do I get through Portland? You know, I was in Washington. They say, well, you go down this street, and they say, when you come to—I think it's 82nd—make a left turn, and that way you miss the city. And it did. I missed the city. Went straight on and ended up in Oregon City.

Good directions.

Yeah, that saved me. If I'd gone into Portland, I would have had a little rough time.

So where did you go to high school?

I didn't go to high school. High school was 23 miles away, and I had chores to do. No, I got through eighth grade, and that's it.

And where was your house after you moved here?

It was down in Salem. And then when I got to be 18 I got a job at Hawley Pulp & Paper, and I come and stayed with my sister in Oregon City. And then I left for the service after a year and a half. Then my mother was moved from Salem to Oregon City later on, while I was in the service.

Melody [videographer]: And he had a family of 10 children

Is that right?

Yeah. And my dad passed away when I was seven, and my mother had 10 children. And at the same time we were well to do because we had $3,000 in the bank, and then the banks closed. She lost everything. And then come along the drought, and then come along the grasshoppers, and things like that. How she ever managed, I don't know.

I think a lot of people came to Oregon during the Depression.

Yeah, but you've got to realize she stayed there for eight years after he passed away, so she struggled through. She got all her kids out here and did what she could. I would have gone crazy. I mean, I couldn't have done that. That's an awful lot of things.

Well, you don't know what you can do until you have to do it.

Roy Paradis

Interview March 2, 2006

So if you could tell us your name and your age?

Well, I'm Roy Paradis, and I'm 81 years old. Just celebrated my 81st birthday at the end of last year. So we're headed towards 82.

Your first job wasn't at the mill? What led you to want to go to the mill from whatever you were doing before?

Well, my first job wasn't at the mill. My family moved out from Minnesota in late 1941, and my first job was spading, hand-spading for a florist…That was my first job. And I thought that wasn't for me. So I decided I'd look for a job, and Crown Zellerbach in West Linn was hiring. Stopped in there, and of course I was only 17, but they were hiring. So I got a job on the paper machines. I stayed there for not too long because the wages there at that time were 78 cents an hour. So as we got a little more knowledgeable of the area…of course the war was on, and shipyards were going full strength—we heard that at the shipyards you could make 90 cents an hour. So my brother and I were both working at the mill so we decided well, why don't we get a job at the shipyards? So we did. Got a big raise; 90 cents an hour.

So we quit the mill and worked in the shipyards until I was drafted in World War II and all that transpired. And when we came home, I didn't go back to the mill right away. I went around the country. We visited friends. We both had a lot of buddies. He was in the paratroopers in the Pacific Theater, and I was in the army and went to the European Theater. And so we had buddies all over the United States. So we decided, why don't we just start traveling, and work and make enough money to travel some more? And so that's what we planned on doing. We got as far as Klamath Falls, and the people worked for—they were farmers— didn't want to let us go, so we stayed until they didn't need us any more. And then we ended up in the Pendleton area because there was work there. So we never got out of Oregon. So anyway, I ended up eventually back in the mill in 1951.

So what was the year that you first went to work at the mill?

Nineteen forty-two was the first year, 1942. That's when my time started, but as you see, it didn't last. But it was in 1951 I went to work and worked in various departments. I worked in what was called the beater department; that was recycling. Most of those people were in the basement under the paper machines. Whenever the paper machines had waste paper, that was down in the basement, and then that paper was all reprocessed, because it wasn't thrown away. And so I was in that department for a while. And I also worked in what they called the supercalender department. And my original supervisor, by the name of Clarence Schearer, saw me in the supercalender department, and he remembered me. And he wanted to know if I didn't want to work for him. So I said yeah, I wouldn't object. So he arranged it so that I was transferred back to the paper machines. And that was my real start on the paper machines. I never left them.

Which machines did you work on?

At that time there were 10 paper machines in operation, and over the years I worked on all 10 of them, starting with No. 1, all the way through to No. 10. The oldest machine, No. 8, made exclusively core stock, the cores that the paper is originally wound on. Of course, it was all made there at the mill. And No. 8 the oldest and of course the slowest machine I think its normal speed, or the only speed, was 48 feet a minute, which was very, very slow. At that time, probably the fastest machine in West Linn was oh, probably 10, 11, 1,200 feet a minute at that time. As time went by, why, the speed increased.

That's a big difference—

Of course, the 48 feet a minute, you can see it go, like this. And when you get a little faster, of course, you can't follow it quite that easily.

We talked with Olaf last week, and he said that No. 8 was so slow that they tried lots of experiments on it, because it was forgiving.

Yeah, yeah, yeah, yeah.

Do you remember making a lot of different kinds on No. 8?

No, I didn't. I didn't experience anything but the core paper on No. 8. The machine right alongside of it, which was No. 7, made the oily paper anyway

The pineapple paper?

The pineapple paper, yes, yes. And that was ooh, that was messy. You just wore one set of clothes when you worked on No. 7, and you didn't wear them anyplace else.

Rosie Schultze's daughter gets a mill tour in 1957.

George gave me a sample of the pineapple paper—it was a second because it didn't get those aluminum stripes on it, so he got to take it home.

Oh, oh, I see. Yeah, yeah, the planting marks.

He said if you ran it, it got so dirty that anything you ran after it just got wrecked.

Oh, yes, oh, yes. That was messy. That was messy stuff.

So your first job was in the beater room?

When I went back in 1951, that was one of my assignments. You know, I didn't have a steady job, so I moved around wherever they assigned me. And yeah, I worked in the beater, which is the mixer department, actually, but in those days the machines that reprocessed the waste paper was called beaters. Nowadays, of course, they have modern machinery, and the beaters are no longer there. I mean, they've been gone for a long time. They have re-pulpers, and they're all big machines.

So what was the first job in 1942, when you were 17?

Paper machine. Yes, on No. 1.

And what were they making on No. 1?

No. 1 at that time was exclusively directory paper, telephone directory. We made both white and yellow directory paper. And an eight-hour shift, if you were unlucky, you had to haul out eight rolls. If you were lucky, you only hauled out seven. So just about an hour for one probably about 38-inch diameter roll. And white and yellow directory paper.

Didn't you have to take the machine down to change color? How much time was wasted?

Not very much. You'd change color on No. 2, for instance. No. 2 made a lot of different grades, and one of the grades was for National Cash Register. And if you remember way back when, they used to have many, many colors: white, blue, orange, red, green. And there they'd make color changes quite rapidly. The stock prep department, of course, had to change the color of the stock, and they knew exactly how much of this color to put in and that color to put in. They did a pretty good job of changing color. And colors were changed quite rapidly. There wasn't a lot of waste.

How did they figure out when they could make more money by changing a machine to a new product?

We weren't involved in that part of it hardly at all, because that was all taken care of, more likely, at headquarters, you know.

But that would take a machine down, wouldn't it, for a while?

Not long. Not long. It would just be a grade change, what we called a grade change, when they decided, and of course the machines were scheduled almost exclusively after the order was taken. So they knew what we needed and how much we needed. And so then they'd schedule a machine to make that product.

I think you said you worked there three times, is that right?

Yes, yes, yes.

So have we gotten to the third time yet?

Not yet. No.

Can you talk about leaving the second time, and what happened?

The time from 1951 until 1964, when I left the second time, we had a safety committee and I was really involved on the safety committee. We had regular meetings, and in 1964 the coast organization had a yearly conference, a safety conference. In my second stint, from 1951 to 1964, I worked on all the machines and I was quite involved in the safety committee. The west coast manufacturers' association had a safety conference once a year, somewhere on the west coast. In 1964 it was held in Portland. I was hourly, of course, at that time, and the conference had a salaried co-chair and an hourly co-chair and I was asked to co-chair the conference that year, which I did. I've got a gavel to prove it. But anyway, during 1964, since I was older in age than many of my co-workers at my level and of course, all promotions were by seniority I could see that I wasn't going to get too far up the line because of my physical age, because a lot of people that were younger than I am would have those higher-paying jobs.

So I decided if I was going to get ahead, maybe I'd better look elsewhere, which is what I did. And another company, another papermaking company, was starting a new mill in northern California, and I decided that maybe that's where I needed to go. So I submitted my résumé, and through a series of interviews, why, I got the job that I applied for. So I terminated West Linn and went to work for this other company, and started their new machine, or helped start it with a crew. And within about a year, why, I started getting, or I should say my wife started getting, calls from West Linn, wondering how I was doing and how we liked northern California, and how she liked California, and how did the kids like school, and all this, that and the other. And of course, I was always at work, because starting a new machine, a new mill, was pretty involved. A lot of time. It ended up I finally talked to the gentleman that was calling, and they wondered if I would come back to the West Linn mill on a supervisory status. So we discussed that and had family meetings and this and that and the other, and I ended up coming back in a salaried position. So that was my third stint.

I started January 1st, 1966, until I retired December 31st, 1983. And in my salaried stint, why, they promised me six months of training for the eventual job that I was to have. So I said okay, I can do that. Well, in three months, why, the man whose place I was supposed to take decided he was going to retire. So he retired, so in three months I had the job, which was a shift supervisor on No. 5 and 6, the coated paper machines on No. 5 and 6. So we did that for a number of years, until my wife, you know, kind of hated shift work. She thought, "Boy, it sure would be nice to have a day job." Well, I finally got my day job, you know, and I was promoted to the day supervisor. And after a short time she was thinking, "It sure would be nice to have him on shift work again."

What was it about the shift work that ended up being appealing?

Well, my day job involved so many more hours. When I was on shift, I always had a relief. I worked my shift, and I had a relief. When I got to be on day shift, I didn't have a relief. I was the day supervisor, and the day supervisor sometimes worked half the night. So I put in a lot of hours

That's management, right?

A lot of hours, and that's what mattered.

So how was it supervising the guys that you used to work with?

I considered that quite a while before I accepted this job, because I knew that I'd be supervising people who were well ahead of me when I was working there, because I worked in the same department. I mean, I left No. 5 paper machine, and I came back as a supervisor on No. 5 paper machine. And I left as the bottom guy on the ladder, you know. On a paper machine there's normally five people, five employees. And I was number five. So then came back, and here I was supervising two machines. It went pretty good. I mean, I didn't have any problems. Always a few because of labor and management; there are always a few problems. But we got along fine, yeah.

Nobody grieved you and said you were a terrible boss or anything like that?

There were probably some. While I was in the mill in California, I was heavily involved in labor there—in fact, I was chairman of the standing committee when I left, and it happened to be a different one that was here. So when I came back, there were some union people that probably didn't like me because I had worked for a different union when I was in California. So that didn't set real good with some of them, I suppose. But it didn't create a lot of problems, you know.

So you said that you were organizing the conference in '64—what else do you remember about that year?

I left before the huge flood. The flood was in the fall, and I left here in October of '64 so we could get down to California before school started. So I left before the big flood.

That's a really good thing.

Yes it was, yeah. They had quite a time. I couldn't believe, we couldn't believe, pictures that we saw, you know, of the area, that flood. We just couldn't believe it.

George was on the Flying Squad. You worked with the Flying Squad?

Well, they worked on emergency breakdown, really, when he was on the Flying Squad. Of course, during that time period I was hourly, so I really wasn't involved in that. When I came back as a supervisor—I don't know if George was a supervisor at that time or not, but shortly he was, because he was the supervisor of the maintenance group in my end of the mill. So if we had a breakdown, it was George and his crew that did the repair work for us. So I was involved with George then, you know, quite closely. But back when I was hourly, of course, those things didn't enter into my thinking. It wasn't one of my worries.

You must have seen some high water down there, though. Do you remember some other years when it was bad?

Yes, yes, I did see some high water. I wasn't involved in any high-water shutdowns. I was involved in shutdowns because of strikes, and that's part of the history. It's not the good part; it's the part that we probably all would rather forget, but it is part of the history. Again, as an hourly person, I was never involved in a strike. I mean I wasn't here. As a supervisor I was, especially the big, bad strike; I mean the one that lasted so long, six, seven months, I believe.

What year?

Strike of '78, '79. At that time there were only three machines left running—5, 6 and 9—and they were all on coated paper. And so we supervisors started those machines, and we ran those machines during that strike period, or part of the strike period, because the strike lasted so long the company was worried that they would lose all their customers. That was the main reason we started them. And as I said, that's a time that we would just as soon forget, but it is part of the history, you know, because we worked awfully hard to get the machines running and I suppose maybe did save some of the company's customers, because the printers had to have paper. And if they buy paper from someone else for three, four, five months, they're not going to want to go back. So that's what the corporation was worried about. It took us quite a while to really get into production, because help came from all over. I mean, supervisory people from St. Louis, New York, Louisiana, Wisconsin, all over they were all Zellerbach employees, all salaried people. You know, some from offices; they had never maybe seen a paper machine, but certainly never worked on one. We had to train those people, and those of us that supposedly knew what we were doing, you know, it was quite a job. And we succeeded.

And when the strike was finally settled, the decision was made not to shut the machines down, just to let the hourly people come in and take over right from there. That was real touchy because here these people had been out of work, you know, for six, seven months at least—not out of work, but had been away from the mill—and here they were coming back and relieving, taking that job from a salaried person. They probably didn't know the person. They knew the West Linn people, and of course some of those were still there. I happened to be there at the time. And there were a few words exchanged during that time period, and we expected that. But the machines kept running. That was the good part. And we were glad to see them back. I was very glad to see them back. I was getting tired of working 12 hours a day 10 days in a row, because that's the schedule we were on.

That was only six or eight years before Crown Z finally sold. So you retired a couple of years before they were taken over.

Yes, I retired from Crown Zellerbach, and James River then took over from Crown Zellerbach. And then James River sold to Simpson Lee. Simpson Lee finally shut the mill down. And then just the one owner, this current one. And I guess they're doing all right. I mean, I talk to somebody once in a while that works down there, and they're apparently doing all right. That's good. You know, I'm glad somebody's working.

Did the strike cause any particularly exciting moments in your life?

Well, yes, there probably were a few. Of course, before the machines were running, all the West Linn supervisors were still on duty because we had to protect the mill from fire or anything—I mean there had to be somebody in the mill. And for some unknown reason to us, light bulbs burned out just by the hundreds. We all thought it was because of the power usage. The mill, of course, wasn't using much power, but the lights were burning out, and that was our main duty: replace light bulbs. And walking down through the basements, of course, no activity, and nutria and rats invaded the place. And walking through the basement, why, here a big nutria would come running across, or rats, or all kinds of critters down there. So that was part of it, and we changed light bulbs I mean, we carried a bag, you know, I mean a box of light bulbs, and every night, I mean every day, everybody was changing light bulbs, changing light bulbs. And we just figured because of the power usage is why they burned out.

That's bizarre.

And of course, you get too many lights burned out, and it would get really dark, you know, really dark down there. Of course, down under No. 7 and 8, which was really bad, that was the first part of the mill. I mean that was the first mill built, and the foundations there was just, you know, a bunch of rocks, with mortar, and that was the foundation. And you get down under there, and of course there's water flowing around here and there. That was a scary place.

You never met the ghost?

I never met the ghost, no. But there were enough sounds down there that we could probably blame it on the ghost.

Did you see people lose lives during the time you worked at the mill?

I was never close to a bad accident. Of course, I knew of several people who lost their life down there, one of them on one of the machines that I worked on, No. 5. I wasn't there at the time, but he got caught in one of those what we called an ingoing nip. I mean a roll with felt or something around it, and of course, that's an ingoing nip. And he somehow got an arm or something caught and went into that system, and of course he ended up in the basement, of course, and he was killed. But I was never involved in a bad accident. Missed some close ones. I remember one time on No. 9 when I was hourly, the web had broken, and we were rethreading the machine. And the final finishing on the machine is what we call a calender stack, a series of steel rolls stacked one on top of another. And the paper goes in and out, in and out, and that puts a finish on it. And on No. 9, which was the biggest machine down there. I was on a footboard, and you had to spear the sheet, you had to cut the sheet, in a narrow tail to start it through the stack. Then as it started through, then you walked across and widened the sheet out.

And at times a lump of paper, or several lumps of paper, would go through and of course would bounce those steel rolls; they would jump. And this one time, somehow the paper got wrapped around one roll, and then it broke loose, and it just really bounced that stack real hard and broke the journal on one roll; one roll was broken on the end. Fortunately it didn't fall out, and here I was standing on that footboard up there. And so of course that machine went down. And I was real impressed with the supervisor at that time. I had to go notify him because he wasn't there; you know, the supervisor at that time in Mill C supervised six or seven machines. So my machine tender or whoever told me to go tell the supervisor, which I did. And I told him what happened, and his first remark was, did anybody get hurt? That impressed me. I mean, he didn't worry about the machine. His first remarks were, did anybody get hurt? And of course, nobody did, so we were fortunate.

So, I don't understand what a footboard is. Is that like scaffolding or a plank, or

It's a plank. When the web comes out of the last dryer —the last dryer is up there quite high—and that's where the web is going to be leaving there into the stack, and it's going to go in way up there at the top of the calender stack. So you have to be up high. At that time we used what we called a spear, a real, real sharp needle-point piece of steel, usually that long or so. And that was used to cut the web across. Now, of course, they don't do that any more. They have a web cutter that's right in the dryers, and you don't need to stand up there any more. But things changed. I mean, they do things differently: more modern, easier, more efficient.

Did you ever spend any time down in the pulp area, in the sawmill area of the mill?

Just visiting.

What was the work like down there?

Oh, that was work. That was work. But when I was first down there, of course, in the grinder room, where they actually ground the softwood—actually, putting those pieces of wood, feeding those grinders, I never did that. Never got assigned that, and I was thankful. That was quite a job, yeah. Even the sawmill you know, I visited the sawmill also, but I never worked there, luckily.

After you became a supervisor, what was it like to be a manager?

Well, it was certainly different, and it took me a while to get sort of used to it. I don't think I probably ever did get really used to it, because some managers are very demanding and hard to please and unable to please, sometimes. So that part of it was bad. But usually we did pretty good at the West Linn mill. As a shift supervisor and also as a day supervisor we had one mill manager, the plant manager, who threw away a lot of our paper, and we thought unnecessarily. But that was his way. I mean, he'd always come in and he'd look at the test reports. Of course, every roll of paper was tested. Every reel of paper was tested by the lab people. And if it didn't quite meet the criteria that was called for, some of that paper we let go, and some of it we didn't. And say if we had a session where there were three or four reels that weren't up to standard, and we rejected those and re-pulped them, and then we'd get the problem fixed and then we were making what we thought was good paper. And maybe later in the day it'd get into the same situation again, many times that manager would come in and he'd throw away all that paper. And of course that looked bad for us so that created a problem. But with our immediate supervisors, I got along fine. I didn't have any problems.

I heard they had some pretty wild parties.

I didn't attend too many of those parties. A lot of those wild parties, those Christmas parties that you've heard talked about—we went to one; we aren't partygoers. That was the end of that one. We just decided that wasn't for us. Yeah, I heard a lot of stories, and you heard Rosie talk about the Horseshoe Club, which was the supervisory club. I went to several of those picnics; those were really picnics. And some of those were a little—what do you want to call them? I didn't care for it.

Was it all men?

It was all men until we finally got a woman supervisor.

What year was that?

I don't remember, and I didn't go to that picnic. But they wondered then, you know, what was going to happen to the Horseshoe Club, because they finally— in later years there were more than one—but when we had the first female supervisor, of course she was invited to the club. And as I said, I didn't go to that because I didn't want to be involved. So I really don't know what happened then.

I was just going to ask you about women's role at the mill and how it gradually started to change from office to working on the machines…

Yeah. I had one little gal when I was a shift supervisor. See, my crew—I supervised 10 people, I mean five on each machine. And she was on the extra board. And so at that time I didn't want a female on my crew. I mean, I just didn't think they could do the job, because it's pretty heavy work. It gets involved; they have to lift quite a bit. And I just didn't think a lady—a girl, a lady, whatever, a female— could do the job. And about the first one I got was a little girl who weighed about 95 pounds, I think, maybe 100, but not over 100 pounds. And I told her later, in later years, she made a believer out of me. She could do the job. And she reminded me of that more than once. More than once. I'd have her for maybe one shift, you know, or maybe one week, and then of course the regular man would be back. And then later, maybe I'd have her again and maybe I wouldn't, so she didn't work for me steadily.

And at one time I had a former Hell's Angel, the Hell's Angels from Los Angeles, California, and he had one earring. And I asked him one time, I said, why do you wear just one earring. And he says, "Roy, you don't want to know; you don't want to know." Apparently he had qualified to wear one earring, and he wouldn't tell me. And anyway, he was big, and so he got to work with this small female, and of course he was quite outspoken. I mean, if he didn't like somebody, you knew about it and the person knew about it. I mean, he didn't pull any bones. But anyway he told me he would work with her—Mary was her name— and he said he didn't mind working with Mary at all. She pulled her end of the job. He didn't have to do her work for her. Mary, of course, eventually ended up working in the production department, I mean the ordering; she made out orders and she ended up doing something else. And I talked to her often, and she reminded me quite a few times of how I didn't want her working for me. And she knew it. But as I said, she made a believer out of me. And so I had more ladies. Of course, then we also had very few blacks. But finally we were getting a few blacks.

So this is in your third stint…

Yes, this was as a supervisor. And I don't remember having a black female working for me, but in the supercalender department, the department that got the paper from us to put the final finish on the product, the supervisor there of course I worked closely with. And he had a female worker, black female, and she wasn't too aggressive. He had to—everybody had to keep after her to do her job. And one time the supervisor told her to do something, and I don't know what it was, and she came right out and told him, that wasn't my job. I don't have to do that; not my job. And so he had quite a time with her.

Was it a union issue?

Oh yes. I mean, they were all union.

Well, wasn't there a department where the paper toweling went, or packaging or shipping wasn't that where most of the women worked in the early days?

Probably, yes, up in the converting department. That was back in my hourly days, early. They closed that down early. I don't remember what year they closed that, but it just you know, unprofitable. And so they shut it down.

When you worked there, pulp was being made on the spot, and then at some point it transitioned to being barged in. Do you remember that?

Well, when they quit making kraft and what we called sulfite at the West Linn mill, and they started buying it, the digesters were shut down—the digesters where they cooked the pulp to make it stronger. You know, ground wood, for instance, in newsprint, it's almost 100 percent ground wood. In fact, in my day it was 100 percent ground wood; it may not be nowadays because they've changed it some. They're using recycled pulp now in newsprint. But when I was first down there, the West Linn mill made their own sulfite, which was a cooked pulp, and it made the sheet stronger.

But when the digesters shut down and the corporation started buying all of their kraft from other places, most of it came from Canada. It was barged in in dry sheets, and then that was re-pulped, see. So it was all bought in—it come in huge bales. And it was re-pulped and added. In the coated paper—what we called coated papers, you know, the magazine stock, "Time", "LIFE", "Newsweek", all of those we sold to all those papers; I mean all those magazines. After you make the sheet of paper and then you put the clay on it, and then when they calender it, they iron it. People call it ironing and it puts a fine glossy finish on it, and from there it goes be cut up in the right sized rolls, whatever the order calls for, and then goes to the consumer.

So one bunch of people just worked a forklift.

Yeah.

Is that shipping, or is it…?

Yeah, most of that's in shipping. When the rolls are wrapped and somewhat sealed, ready for shipping, then the forklift driver, he'll stack it out there on the dock. They used to haul it out in barges; then they'd load it into barges, and out it would go. In later years most of it went out in trucks. Again, modern.

Do you have any stories about about the locks or the PGE plant, or fish ladder?

I don't remember anything like that. Of course, George, you know, he was all over the mill. I mean, basically, I mean in his earlier years, and then even when he was a supervisor of the maintenance crew at my end of the mill. They went to other parts of the mill also. I mean, they weren't exclusively in Mill D. When another machine was down, why, if they needed extra maintenance people, why, that's where they'd go.

Could you describe what it was like when the locks were running all the time? All the logs didn't go into the mill…

No, no. Logs went up yeah, rafts. Raft after raft after raft of logs went up to the mill. Or up and down; I think some of them went up, some went down, I think. But early, you know, all the paper went out on barges. I mean, it all went out on barges. And everything that came in came in on a barge. Like when the digesters were still running, barges of yellow sulfur would come in, and then they'd unload it, and a clamshell would haul it, take it out of the barge and put it in a bin, and that's what fired the digesters.

I've seen pictures on the Web of people that lost fingers in mills.

I didn't lose fingers until I retired. Then I lost those three. But anyway. I was really involved with the safety committee.

When did they start caring about it?

Oh, that I couldn't tell you. I don't know, because I know there came a time when we well, we always had to wear steel-toed shoes. Maybe not the first time I was down there, in '42, no, but in '51 I'm sure I had to wear steel-toed shoes then. And then, of course, next was hearing. You know, nobody wore any sort of anything in their ears to protect their hearing, and that finally got started. And of course, when I was on the supervisory level, why, that was one of the things that we had to look for all the time, was to make sure that people had steel-toed shoes, had hearing protection.

And when the long hair style came in for everybody—at that time it was just men because we didn't have any ladies working then—and these young guys, you know, would let their hair grow, let their hair grow. I mentioned ingoing nips before and, especially around paper, there's a lot of static. And they'd get under the web, under that sheet of paper going, and you know, the static would just pull their hair right up. I mean, hair like her hair, for instance, would be standing straight up, you know. And so it was dangerous. There were a few that did lose a lock of hair. So we finally got to the point where we insisted on hairnets. That didn't go over very good. That was a big problem. That was a problem because, you know, these guys just didn't want to wear a hairnet. But it was either get their hair cut or wear a hairnet.

I had this one employee; he was the son of an office worker, a salaried office worker. And a good worker, real good worker. But he had long, long hair. And I kept after him. You know, you've got to get a haircut; you've got to get a haircut.

And finally I told him on the last day of day shift or swing shift or whatever, I said, okay, if you don't get your hair cut on your days off, when you come in the next time I'm going to send you home. I said, you have to get your hair cut. Well, he came in, you know, and started a new week, and I almost sent him home because I was sure that he hadn't got his hair cut. Well, he insisted his girlfriend cut some off his hair. Well, how much I don't know. And softie me, I let him work. But he finally did get his hair cut so he didn't have to wear a hairnet, because he absolutely didn't want to wear a hairnet.

It was long enough to be unsafe, but it wasn't long enough to put in a ponytail?

Well, they weren't wearing it in ponytails then; it was just long hair. That changed later. Maybe the ponytail came in place of a hairnet. I don't know.

Talking to Olaf about his job, he said those last seven years were the only time he really liked his job. So of all your experiences at the mill, did you have favorite times?

What I liked or didn't like? Right, getting back to Ole. Ole, his last seven years, he was a winder specialist, and if I had a problem with one of the winders in my department, you know, call Ole. Ole would fix it. He was good at his job. And that's why he was really his own boss. I mean, the boss would ask him, what are you going to do today? Well, he'd tell him. But he was very good at analyzing. You know, if something wasn't working just right maybe there wasn't anything broken; it just maybe didn't work quite right and he'd come over and he'd watch it run for a while, and he'd analyze it. He was just a good maintenance man.

As far as my time, after I got on day shift, after I was the day supervisor, I called on quite a few customers, especially in Los Angeles, usually when they were having a problem. Pacific Press, for instance, was a big customer of ours. And if they were having a problem, you know, experiencing too many web breaks, or it wasn't printing like it should, a group of us from the mill would go down. And I went as part of that group several times, quite a few times. Always somebody from the technical department and somebody from production, which was me, and sometimes somebody from the main office. Sometimes I'd end up down there by myself; I mean it would just be me. And hopefully I satisfied the customers while I was there, you know, gave them some answers. If I could see that they were doing something that maybe wasn't what it should be, it affected the paper the wrong way, maybe I could help them. That gave me some satisfaction.

And as far as the mill is concerned, I don't know; I had some good times down there. The mill was good for me overall. I mean, a lot more good times than there were bad. There were some bad, yeah, there were some. Especially when I was on straight days and I'd go in at seven o'clock in the morning, and lots and lots of times I didn't get home till seven, eight, nine o'clock at night. Ask her over there.

That's hard on the kids too.

Yes. Then, of course, the kids were gone. See, our youngest one is 47 now, so they were all gone. It was just she and I. She's sitting here waiting for me, you know, and that gets old. So when I retired, I'm going to tell you about that. In '83, I always wanted—I mean I thought that I was going to be a woodworker, so I had just started that, and so she gave me pretty strict orders. Dinner at that time was at six o'clock, and I'd better be in the house at six o'clock, ready to eat.

You were out working in your shop?

Yeah, I was out there working, out there doing woodworking. And I—right now it's slacked way off, but in my retirement years the first 10 years were really enjoyable. I mean, I really enjoyed what I did, and retirement was good for me.

So how old were you when you retired?

I was 58 when I retired. They offered early retirement. This was when corporations were downsizing and the first group I believe they had to be 60. And at that time we had already planned I was going to retire at 62. Of course, most everyone was retiring at 62 then. And we had planned that that's when I was going to retire, at 62. And then they came out with this early retirement, but you had to be 60, and of course I didn't qualify. They didn't get as many as they thought they would, apparently. So the next year, I think it was, they dropped the age down to 58, and I was 58 in mid-November. I retired January 1.

That's young!

Yeah. It's been enjoyable. I've, well, been retired since '83, so quite a while.

The West Linn Inn was there while you were there, wasn't it?

Yes, it was. When I first started, the West Linn Inn was there. Let's see when did they go out? It was there when I was working in '51, because when I started work in '51, I lived in Woodburn, or we lived in Woodburn. So I drove from Woodburn, and we always walked through the West Linn Inn on our way down. I don't remember when they closed the Inn, I remember when they did, but I don't remember what year. Yeah, that was quite a place.

Now, you say you walked through there every morning? What…to get coffee or…

Well, probably just to see who was there. I don't know. And a lot of people, you know, would come early, and they'd wait in the lobby. You know, they had a huge lobby with a lot of chairs, and a huge lobby. And a lot of people would wait there until it was time to go to work. Of course, being as I drove from Woodburn, I usually didn't have a lot of time because I timed it as close as I could. I mean, I didn't want to leave that early.

What car were you driving?

Oh, dear. Was that after the 1938 Chevrolet?

    Mrs. Paradis: It was a Ford.

Yes, a 1949 Ford. That's what it was.

And was the highway between Woodburn and West Linn paved?

Oh, yes. Oh, yes. I did get stopped once. I had a rider with me one time coming to work midnight shift, so it was 11 o'clock or so. All of a sudden here's this red light behind me, you know. And so I pulled over and stopped, and the policeman gets out, and you know, he shines his flashlight in the car and talks to me and asks me, you know, where I was going and where I was coming from. I finally asked him, you know, what was—why I was being stopped and he was so interested in my car. And there'd been a robbery back I don't know someplace in Salem or someplace, and I kind of fit the description, and so he stopped me.

So you were late that day.

I was probably a little bit late that day, yes.

How did you like working at Willamette Falls? Did you guys ever get to appreciate the fact that this incredible waterfall was out there?

I don't think I thought about that part of it or not. I mean, I don't think I did.

So it was just part of the business.

Yeah, just part of the business.

The water comes in, and the water goes out.

We were here. The mill is here, and we were doing what we were supposed to be doing. And I just didn't think of that part of it.

Okay, did any of your kids or anybody else in your family work at the mill?

Well, yes. He's over there shaking his head because I wouldn't let him work at the mill [son is observing]. And our second son, he was a little more determined, and he went to work for Crown Zellerbach, but in Camas. So he drove to Camas. But I wouldn't let him work in the mill because there was always controversy if a boss's son was working there. Some people were always saying that, you know, well, he's the boss's son; that's why he gets away with this, or that's why or that's why he got the job. Always had some reason. So because of that, I wouldn't let my sons work at the West Linn mill.

So were the people who had generations working there were all hourly then?

Not all supervisors felt the same way. Like this young man that I talked about previously with the long hair. His father was a prominent salaried person in the office. And there were other salaried supervisors that let their sons work there.

The Bonn family, for instance. The senior Bonn, of course, was hourly. The son was hourly, who became a supervisor, and he had two sons that worked there. In fact, I've got a picture of those four people in a company publication, where the senior, the son and his two sons worked at the mill.

Well, did you know the Freemans?

The Freeman I know of was paper machine hourly man. And I think I only knew one, and he was in the other end of the mill. I knew who he was and I knew the name, but I didn't know him personally. I think when I first became aware, I think there were 1,400 and some employees there. That's when all 10 machines were working. Fourteen hundred and some employees. Of course, we ended up with a lot fewer than that. I mean, as the machines kept shutting down and shutting down, and I don't remember how many we had when I retired. I don't know; probably 450, 500, probably.

Would you like to tour the mill today and see what's it's like, see how it's changed?

After I retired I went down several times. Was called one time when—I think it was when this company [West Linn Paper Company] took over. They wanted to make a certain grade of paper that we had made when I was there, and they couldn't find any records, and they wanted to talk to some of the people who had been working there when the grade was made. So they had called a number of us. There was half a dozen of us that went in and had a little session where they tried to pull information out of us, you know. Whether that helped them or not I don't know.

So you were the brain trust, huh? They thought you might remember the recipe?

That's what they were looking for, something like a recipe, yes. And you know, people in the stock prep area might have remembered. I knew generally what was in it, but I didn't know how much of this or how much of that. I knew what was in there, but the stock prep department had to come up with the right combination.

Olaf talked about the department where they cut the paper—they had the big cutters.

Oh, oh. Converting.

When you talk about the nips on the machines, I know what a break is, or a shear, but does a nip work like a scissors?

What's a nip? A nip is when a roll and either a fabric roll come together, or in what we called a calender stack, a series of steel rolls stacked one on top of the other. So wherever those two rolls come together, that's a nip. That's an ingoing if the paper's going in here, that's an ingoing nip. For instance, in a dryer section, where a felt—when the paper goes through the dryers, there's a felt holding the sheet of paper against that steam dryer. So every roll that felt is rolling on, and wherever that fabric touches a roll, that's a nip, because if you get in there it'll pull you in, see? When you're feeding something into a—what can compare to that?

A wringer washer.

Oh, a wringer washer. Okay. You got it. That's a nip. And…

It's not actually a cutting process?

Not that part of it, no. Usually, in a nip where a felt is concerned, it'll pull the person, the arm or whatever, in a stack or a series of steel rolls. You talk about it's not a cutting thing, but it'll cut a finger right off. I mean, it would pull the arm in; it would just cut that finger off. It would, you know, smash it, and that's what would happen there. So it would cut that way, but it's not like a scissors.

So it's really every point where there's an opportunity to be sucked in. A dangerous point.

Right. That's the danger point. That's a nip. And on a machine there are many, many, many nips. I mean, some places aren't accessible to the man unless he reaches, you know. If he was just walking by, there's no danger. But you don't want to reach in there, because it could grab you.

The other thing about the machines that we've heard is, well, it's hot. The dryers, the heat. It wouldn't be a very pleasant place to work in the summertime, for sure.

Right. Yeah, it's hot. Summertime and because the dryers create a lot of heat, a lot of heat. And especially where there's two machines side by side, like there were in Mill D, No. 5 and No. 6 side by side with an alleyway in between. But those dryers are hot, steam-heated, and summertime I'm sure it was 120, 140 degrees in there. I mean, it was hot.

And you had some fires, didn't you?

Once in a while, but in my days we didn't have any serious fires, no. Never experienced any serious fires. They had a fire in No. 9 in the electrical part of it. No. 9 was driven totally with electric motors, where all the other machines were driven with a line shaft. There's a big room where all the circuit breakers were and all that. So they did have a fire there one time. Totally electric.

And electricians were in charge of the electric, is that right?

Yeah, yeah, yeah, yeah.

Not just for union reasons.

Yeah. But all the other machines were driven with a line shaft, and in most cases the line shaft's in the basement, and then a belt goes from there to each section.

Are those the motors people talk about moving up out of the water, that drive the belt on the shaft?

No, the motors they're talking about are electric motors that drive pumps, because there's a lot of pumps for many different things. You know, the solution that comes out of the head box when the paper first starts to form, I mean it's about 99.6 percent water. There's very little pulp there. But the water starts to drain right away, and of course it ends up as a sheet of paper. So there's a lot of pumps, and not only for the water, but for the stock and for the there's pumps to create vacuum because there's a lot of vacuum being used on a machine. At the wet end there's many—every press has a huge vacuum pump, because that's how the water is basically taken out after it leaves the forming fabric. The forming fabric is drained, it drains, and…

And that's the mesh…?

Right, right.

We're done.

Ed Witherspoon

Interview March 14, 2006

Would you mind telling me what your name is and how old you are?

My name is Ed, and I'm 79 years old.

What year did you go to work at the mill? Did you just work there one time?

No. I worked there the first time in 1943, between my junior and senior years of high school. And then they kept me working on the weekends during my senior year, and after I graduated I worked there 'til I think it was sometime in October. And then I left for the service.

So you left for the service in

Nineteen forty-four.

Right at the end of the war. Did you go to Europe?

Mm-hmm.

When you came back, what happened? Did you go right back to the mill when you got discharged?

No, I didn't go right back to the mill. I never went back steady until I think it was about 1950. And then I was there from then until I retired.

What year did you retire?

I retired in 1984.

So how many total years was that?

That I retired? Well, they credited me with a little over 35 years' service altogether. That's as long as I could stand it.

So did you take early retirement like some of the guys did?

Yeah. I wasn't quite 58 when I retired, so that's a little over well, it'll be 22 years in May.

Some people took early because there was a strike happening or something.

There was a bunch of us went out just before the last big strike they had down there. I think we went out in May, and the strike went in in June. They were out a year.

Did you get severance packages?

They gave me some severance, yeah.

When you did your first job, when you were in high school, what did they have you do?

Working on the old No. 6 paper machine. I was a roll buck, taking finished rolls of paper out to what they called the finish room, and put them on some horses. Then they put the end wrappers in and wrapped them up and shipped them out. But that was my job, just to take it off the rewinder and take them out and deposit it on those horses. They were made out of wood.

Like a sawhorse?

Yeah, only they were real low. They were the same height as the dolly that you took the roll out on, and then there was a couple of straps on the side. You grabbed hold of them and you pulled up on them, and it would make that big roll of paper go out on these what they called the horses. You had to be real careful you got them balanced on there right, or they would tip one way or the other. Then you'd have a few people mad at you.

How much would they weigh?

Oh, I would say probably close to three-quarters of a ton, I imagine. Fifteen hundred pounds, something like that.

And what kind of paper was that in '43?

Newsprint. At that time, in '43, that's all they ran down there on those two machines. Five and six were almost identical, and they ran strictly newsprint on them two. Later on they converted them over to run the coated papers.

They were making their own pulp for the newsprint, right? They didn't import pulp…

No, they basically made their own. The wood would come down a flume from the wood mill. Before it went in the digesters, they chipped it up into small wood chips. They would cook this in their process up there in the digesters, and that's what made a lot of the pulp. But there wasn't any brought in in them bales like there was later on.

So in the early days the log feed was from the top end of the mill?

Yeah, they would bring logs down on rafts, and they had a place there where they could run them in between so that they couldn't get away. There was a big chain, and it would grab these logs and it bring it up into the mill. And then of course there was a cutoff saw right where they came in, and they cut them to certain lengths. Then it took it and put it on another conveyor, and it would kick it off on that onto some more chains. There were saws that were, oh, probably about that far apart that would cut into those big wood blocks. If there was any bark on there, there was a bunch of guys standing down there in certain areas, and they'd have to pick these off and grind the bark off that was on them. And then it would go on down this flume down to the chipper room.

So they wouldn't put them in the debarkers?

At one time they did it all by hand. There was quite a few people working there, picking these things out of there. Some of them didn't need debarking, but there was quite a bit that did.

I've got an old picture that shows that the logs went through a tunnel thing with really strong jets of water

Well, that was called the ring barker. It was run by hydraulic water pressure. That's what peeled the bark off, and of course that put all them other guys out of work.

So when they started blasting it off with water, a bunch of jobs went away.

Mm-hmm. Then the bark would go down to the boiler room eventually, and they would burn it.

Did they call that hog?

Hog fuel. There was no wasting anything.

That's kind of modern, that they used every single bit of the tree...

Yeah.

So, it sounds like very hard work.

Some of it was. Of course, then I spent probably about almost five years in the converting plant. Her and I worked at the same shifts all the time. And then there was a job opening came up in the maintenance, and I bid on it, and I got lucky enough to get one.

So that got you a raise?

Eventually.

Ring barker, mill pond and upper canal in background as log comes into the sawmill end of the mill.

Do you remember how much they were paying you when you first worked there?

When I first went to work in '43, I think it was something like 80, 85 cents an hour. Of course, when I left I was getting over $16, so it made quite a difference.

That is quite a bit of difference.

After you got that job you had to take a test on blueprint reading and some mathematics and that stuff in order to keep that job. And you had to pass I think it was two or three of them. Then they gave you your journeyman's rate. And then later on I got what they called a plus rate, so that made a little more money. It wasn't by seniority. You had to basically earn that.

It was a bonus of some kind?

Yeah. I guess you were supposed to know a little bit more than what just a plain journeyman knew. But basically we had a pretty good crew; they could jump in on just about anything and fill in. You didn't work with the same guys all the time in the maintenance. You had—there was quite a number of people in there, and certain jobs you would work with guys, and then during the week I was in what they called Mill C. And I was in there every day. But then if a machine went down, they would bring guys in out of the other group, and that's the ones you would work with on some of these repair jobs.

Was that the Flying Squad?

Flying Squad was separate. They were just a group, I think about five or six guys. I was in there for a short time, but then I went into Mill C and I stayed there until I retired.

The mill was union then?

Right. I think during the time I was there, I was out on three strikes. The first one was 10 days about, and I think the second one was a little over two weeks. The third one was nine months, so that was a long time without pay, you know. But at the time I didn't something I shouldn't have. I went up into the national forest; I cut a load of wood every day, and I come home, and I had enough I could sell it. So this brought us in some money. And then there was a foundry over in pretty close to the Jantzen Beach area needed somebody to set in some machinery for them. And they called the union up here, and I went over and talked to them, and they put me to work. And they kept me going all that winter, up until a couple days before I went back to the mill when the strike was over. So I really wasn't hurt too bad.

That's a long time. A lot of guys must have just gone off and never come back, I'd think.

There were some that left, yeah. And then others, they stuck it out too; tried to find what work they could here and there. You didn't get anything from the union at that time. And I understand on the one that happened after I retired that was for a year they gave them so much money every week. So that kept them going. But that's a long time too to be…

Do you remember the day you went back to work at the end of the strike?

No, I don't. I don't even remember when that strike hit. That's when you were working up at the bowling alley, I believe. That was what? About '79, '80?

Margaret Witherspoon: Eighty, '81.

Somewhere around there.

When we were talking to Roy we got story about the light bulbs in the plant going crazy. So these supervisors would be running around in the basement with a box of light bulbs, and nutria swimming around their feet while the machines were down.

I didn't hear about this light bulb business. That's the first I knew of that. That kind of amuses me.

He said the basement in the oldest section of the mill, where the water still came through, had nutria and rats swimming around in the dark down there.

I seen a couple of them later on. They were pretty good sized.

In the basement?

Well, it was outside by the old boiler plant. They had an area there that didn't have any concrete or anything over it; it was ground. And that's the only place I ever seen any. Except one time a beaver got in the steam plant. I still remember that. They had to call in a guy from the game commission, and he came in and got hold of him somehow and put him in a sack and took him out. But that's the only time I ever heard about that. But I did see that.

Were you ever down there when there were fish in Mill A?

The only time I ever seen any salmon was you could look down through a grating. The water level was below the floor, but you could see some of them big salmon laying in there because the current was coming through, and they would go as far as they could to the steel gratings. They couldn't go any further. And then they would just lay there. Then sometimes they would shut the flow of water off, and then these salmon would back out and some of them go up the old fish ladder. But I don't suppose all of them made it. You would be surprised at the size of some of them fish that were in there: big.

Somebody, I think Olaf, said that sometimes a salmon would come up through a drainpipe, but would leave in a lunch box.

Well, I imagine there's some that went up over the stairway, too, and up out through the filter plant, and went home that way, but all in one piece.

So was there another way where the fish could have been deluded and thought it was the fish ladder?

No, basically it was always up there at the grinder room area, but I don't recall anyplace else that they could come in. Of course, they hung out over there too where the PGE turbines were, because the water came down through there. And they would lay around in big tubes, where the water, after it come out of the turbines, would go back into the river.

Did you ever work with the guys from PGE?

No, we had nothing to do with them whatsoever.

Or the locks people?

No. No.

Do you remember when they started bringing the barges with the pulp, after they shut the sawmill and the grinder room down?

No. They've brought barges up there ever since, well, that I went to work there, I guess. Because the only way that they could move the paper out was either by barge or truck. Yeah, that's the only way they could do it. Wymore had the warehouse over there across the locks. Certain rolls of paper they took out to there, and he trucked them out. And everything else went into the barges, and they took down to Portland.

There's some photos of a landslide that happened when they used too much dynamite to blast that road and make it bigger.

I guess that dented up quite a few cars too, from what I heard.

Wymore wasn't part of the mill?

No, I think they had a contract to truck the paper out, because they had their own trucks and everything. They had a warehouse over in Oregon City. I imagine they took some of the paper there part of the time, and I don't know if they would reload it and ship it out or what, but they loaded up these big vans and then hauled it out that way.

Now, do you remember much about how they brought the clay in when they started making the coated paper?

That was brought in by barge. They had what they called an airvator, and basically it was just like a big vacuum cleaner: it would suck the clay out of the barges and put it up into what they called the silo they had there. From there they would use it in mixers. Originally them mixers, I think, were basically out of a bakery someplace; had big paddles in there; and they would add their chemicals and so much clay to it, and then of course these big paddles, they would mix it up real good. From there it went down to five and six coated machines. But somebody had to be on those barges when they were sucking that up, because they had to keep moving a big hose around in order to keep hoeing; if it didn't go in there very good, well then this guy would have to rake some down.

The clay barges came from some landing upriver on the Canemah side?

Well, they came up from Portland. But then later on Crown had a place close to Oswego where they brought some chips in on trucks, I believe, and loaded them. I'm not sure about that clay stuff, because I think that basically came up from Portland someplace; I don't know. That's where this Henry Herwig could probably tell you quite a lot about that.

I know that your wife, Margaret and you worked side by side. Is that where you met?

Margaret: No.

I asked her if you guys met up in the converting.

No, no, no, no.

So the mill had a policy that you guys could be married and still both work there?

As far as I know, they never had anything like that as long as I was there, because there was actually quite a few couples there working there. There was some others in the converting, and then of course there was some other guys' wives that worked up there, maybe down on the machines or some other department.

Did anybody else, like your brothers or your father or your kids, end up working for the mill?

My dad, he worked there. And we had kind of a standing joke. You know, I had a lot of relatives. They made the remark that if we all took the day off they'd have to shut the mill down. Some were Baisches, some were Eckerts, and of course my dad and I, we was the only ones with the Witherspoon name, but there was actually a pretty big bunch there at one time.

But your own kids—did they work there?

We had three boys, but none of them worked there. One of them, he's a setup man for Leopold Stevens. And another one works for a tree care outfit; he's a maintenance supervisor or head guy or whatever you want to call him. And of course Bruce he's the middle one he works for an electronics group, and then he does photography on the side—weddings and senior pictures or whatever.

Roy said it was really hard if your parent was a supervisor and you were hourly…Did they kind of think that they were special, or?

A guy by the name of Ed Baisch, he finally got to be yard supervisor, but he didn't have any control over me. Once in a while I would snitch one of their hand trucks, and he'd come and talk to me, and he says, "You're not supposed to do that, you know." If I needed one, I went back and got another one.

A little bit of an outlaw.

No, I was—just when I had to be.

I heard that you had to buy your own boots at the mill, and you bought them from the mill.

Shoes? Yeah. We had to buy them for some time, and then the union got something through that they furnished you, I think it was with one or two pair a year. But I didn't figure that was any great big thing, because normally I figured I could buy my own shoes.

I saw a website on Crown Zellerbach history, and they had pictures of hands missing fingers.

Sometimes them fellows that had fingers missing, it would happen on what they called the stack. It's where the paper would come in, and there was a series of rolls on top of each other. The bottom one was a big one in diameter, and then of course as they went up they got smaller. The paper would come in and keep going around like this until it got down to the bottom one, and this is where some of these guys lost some fingers, because they had in there what they called a doctor. Now, a doctor was when the paper came in, if there was any scabs on the roll or anything, this doctor had a blade in it, and it got real sharp from rubbing on the steel roll. And this is where some guys had lost fingers. They would get them caught either in the stack rolls or get them cut on that steel blade. Actually, they were bronze.

Bronze blades?

Bronze blade. They were sometimes, oh, six, eight feet long. Not eight feet, but about six anyway, seven. One time I cut myself on one of them, and I went up to our medical clinic, and they sewed that up for me. It wasn't a big cut or anything. And they said, how did you do that? I said I cut it on a doctor blade, and he looked at me, you know—got quite a kick out of that after I told him what it was, you know. [grinning]

Roy said a nip is a place where something comes together where there's a chance to get hurt. The doctor blade would be in a nip, is that right?

Would be just below the nip. The paper would come out, but the doctor blade had to be below that; otherwise the paper wouldn't have come out of there right. This nip is when there are two rolls set one on top of the other, where they touch, that was called a nip. And every other one would be going a different direction. So this is where they would get caught in there sometimes. If the paper broke, sometimes it would start wrapping around the roll and keep building up. Well, you had to cut that off in a big hurry. Otherwise I don't know what would have happened.

I try to imagine these big rolls and think of the wringer washing machines.

Well, basically, yeah, you could call it that. But they were all made out of steel, and a couple of them were constructed so they could put steam in them. So of course in this stack of rolls there was probably seven, eight rolls, one on top of the other.

So was steam and water dripping out of the screen or mesh?

There was no water.

No water?

Not on the stack. That's where the paper was already dried. It come out of the dryers through the stack, and then it was completely dry. They took the paper off the machine in a big roll. They had big transporters there, and they would take them out to the supercalenders and put them up in there. And then by hydraulic pressure squeezing on the rolls, and steam, they could it did something to the coating on the paper. But outside of changing some of the rolls when they would get damaged, that's all I had to do up there. But I didn't do that very much.

So what did you mostly do in the machine area?

Anything that had to be done on a paper machine: change rolls, stack rolls, repair pumps, belts. There was a fellow by the name of Boyd Ringo; he was a belt man, and he had to go over the whole mill. Then they picked out a helper for him, and I happened to be the one that helped him for quite a long time. If he wasn't there, I had to do it by myself. So I got acquainted with belts.

Well now, where do you get belts from? Do you have to order in bulk quantities of belts, or did you make the belts, or…?

You knew what the lengths were, and they allowed so much taken out of each belt to get the right tension to make it work. Well, when it was leather, those would stretch out so bad, and then you had to take the splice apart and put some clamps on there and shorten them up and redo the thing. But it was forever a big job keeping up with them things, and then they finally went to the plastic. And once you put one of them on, they would run for years sometimes without doing anything. And if they got to the point where they were too loose, you just ordered another belt out, and they knew the length that it had to be. And you had to splice it together on the line shafts.

Splicing plastic?

Yeah, it had plastic inside and a leather covering on the outside. On the inside surface and the outside it just had a little kind of a fabric over the top of it. But the plastic was probably three-sixteenths to a quarter inch thick, and depending on how rough a service that had to pull—like presses—and that it would govern the width. Sometimes they were only three inches wide; sometimes they might be six, depending on if it needed a lot of power, you know, you couldn't have them slip. If they slip, the paper on the presses would break, and then of course we had to wash everything back up and start in again, and that cost them money, you know, besides a little production.

So how long would a belt be? I mean, five feet, 12 feet?

Oh, it had to go all the way from the basement up through the next floor, so they had to be—I can't remember lengths, but I would say some of them were at least 50 feet long. You could get it in big rolls, but if you got it a big roll, they had a machine down there, and you could run it through that and you could taper the ends. They all had to be tapered, and the little platform that you put them on when you run it through that little machine, it would put that taper on that splice. The splice itself wasn't over probably about that long. And then they had a chemical that you put on there, and then of course you had to cook it.

Oh, like a catalyst or something?

Yeah. And the heat in this press, you had to leave it on so long in order to bond these two pieces together. And if you didn't do that right, well, the thing would come flying off, and then you'd hear a nasty report the next day, you know.

So you had to strike a balance between doing things as fast as you could to keep things moving, and not doing it so fast that it broke right away.

Right.

It had to be perfect.

Yeah, if, say, a belt did break, you could get by by putting on kind of a leather, but it had to be real wide, and then they would get slack in them before very long because they couldn't stand the tension that was on them.

They must have had a tannery somewhere close by.

Well, somebody had one. But after that plastic started in, well then they done away with almost all of the leather.

So did you have certain things you liked and certain things you didn't like to do very much?

Well, there were some jobs that weren't very good. Before I got on the paper machines I worked with a group of guys all the time, and they got the dirtiest and the hardest work in the whole mill. I was with them for some years. But we did riveting, a lot of concrete demolition and different things like that. Of course, that riveting, that got to be quite a chore because them rivet guns, I think they weighed 90 pounds, and you had to pick them up. And of course you had to hold them on a horizontal. And when they put these rivets in through the pieces of metal, well, one side of this rivet already had a head on it, a rounded head. But with this big machine it took two of us to get that up there, and there was two guys on the other side pushing to keep that in the holes. The guys on the inside—that's where I always was—we had to take this gun and kind of round that head in order to get that rivet in there tight so that it wouldn't keep working. So when you used a two-inch rivet, by the time you put that round head on one side, well, you had to pound on that thing pretty good with that gun until it got that rounded head on it.

Of course, they were red-hot, and sometimes when you'd put this gun on there the sparks would fly, and you had to wear a shield and a leather cape affair around your shoulders, and it would cover the front of you in order to keep from getting burned. And then your gloves, they were real heavy leather. And you could take quite a lot of heat through them things before you could even feel it. But that was pretty hard work.

It sounds like a lot of places in the mill were really hot places to work.

Well, even on the paper machines in the wintertime, if you had to do a repair job on a dryer or something, that was real hot, too, because when they shut them down that steam had been in there all this time, and they were hot. And maybe by the time if the job took long enough, they'd cool down some. But that usually took quite a while.

So the riveting, it must have been loud too.

Mm-hmm. You didn't have any protection at first, like they used to. You know, later on they got earmuffs and earplugs. We did use some earplugs. It deadened the sound some, but you still got a lot of it.

George Droz told us about some really nasty jobs in—I want to say the sulfur plant, but I don't know if that's right. But like these big mixers, these bowls, and they had to lower men down inside of them to redo the brickwork inside.

Oh, that was digesters. That was in an area they called Mill B. And these digesters, they had big tiles, and it lined the whole part of that digester from the bottom opening where the well, the big valve was on the bottom clean to the top. And what maintenance had to do a lot of times, when these things would start loosening up and they would re-tile the whole inside, you had to go in there and chip them things out. And everything had to fall down through the bottom, of course, in order to get it out. They took the big valve off on the bottom, and that left an opening down there probably about that big. And while you'd chip these off with these chipping guns, the pieces would fall down there and then go out through the bottom. And somebody had to be down there to rake it away. And of course then tile-setters, they always came in from the outside and re-tiled them.

And then too, they had what they called the blow pits. That's where, after the wood chips were in this digester and the acid came in and finally worked it up into a pulp, when that got ready to blow, they would blow a whistle. And everybody around that area, you had to get out of there in case one of these lines would break or something. And they would pump it out of that digester into what they called the blow pits. And that's basically where the stuff cooled. And then they would kind of wash it out of there with a big—you might say a fire nozzle—a lot of water pressure on it. And those weren't very nice to redo either, because you'd get all this acid in there, and it would get into the the bottom of it was made out of lumber; I think they were about four inches wide and then, of course, the full length of that pit. And they were about eight inches wide. Well, it wasn't very nice to get in there and have to cut that out. You'd get all that smell from that acid; it got pretty strong sometimes.

Sulfuric acid?

Yeah, right.

And then where did it go when they washed it? What happened to the sulfur?

It went into the tanks. Down below there was tanks, and it would go into them. And then they could pump it from there to wherever they wanted to, different areas. A lot of it went to the mixer room. And of course, the mixers, they mixed up the pulp for every machine in that mill. See, at one time, when I first went to work there, they had 10 machines. And then they kept cutting back here and there because some of them were so slow, not very wide, not fast. I think when I left there, there was probably about five, I think; four or five left. All of Mill C was shut down, and they had No. 9, 5 and 6. Four, I believe, so that it was probably down to about four machines. When I first worked there, I think there was two thousand or 1,200 I'm not certain; don't quote me on that. There was a good bunch of them.

Why did people stay at the mill for 35 years…what was the best part of your job?

Oh, one of the easiest ones I had, I think, probably was doing the table rolls. These were rolls that stretched on the wet end of the machine, and the wire went over the top of them. Of course, it would keep running around all the time, and this is what formed the paper when it come out of the head box to these rolls. And of course, they carried that wire. And the job I had for quite a few years was keeping care of them rolls, changing bearings and so forth. And I had to see to it that the machines I was assigned to always had some spares, because they were losing bearings all the time due to the heat, water and side tension. When a bearing went out, well then they would change all these rolls, and of course then you had to get them all ready for the next shutdown. But I could work by myself; I didn't have to have anybody with me or anything like that, and they didn't bother me any. As long as I had spare rolls to go, nobody even come around there. But I had my own little work area where these rolls were. And then there was racks where they would fit into, and you could fit them up along the wall and keep your spare rolls in there. But that was probably about the best job I had in there.

Because you had control over your job?

Well, I didn't need anybody standing there watching me or telling me to do this or do that. Of course, if I made a mistake, I heard about that too. As long as you did it all right, you didn't hear nothing. But one little mistake and you'd have everybody on your ear.

How big were the bearings?

On those particular rolls they probably weren't only about that big around.

Golf ball size, or…

Not any bigger than that. Well, maybe a little bit, but not much. Biggest bearing job I ever got in on changing was down in the supercalenders, and they were huge. You had to have a chain block to even lift them up. On the big bottom roll, them bearings, they were probably that big around. And we'd go up to the old finish room, and they had some old conveyor rollers up there. They were probably that wide. And then they had these rollers that were about that big around. Well, you'd get a section of that, and you'd put sheet metal down on top of it. You would lift this bearing up and put it on top of that sheet metal, and then you'd have two torches, and you'd light them up and turn them on full blast in order to put that heat on these bearings to get them a certain temperature. You had a stick about as big around as my finger, and they were about that long. And you could get them different degrees.

A bearing had to be 200-some degrees before you could slip it onto the journal of this big roll. Well, you'd keep that heat on there long enough, and you'd take this temperature stick, and you could touch on the outside of that bearing. And when it got hot enough to be put on, it would melt that stick. Then you knew that you had to get it on. But you had to get that bearing all the way on in one shot, so that's why we used this sheet metal and them rollers. You could just—and it had to be at the right height. And you could slam it in on the journal. But that was kind of interesting too, you know. Most guys didn't get in on that. But that was one of the—when I worked with that one bunch of guys, we changed a couple of them down there one day like that. But sometimes it would get pretty interesting when you run into something like that.

What does a bad bearing do?

Run out of grease is one bad thing. They start overheating, and you could hear them when it would start going bad. And of course, if it got bad enough it could even conceivably stop that roll from turning. But some bearings, the roller bearings, if you had a used one that was off of a roll, you could tell by the color of the inner race and the markings on that race whether this was starting to wear quite a lot. And you could take a feeler gauge, and you could put it around a little tubular bearing inside them rollers, and if it was getting too much clearance in there, well then that would start to kind of hammer a little bit. Then I seen some of them get so hot that they were just smoking. And of course you had to shut them down, not just automatic. And you'd either have to change a bearing in place, if it didn't hurt the journal, or you'd have to replace the whole roll. So sometimes it would take several hours to do that.

So did you work shifts, or you just had to stay and work until the job was done?

Yeah. A lot of times I was supposed to get off at quarter to five. I didn't know when I went down there in the morning if I was coming home that night at five o'clock or whether I was coming home at midnight. I did work an awful lot of overtime. And them times I would get home, there was a quite a few times they would call up, have to go back down and do some work again where something went wrong. And then the union finally got something in there where they couldn't work you over 16 hours at a time, which was a big help to us. There was a few times I would get home at two o'clock, three in the morning, and they'd want you in there at eight the next morning, you know. But then when they got this deal in they could only work you 16 hours, they either had to have some other guys in there at midnight to replace you, or the job was done at that time. And too, it got to the point where if you worked a double shift one day, if you wanted to take the next day off you could do it, and they didn't say nothing about that. But then you lost a lot of your premium time if you took an extra day off. But I got the point where if I worked 'til midnight, I took the next day off, because I figured I made enough on my time and a half to cover that next day's wages.

You have to sleep sometime, and see your family. So about the gigantic bearings. I'm trying to imagine where they got those. There must have been companies that manufactured these huge cannonball things…

Yeah. If I remember right, Fafnir was one of the bearing outfits, and oh, I can't remember; there were several of them. They got most of them out of Portland at warehouses. The mill stocked some of the smaller ones, more to the vital areas like certain rolls or whatever they used them on pump bearings and that stuff. If it was quite a critical pump or roll, they would carry spares. The storeroom a lot of times would stock a few extra bearings, and you could just go up there and get one. However, if they didn't have one, they would have them sometimes brought out from Portland by taxi or whatever, and then as soon as it got here you would put it on and get a job done anyway.

So what would you do when the floods came? Did anybody try to move the spare parts up out of the water?

There was what they called the basement, and then the sub-basement, in Mill C in particular. We didn't keep any spare rolls or anything like that in the sub-basement, but there was pump rooms down there, and they had watertight doors. Well, when the water would start to come into the basement, you worked in boots, hip boots, all the time. And you would have to go down there and close that watertight door. And then they had bolts made up, and there was brackets welded to the inside of that steel door, and then it had a rubber gasket where it touched the concrete door opening. And they had pieces made out of metal with slots so that these bolts could go through, and they were wide enough to cover the opening on the door. And when you slipped this bolt into that position on the door and started tightening it up, it would pull that door tight against the concrete wall. And there was maybe, oh, six to 10 bolts on a door. By keeping these good and tight you could seal that; water wouldn't come in. That way they could keep running the machines until the water got up onto the next level.

And if the water got up to the next level, they would keep them running until the water got up on the base of the motor, before it would get into the motor. Then they would either shut down ahead of time or wait till that point and then cut off whatever piece of equipment went to what machine, and then they would shut the machine down. And of course you had to pull the motor out and take it up on the top floor. In order to get them upon the top floor, you had to do it by elevator. Well, when the water got down in that sub-basement far enough that if you had to get close to where that water level was before you could get off on the elevator, you had to add counterweights like a steel roll that was pretty heavy inside and leave it in that elevator car—when it would hit that water—otherwise it would knock the cables off of the drums, and then you were down completely. But if you kept enough weight on that elevator cage and when it would start to hit the water you would you run these yourself.

It was like a dumbwaiter with pulleys kind of thing?

Yeah, similar. And then by hitting that water real gently, you could get to where you wanted to take this roll off, or motor, or whatever you had, see? Otherwise, it was a job to change cables on an elevator, because there was drums that was probably anywhere from four to six cables on it, just one car. And of course, you didn't want that to happen. Water was in there.

Right.

I think one time, the worst flood I was in, we actually worked in water for, I don't know, a week or so. But to get that place running again we worked six days a week, 10 hours a day until we finally got everything back in and running. That was the worst flood they had. I think that was in '64. It almost got up to the main floor, and that was bad.

We had one story—maybe it was George—they were down in the welding shop during the '64 flood, when way up above their head a log came through the windows—did anybody ever drown there?

No, we didn't lose anybody. I can remember one amusing thing that happened one time. Down in the sub-basement they had ditches—you might call it a ditch, but it was a place for water to run through. It wouldn't be on the floor, but anyway, when the water got up high enough, some of these had a wooden kind of a grating affair over the ditch. And this one fellow, he was walking along there, and of course he didn't know where he was going, and he stepped off into one of them ditches. Here his hat was floating down through the water and everything. Of course, if you laughed at somebody like that when they was a supervisor, you got in trouble. I couldn't keep from laughing. That one guy looked at me, he got some stuff all over his shirt and over his glasses. The pulp was hanging off his glasses. I looked at him, and I had to laugh. And he looked at me and he said, "What's so funny?" he said. "I've got to wash them myself." That was old Ed Karbonski. I never have forgot that. He was my supervisor. He was a good-sized man. They called him Big Ed, and of course, naturally I was Little Ed, you know.

I know Rosie was "Rosie." He said if anybody called him Bill or William, he wouldn't know who he was talking about.

He was machine superintendent over in Mill C, and I did a lot of work for him, of course, on the machines.

He had a pretty good handle on how to tune machines, since they sent him off to Holland.

He was one of the better ones because he worked his way up from the bottom. He went right on up. He knew every job that was in there. But for some of them they just hired from the outside, and you didn't have any way of knowing whether they had much paper machine experience or they learned it out of a book. No, no. You could tell if a guy had worked his way up through the ranks. Some of them old-timers down there, they could walk into a machine room in the morning—some of them the guys could tell you whether the dryer had a lot of water in it or no water, just from the sound.

That's a relationship!

Yeah. Of course, well, they some of them guys worked there from when they were little kids. Or not little kids, but I mean working age on up 'til at that time they worked 'til 65 before they'd get out. So some of them got to know them pretty doggone good.

How many different kinds of paper can you tell? Like what machines made what paper?

Five, six, they ran coated. Nine was finally converted over to coated. At one time No. 9 ran nothing but "Oregonian" paper. It was almost a steady run. Of course, well, they'd use it more for other newspapers too, but it was mostly newsprint. And then some of your smaller machines, they would run specialty papers. I don't remember just exactly the names of them and everything. But those smaller machines, they were a lot slower and weren't very wide. They ran toweling down there on No. 4. But the blue windshield toweling, that used to be sent in in big rolls. And then where she worked, up there in converting, the machines up there would cut them into those paper towels like, you know, you see in a service station. I'm not familiar with all the different types of paper. I know that one time they made paper for the "Denver Post" and the "Salt Lake Tribune", but that was years ago.

Well, I've got a sample of Crezon. I think that's what they called it. It was a construction paper.

That was real heavy. It was made on No. 7, I believe. Maybe 10, too; I don't remember now. But it was probably twice as thick as – [Margaret brings in a towel sample] yeah, there's a blue towel.

Wow, neat!

Well, these rolls would come in; they were that width. But they were probably that big in diameter.

Ah-hah. That's a skinny roll.

Yeah, well, when they made these, I think there was four sets of rolls to one machine, and there's two to each one of them there, isn't there? Well, that's what they call interfold. And by having these four rolls, the two bottom ones would make one side of this. See, when this would come out, there would be another one of these, but it would be folded in here and go this way. The bottom rolls would be one-half of it, and the two top rolls would make the other one. And there was a counter of some kind, I believe. When these women were sitting at the machine, banding them up, they got about the same amount each time. They put this paper band around it, and then of course, so many of those bundles went into one box.

After the box got full, it would go down a belt to what they called the sealer. It was a machine that took this box up in there, and as it passed through it, it would put glue on the bottom flaps of this cardboard box, and it went a little bit further, and then it would close them up. And then I think it even had some heat there, a little bit. Not much, but to kind of seal that together. And then it would go down a big chute three flights of stairs, or three floors, down to the main dock. And then there would be two guys and a mule driver down there. Two guys would pile the boxes as they came down this big chute onto these rollers, and when the board got full of one certain kind of toweling, he would haul it away, and then you'd have another empty board there, and of course you'd start filling it up. I did that for a while, too.

When you worked in converting, is what you just described part of what you were doing?

Yeah, and then part of the time I worked on what they called the can machines. You know in some of the restrooms when you grab hold of that little lever and you pull it down, it puts a piece of paper out there so far and you tear it off? We converted the big rolls into these small ones. They were about as wide as that. And then they would be about that big around. Well, this is what you did all day long. That was a boring job, you know. Start that thing up, and it would run for just a little bit. You'd shut it down, pull the shaft. The guy back there, he'd have to put a plug in one end and put them in a box. The box got full, he'd push it off on a conveyor and then it went up through that sealer. Well, if you had a job like that, you knew what you was going to do from one day 'til six months later, you know.

Not much stimulating…

No.

I know Olaf worked, I think, at Hawley before he went to the war.

Olaf and I worked together quite a while.

He said his first job was at Hawley packaging toilet paper. He said it was awful.

Olaf, he could take care of himself. They gave him a job—basically he was taking care of the rewinders, and that was his job—all the rewinders on most of the machines. He did a good job. Then sometimes if he needed some help, he would come and see me or one of the other guys down there, and we would go up and help him change something if we had to. But he was always good to work with.

I think he said he liked the last seven years…

Yeah. I think that was when he was doing them rewinders.

So, like you said, it was more fun for you to have your own little space.

Yeah. Later on it wasn't so much fun any more. The crew leader—I forget what they called him—anyway, if he was gone on vacation or something, I filled in. He had a radio on, and they'd furnish you with a cart sometimes. If something went wrong somewhere, all they had to do was call on that radio, and then of course you had to jump on this cart and run down there. That took a lot of the fun out of it.

Because you were on call to the radio?

Yeah. They could talk to you. Of course, you could talk to them too, but you didn't do that unless you had to.

Did you go to meetings or trainings at the West Linn Inn?

Not really. When we worked in converting, sometimes we would stop in there if you was on swing. Some of them would stop in there and have coffee, but basically I didn't have much to do with that up there. I guess we did have a couple of banquets in there.

You didn't go into management? You didn't supervise?

No, no, no. No, a lot of times I would have a crew. I would run that, but no further.

When you were lead worker?

Yeah, yeah.

Okay. And you were never hurt at the mill?

Not really. Got banged up a little bit, but I mean not enough to lose any time. Of course, when you've got three kids, you know, and you would sit down at the breakfast table, you might not have felt good that morning, but when you'd see them groceries disappear, you'd reach back and get your lunch bucket and pick it up and go out the door, you know.

So you retired at 57, and what did you do? Did you miss it? Like the next week, when you didn't go to work, what did you do in the real world?

Margaret: Drove me crazy.

That's not true [smiling]. Let's see. I think wintertimes, when I could get up in the national forest, I did cut a lot of wood. And I wasn't supposed to do it, but I did sell a lot of that. And then Charlie Calhoun had a place out there by Fischer Mill, and I think there was 20 acres of timber, and I got all the alder off of it. And this guy that cut all the fir and everything else, that's what he took. He delimbed all this alder and stacked it. Well, being as I knew Charlie pretty well, I got all that alder wood from him, and of course I cut it up into fireplace length, and I sold all that. It took me quite a long time to get all that done. She helped me a lot on the hydraulic wood splitter we had built. She would run the cylinder deal, and then I of course would handle the logs, and we'd split them up that way. I had a good part of this backyard here full of wood one time.

So you kept working.

Oh yeah. And then a guy by the name of Bob Stein, he has a bunch of service stations around here. He called me up one day and wanted to know if I would work for him. So that worked out real well because if we wanted to go down south in the wintertime, he didn't care whether I left or not, and then when I'd get back I'd go back to work for him again. So I kept busy that way too.

Margaret: Part-time.

And then about 12 or 13 years on what we called the wood toys. I got a little shed out over here on the back, and I got my equipment in there. Well, we made a lot of fire trucks and little wood cars. And she had napkin holders, stuff like that. And then in the fall, when the schools would have their fall bazaars, we would go to them. Oh those, they were made out of the scraps. And we would go starting in probably the first part of September, whenever they started having them, up until Christmastime. Sometimes we'd go every weekend, sometimes we'd skip one. And then we would sell the stuff that we had made to the people who would come through the bazaars and wanted anything. This kept us pretty busy up until last year. Well, last year we our sales are starting to fall off because there's too much plastic, and I can't make something out of wood that they can do with plastic.

If you had the chance, would you like to go back down to the mill today and see what they're doing?

You mean just to tour it? Yeah, I wouldn't mind doing that. I think the last time I was down there was when Crown Zellerbach had their hundred-year celebration down there and we went down. But I think that's the one and only time that I've been back in there. I really didn't have too much desire to go back, but I did go to that. But it would probably be kind of interesting to see what they've done down there now, because I understand they've done a lot of changing and everything, and I think there's—I don't know; they've probably only got two or three machines, I suppose.

Paul Miken

Interview March 29, 2006.

So what is your name and your age?

My name is Paul Miken. I will be 80 years old this year.

And what year were you hired at Crown Zellerbach?

I started at Crown Zellerbach in 1946, and I retired in 1983.

Do you remember your first day of work?

Yes. My first day of work? Naturally, it was quite scary. I didn't know what to expect. But the biggest factor was, when I got inside, how noisy the machines were—were just roaring, and of course I wasn't used to that, and it was very, very loud. But on the other hand, the company always provided us with ear protection, which was outstanding.

Even in 1946?

Yes, when I started in 1946 ear protection and hard-toed shoes, it was a must.

So, I heard they only gave you one pair of boots, and you had to buy the rest.

We had a choice of buying our shoes there or another place, but they had a storeroom that was loaded with all different type of shoes, and the shoes had to be bought. Even the first pair, they took it out of your pay, which was very nice, at that time.

What was your first job title?

My first job, I parked cars and I worked as a gardener. And then later on, I was a guide, taking people through the mill. It took about two hours to take the people from one end to the other. And then we done that twice a day. We would start the 10 o'clock of the morning tour, and then the afternoon was at 2 o'clock. But we had busloads of youngsters from the grade schools, and we had people from all over the world stopping in and going through and seeing how paper is made. It's really amazing, an amazing process of making paper, yes. Well, we would always have 10 people would be in a group, or sometimes 12 adults, but, about 10 youngsters in a group. That was a safety precaution, you know, that we wouldn't let them run here or there.

Did you have to study so you could talk to them, if you hadn't worked in the mill yet?

Oh, well, that's very, very true, yes. They had information about amount of tons of paper produced a day. And of course, the youngsters, I think they had to be at least 12 years old to go through the trips. But they would watch the forklift drivers driving up and down the dock; they thought that was really neat. And of course, the most fascinating part was the sawmill, where the logs would come in singly, and in about one minute they would be cut up into two-foot blocks long and one foot square. Then the blocks would drop down into the basement where you were talking about, and they were ground up into pulp. That was the start of papermaking right there.

I heard when the logs came in, they had to be debarked in a big debarking machine.

Actually, barking the logs, we had about three different ways. When the logs were lifted up on the deck, it'd be a single log went through high-pressure water, and then when they were cut up into blocks of two foot long and a foot square, they were also put into a tumbler, like you're indicating. So it tumbled the blocks to get the bark off. You could tell in the finished product of the paper—you see little specks in there, that was specks of bark. So it had to be 100 percent bark-free to make top-quality paper.

Then they took the bark and took it down to the hog fuel burner to run the mill?

Any log, and part of the log, that would not be used for pulp purposes would go through a hog, and this hog had 36 knives on it, and it would be going very fast, and it would chip up the part of the log so the firemen would be able to burn the wood to make steam, and that steam in turn would be used to dry the paper on the paper machines. So it was just like butchering—the butcher uses everything but the squeal of the hog. Every bit of it was used: the bark, the bad logs and everything was utilized. If we couldn't use it for pulp, we used it in the firemen's bin, and they would burn it up for steam.

Roy said that the mill actually had its own fire chief?

Yeah, we did have our own fire chief. Matter of fact, a good friend of mine, Richard Buse was a fireman. But the first one was Mr. Nixon. He worked there for 50 years, and he was the head fireman. I was actually a volunteer fireman for 30 years. We would meet every month. We'd have exercises of putting out fire—gasoline fires, chemical fires, and also working the equipment so we could get the equipment onto the fire hydrants. It was quite a strenuous procedure being a fireman.

This was just for the mill?

Absolutely. There was probably 50 or so firemen on the force, because, see, we worked three shifts, so therefore they had to have firemen on each shift, plus the day shift, so it would be covered at all times.

So how did the firemen relate to the hog burner? You said 'the firemen'

Okay, the title "firemen"—they were the people that operated the fire in the furnaces, you might say. We called them the firemen. And of course, the volunteer firemen, they were different people, yes.

I see. I saw one picture of the mill and there were two huge perfectly square piles, and I thought they were stacked wood on either side of the locks on the end of the mill.

What I remember, it wasn't wood; it was pulp. Pulp was stacked very high in the areas, and when the water was high, you see, then they would do a lot of grinding, grinding up the pulp logs. But in the summertime there wasn't much water flow, so they could not grind because they didn't have no power, so therefore they would take these laps of pulp and throw them into the digester to beat them up and to make the base for making paper. So they stacked a lot of pulp outside to get ready for the summer drought.

That's very interesting. How many years did you work at the mill?

I worked for 36 and a half years for Crown Zellerbach.

The whole time it was Crown Z.

Yes. When I retired in the mid-80s, a gentleman from Great Britain came over, Jim Goldsmith, and he brought a bunch of accountants with him. And he bought out 51 percent of our Crown Zellerbach stock, and that's when Crown Zellerbach was no longer a paper-producing company. And Mr. Goldsmith sold off all the papermaking machines. He kept the timberlands, and he went home with $400 million in his hip pocket. So that was the end of Crown Zellerbach after—we were over a hundred years old. That was the end of Crown Zellerbach making paper, yes.

Who owns it now?

It's Canada, yeah.

Have you been back down there since you retired?

Once.

Tell me about it.

Well, a young man that I used to work with, Fred Smith, retired, and they invited me to come down to celebrate his leaving. It was very sad about Fred. He was very athletic, and it was about two years after he retired he went to Hawaii. He loved to—what do you call it? skiing on the water. Anyhow, there was a bad wave that came in, and he lost his life in Hawaii, on vacation. Yes, that was Fred, Fred Smith.

Did it seem different, or the same, or...

Well, there was less people, but I think it was actually the same process.

Yeah, somebody told me it was close to 2,000 workers at the peak, on all shifts.

The figure that I had was around 1,700 people. We had 10 paper machines that were operating, making all different types of paper. We made the paper for the white and the yellow pages of the telephone book. We made Crezon paper. That was a little thicker paper that was put on plywood under extreme pressure and heat, and this made the top finish very, very smooth, and it was very good for building tables and so on. And we also made paper for the pineapple plantations in Hawaii. This was a darker paper, but it was marked for planting pineapple, and this would suppress the weeds. So we made a lot of paper for that, yes.

And you appreciate that now that you're a master gardener!

Yeah, that's very true, very true.

George Droz said that he and his wife had gone to Hawaii and they saw it out in the field. They just take a machete and slash that X and stick the pineapple through it.

That's right. Yes. And we also, in 1948, we started out with a brand new paper machine. This paper machine made—they called it coated paper. This clay that was on the paper came from Georgia, the state of Georgia. And it was mixed up with the other material that made this paper very slick, and we made paper for "Sports Illustrated," "Sunset Magazine" and "LIFE Magazine," and many brochures were made with this very slick paper. It was really a brand new process. It was the first mill west of the Mississippi that made this particular paper. That was in 1948.

So you went on in '46; you did the tour guide thing, parking, your other duties…how did you make that transition into the mill proper?

We had a good system that you could apply for these jobs. These jobs were placed on a bulletin board, and you could apply for them. And that's how I went on and worked on the river for eight years. I thought it would be a tremendous job, because we had four boats. We had the Crown Z, which was the largest one, 48 feet long. And we had the Mary Jane and the Viola and the Billy K. We would move and sort and move logs, and we would move barges. The barges were bringing raw material through the locks, and they would be unloaded. But every day we'd make 600 tons of paper, and they'd be loaded out on the barges. Actually, we moved the paper to the customers by water, by rail and by truck. So the finished product was always moving.

I've heard there was a clay siding upriver and clay was barged down.

The clay came from Georgia, the state of Georgia, on railroad cars. And therefore, we didn't have railroad spur at our mill. But Pulp Siding—probably three miles from the mill. The railroad cars were unloaded, or the clay was unloaded onto a barge, then floated down through the locks, down to the lower river, where it was unloaded at the mill to mix with the other material for making this coated paper. It ran through a bunch of rollers, just like an old wringer in the old washing machine, but these wringers were probably 12 feet tall, and these rollers were probably about a foot in diameter, and it would go through a series of these rollers, and it would shine up this paper and make it nice and slick, and it would really gloss and make top-quality paper.

Is that the super calender machine?

Yes, supercalenders is what they were known as.

So Roy said to ask you if you remember when he yelled at you to rescue a loose barge.

Well, during the '70s we had a dispute with the union, and therefore the company went on strike for eight months. And we did not employee any hourly workers, because they had their union, and the management had contracts that they had to fulfill. So therefore we had people from Molalla; they were loggers, and we had them working inside the mill. We had people from New York and all over, Georgia, Louisiana. Anyhow, we had a skeleton crew, and of course some of the people were not well versed. We were moving a barge from Pulp Siding down through the locks, and somehow the pilot—we had the captain and we had a pilot on each side of the barge to tell the captain if he should go right or left or whatever—but anyhow, we got over and we were hanging over the falls because we got the wrong movement. But anyhow, we didn't hit no rocks or anything, so we backed off and got the barge down to lower, where it belonged. But it was quite scary. We had a lot of people on the bluff, on the banks, watching the maneuver, because it took about an hour to do this. So that's the story of our barge.

Hanging over the falls?

Well, it was very close. And of course, the water was down. See, if the water was high, the draft of the water would just take us right over the falls.

It was summer time.

Yeah, it was—it was very sad, though, that we worked for eight months, and you know, the hourly people—it was very difficult for a lot of people.

Roy talked about having skeleton crews, supervisors who might not know one end of a paper machine from the other.

No, that's very true; that's very true. Actually, at that time we had the top attorney working for us in the sawmill, from San Francisco. He was the head what do you call it when they were…?

Like a negotiator?

A negotiator. He was the head person for Crown Zellerbach, yeah, yeah.

Roy said that for him the most awkward thing was the hourly guys came back into their job and had to take it from a supervisor…

Yeah, probably, but we didn't witness that at the sawmill, for some reason. They were very glad to get back, number one. They were very glad to get back on the payroll again, and do their jobs, the jobs that they had done. But on the other hand, they found out that some of the jobs that the union people are doing, they eliminated some of those jobs because they felt that the people were not working the eight hours.

So they walked in their shoes and decided that it wasn't that bad a job?

Well, like wrapping up these rolls: They had three people, and they found out two people could do the same job as three.

Even two supervisors.

That's right. It was tough times, yeah, for those people, really.

People say, "…it's the strike that lasted seven months," "the strike that lasted nine months," "the strike that lasted eight months…"—that's a long time.

That's very true. And we used to have a bowling league with the hourly and the management people. And of course there was a lot of bitterness, so they cut that out. They bowled, I think, the Gladstone alleys, the Gladstone bowling alleys.

I had heard that the West Linn Inn was built for an earlier strike, way early, and that it was built in six weeks. What year?

Probably 1920s, something like that.

And they actually built it to hold 400 scabs or something inside of the picket lines?

Yeah, I'm not sure it would hold 400, but it was probably a good 100 or 150 people, because the top side of the hotel was made for people to sleep in there, you know.

You were saying they posted things on the bulletin board.

Well, you applied for the job, and then of course it was up to management to pick their people. They tried to pick their best people that they could. But this was a tremendous promotion for all the people in the mill. And you could go to any school, but you had to have a C or better grade. You could take any course you wanted to, and the company would pay the bill for that. So you could really learn what you wanted to learn. But 99 percent of the people would not do that; and actually, working shift work, it was very difficult to go to any school because, you know, they worked five days on day shift, five on swing shift—on swing shift, you know, the colleges, a lot of them were closed down and graveyard shift, so they just didn't have the opportunity to go there. But they were very good at educating people. Matter of fact, I took a lot of courses in accounting, and in tax work—and I still have my tax license. After I started in '59, there was another fellow, Bob Lane, and myself we were about the same interest in accounting and helping people, so we more or less worked together. But it was unfortunate Bob passed away as a young man. And he was accountant for Crown Zellerbach for a number of years. But he had a heart attack, and he left us.

So what was the first job you bid on?

I bid on the river job, yes.

Does that mean you were the pilot on one of these boats?

Yes, but you didn't start as a pilot, you know. You worked on sorting the logs. This was the main thing, because we had logs come in by rail and also come in by the river. Crown Zellerbach had a lot of land in the Molalla area, so they built their own road so they didn't have to watch the weigh limit on the trucks. And then they would dump into the Willamette River at Canby, Oregon, which was probably about six miles from our plant in West Linn.

Six river miles.

Yes.

When you say sorted logs…the river is full of everybody's logs, logs that are going to go through the locks and logs that are coming to you—how do you tell the difference?

Yes. Well, we knew our logs, and other logs were in big rafts where they identified them, and they would bring the logs through from Corvallis, Albany and Salem area, Willamette River. There was a lot of logging going on, and there was a lot of lumber mills in the Portland area. As a matter of fact, there was a mill down the Portland area right off of Macadam Avenue that made furniture. They would want logs, and there was probably, I would say, 10 or 15 mills that required logs, and they'd get their logs from the rivers of course. Every log had to go through the locks—they'd have to have a boat up on the upper river and then a boat down on the lower river. This was done by private people. They rented their little boats from the Bernerts' Company, and anyhow they made a living, and they kept things moving.

So your logs would go in at Canby from the Crown Z farms, and they would come down with just one boat; you didn't have to worry about the lower river because your logs went right into the sawmill? And you had two pilots and captain on each, or—?

Yeah, on the larger boats we had three people, yeah, the captain and two pilots. And on the other boats there was a captain and a pilot, or a deckhand, we'd call him. Yes.

Zellerbach tugs maneuvered around heavy log handling on the upper river and at Lake Oswego's log hoist.

And did you ever have to work when the water was really high?

Every day. Oh, yes. That's when it was really important to be sure that these rafts were tied securely to the bank. We always had a big tree or something; we would have an inch cable. That would be the main cable that would be holding these rafts. And there was probably three different cables on there to hold the logs in towards the shore. And then when the water would rise, we would have to move these rafts in towards the bank to keep them out of the swift current, out of the middle of the river.

You did that for three years?

Eight. I thought that was a good job, because you worked and you had a progression ladder that you know, you sorted logs, and then you worked on the barges. You see, every day we had to have people moving the barges through the locks. And then eventually, with the four boats, you'd get to be inside with a little stove to run the boats. And so there was a real incentive to work hard and to be able to get promoted to run one of those boats. And you had to be a licensed pilot to operate these boats.

A river pilot. So you're an accountant and a river pilot.

No, that expired. But the accountant business, we had to apply for that. We needed 30 hours to keep our license.

So it took a year to get your river pilot license, or…?

Normally it would take about a year, but they were real good. You went up to the Seattle area, where the district Coast Guard people were. And then you had an examination you took, and you'd pass it, and then they'd give you a license.

When you were on the river so much, did you feel more like you were part of the river culture, like the Bernerts and the other tug people, or more like a mill employee?

Oh yeah, you felt like you were independent, because you got your orders during the day. At that time we didn't have radios. But then later on, each boat had a radio with which we'd communicate, which made it a lot safer and a lot nicer when we went on the water. Actually, we had high water in '55; we had a flood. And then there was one in '94. And of course I was retired, but there was the big flood in '96, so there was it was quite devastating; we lost a lot of logs in all those.

When did the pulp operation stop?

I believe it was '67 or '68, because that's when I applied for the job in Oswego. We would eliminate the digesters in West Linn, but then we would have the chips. We went into wood chips, and the wood chips would go into the digesters. They would cook the chips for eight hours and make pulp out of these chips. And therefore, these little wigwams—they were all over the Willamette Valley— they were burning the side of the log, you know, because the tree is round, and then the sawmills are made for square logs. So you had to trim the sides of the round log to make it a square log, and the sides would be taken into the wigwams and burned. They found out that there was a lot of pulp, or a lot of wood chips could be made, so they decided they would take the bark off, and then they'd run the side of the log into a chipper and make chips. These chips were about 7/8 inch long, which makes the ideal paper. And we selected Oswego; we had eight acres that we rented from the Port of Portland, and we made underground conveyors. The barge—it would hold 600 cubic feet of chips. It would hold 60 truckloads of chips, and it would be sent to Wauna, our mill in Wauna, and also to Camas. So therefore, that's why we had to shut down this. So therefore we got the pulp from the different mills, and we did not have to have the digesters at West Linn to pollute the water and the air. And that's how we started with the Oswego chip exchange.

I didn't know that. Never heard of it before.

Well, it didn't last—matter of fact, even some of the vice presidents didn't know about it either. Actually, our president, he would stop, he would drive by—I forget his name right off the bat right now—but anyhow, he would stop from the airport. He'd stop at Oswego. And then from Oswego we'd phone West Linn: "In 15 minutes you're going to have the head man coming." So he would go to the other area. But the reason we had it in Oswego, railcars would go into Brooklyn Yards, and to shift them out of there, it would cost $50 a car to run them any other place. So therefore, we would unload about 50 cars, times 50—that was pretty good.

Big money then?

Big money at that time. And we had a lot of trucks. We would get anywhere up to 50, 60 truckloads also coming in from Warm Springs, from the Eugene area, the coast area, to unload their chips.

You said the chip plant didn't last very long, that even some of the vice presidents didn't know about it?

Very true, very true.

And you got 15 minutes' warning to the people at the mill...

West Linn, that's right. But anyhow, we'd save money by transferring from Brooklyn Yards to out there.

So how many years did that last?

I'm trying to figure out. Okay, let's see. I started there in '66...'86 it was probably about 20 years. Yeah, because when I left, there was somebody else that took my place. I did for a short while, but then I got the sawmill and the Lake Oswego wood chip job. Well, actually I had the river job, the sawmill in Oswego.

So you bid on all of them?

Yes, I was superintendent. That was my job—three, yes.

So you had off-site places to go and supervise, as well as on the island.

I had lead men; they were very good. So therefore we worked together on that.

What about maintenance? Who maintained the Oswego place? A separate crew entirely?

No, it was from the mill. They would come to the mill, yes. It was sort of bad for us, because anytime we wanted anything done electrical, we were not licensed to do anything electrical, so therefore we had to wait. And if they had a paper machine that was down, the paper machine always had priority, because that was the finished product; that's where our money came from, is paper. And so therefore, we had to sometimes wait for three or four hours before we could get going again. But that was one of the things that was part of the game, you know.

At some point you made the move from hourly to supervisor.

I was an hourly person. When I bid on the Oswego job is when I became salaried.

So what does a bid look like?

Actually, to work for Crown Zellerbach you had to have a high school education. That was number one. And then they'd bid a job up; they'd put it on the bulletin board that there's a job opening in Lake Oswego. And then if they picked you, you interviewed with the personnel people, and they told you what it consisted of. And then you had a choice to either take it, or maybe they didn't want you either. So that's when the decision was made with the employer and employee.

So you might have 20 people apply for that job, bid on it?

Or more, or less, yes. But you know, the people, when they'd get started with a job, they would very seldom want to move. We had people down in the grinder that worked down there for 45 years, and like you say, it was just like a dungeon. They handled those blocks of wood—some of those blocks of wood were up to 60, 70 pounds, and they'd have to put in three of these blocks into a stone. The stone was probably four foot in diameter, and probably two feet wide, and this was run by waterpower. They had a shaft on them, about an 8-inch shaft, that was turning this stone. It was going quite rapidly. And there was a jiggerman and he would—depending on the quality of the pulp—sometimes he would have an iron, a piece of roller iron [jigging burr], that he would run on this stone to make it coarser, so the stone would grind faster, but also it would grind with longer fibers so they could make good paper. But some of that paper was made for newsprint, so it really wasn't not top-quality paper at all.

So some people chose to do that hard work?

Yes, because, okay, you kept those stone, you kept the pockets under pressure with wood in them, and you did that for eight hours, and you ate whatever you wanted to eat, and nobody bothered you. And you see, when you got on a paper machine, there was a lot of people who was hollering at you—do this, do that. You know, the paper's broken, and you've got to get the paper out of there let's get going. Actually, when I got off the paper machines, that was one thing—I went to personnel and said I don't like this; it was shift work. We worked, day, swing and graveyard. And every week it was a different shift. And this was about 1951, when we were planning to get married, and my wife didn't like it, and neither did I. So I changed jobs. But anyhow, some of the people just liked being…like the millwrights, you know. That was a greasy, dirty job, but then the pay was good, and it was day work, and they didn't complain too much. And actually, the people that would start, and a lot of times they'd go on to other jobs. And I had a chance to go to work for a safety company, Sanderson Safety Supply, and I chose not to. I only had three miles to go to work, and parking was good, and you know what? I never had a check that bounced from Crown Zellerbach. It was always good money, so why move?

So how did they choose between you guys?

Well, it's just like any other job; they'd see who's qualified the most. I started to be manager when I was 14 years old; that's when my dad died. So therefore, we farmed, and I guess maybe that had something to do with it; I don't know. I was active in sports, so I managed a softball team. I felt that I had also done supervisory—I thought if I didn't stay there, I'd at least have some credentials of being a supervisor, which I have. And so therefore I think probably that's why they chose me, I guess. I don't know. A lot of it was just verbal with the personnel person.

So after the river you went to Lake Oswego as a supervisor of three different units? And that's pretty much where you stayed.

Ended. Yes, very true.

How much did you feel you were watched when you were supervising?

Oh, everybody watched you. Union people watched you. Matter of fact, I even got called in the office once. I was helping at the sawmill; we hired students, you know, for in the summer time, when it's during vacation time. And this student was supposed to take care of the part of the conveyor where the blocks dropped down. Some of the blocks were too large; they had to cut them up with a chainsaw and split them and drop them down on the conveyor so they'd run through the hog, like we talked about, and then go to the steam plant. And I was helping him and showing him the safety factors of the chainsaw and how to spread your feet apart when you're chopping wood. And the union person seen that and said, "Well, are you working at an hourly rate?" I said, "No, I'm just showing him." But anyhow, he went to my boss, and of course my boss said, "You don't do," he said "you show the people, but you don't do it to take somebody's place." which I didn't. A young fellow, just going to college. I wanted him not to get hurt so he could go back to school.

So everybody's watching you, you know, and of course production was number one, production and safety. Actually, we had 12 years without a lost-time injury in the sawmill, which is unheard of. So anyhow, we had good records. And every morning I would go out and check with each member of the crew, the first thing when we started. I would walk through the whole mill. And we had three stories, so we'd walk those and then talk to the person every day, because the reputation was before that some of the people drank quite a bit, and if I smelled alcohol on their breaths, they'd either drink a bunch of coffee or I would sit them down for an hour or so, or give them a job sweeping the floor instead of running the machinery. And they found out that I was that strict, so therefore they'd come in very good shape.

Are you the only person in your family who worked at the mill?

I had an uncle that worked down at the mill. And you know that John Raines was the original personnel man, and I think that he would hire people from different clans. My parents—see, I'm the first one that was born here in the United States. My family came from Lithuania, and therefore they felt that those Lithuanian people—they had a lot of them working down at the mill—they were good workers. And anyhow, my uncle was down there, and this was when I was out of high school, and he says, "Come on down and I'll get you a job." And that's where we started, with his prominence…you could make some money, and there we went.

So you went right out of high school?

Well, I bummed around for a while and worked up at some dairies and stuff. I think I was 20 years old when I started down there, or a little bit less. Nineteen, maybe.

Ed said if everybody that he was related to stayed home, the mill would have to close.

There was a lot of them; that's very true. For instance, there was a Benski family. One of the boys is still working there. But the grandfather, his father, and now the boy's working there, so that's three generations. Like I say, it's still a good place to work.

Anyway, lots of clans down at the mill. And on the river.

Yeah, that's very true. Yeah, there was three brothers on the river, and you know, you had to get along with them, or you didn't work there very long. Planton boys. Henry, Stanley; there's two of them. But anyhow, yeah, there were three brothers down there. I think it was Joe, Joe Planton, Stan or Henry. Yeah. They did not want to go on the tugs, so therefore they worked sorting logs the whole time. Yeah. And there was a Zaniker. He was a pilot. Yeah, there was quite a few old-timers. And they all worked 30, 40, 45 years for the mill.

You said if you didn't get along with them, you were in trouble, and I was trying to figure out what you meant. I thought maybe they were independents…

No. They all worked together. Yeah. They all drank together too, you know, and I think that—there was a lot of it on the river.

Any stories about the people at Sullivan or at the locks?

Oh, they were very helpful. You know, we always needed help going through. Sometimes, you know, you had to catch a hook with a towline. We had a towline that was an inch and a quarter in diameter to slow their boats down. You see, a lot of times we had more than one barge, so therefore one barge would go up by itself, and then when they let the water, you know, there was about 40 feet difference between the upper river and lower, and there's four lifts. And so each lift, you had to catch a stopping line. There was a hook there you had to hook onto your line to slow the barge down slowly, because you couldn't just snub it down real quick, or they would break the line. You had to slow it down slowly, and of course you didn't want to go into the gate. You had about 10, 12 feet to go to the end of the locks, but you didn't want to hit those gates, because it would ruin those gates, and then they'd have to shut down and replace them and so on. It would be a tremendous job, and so we had to be very, very careful not to do that.

I never thought about timing it so they could get all that momentum slowed down so it would stop exactly where they wanted it to stop. That's quite an art.

Yes. Oh, absolutely. Very, very true.

Do you remember how much the first job you got at the mill paid?

I think it was around 80 cents an hour when we started; something like that—82, 80 cents.

That was for the tour guide?

Yes. Well, a lot of the jobs are just base rate. And if there's advancement, then they were a little bit different. Just like in a crew, there was somebody who might be leading the crew; they would probably get 10, 15 cents an hour more; something like that.

And by the time you quit, you had a pretty good retirement package?

Still do. But, talking to some of the other people, they differed in this because, you know, they'd always get—like our Social Security, you know—you get a 2 percent, 4 percent raise, I guess.

Cost of living?

Or whatever, yeah. And we never did. We just got our package deal, and that was it. But I can't complain. And we also, with the salary, we got stock into the company, which made a tremendous difference, because you felt that you were part of the company, and I know that times there I worked six weeks straight, Saturday and Sunday, because the water was up high, so therefore we had a lot of problems with securing things. And I worked six weeks. But on the other hand, I got to retire when I was 56. So it's been over 23 years retired, yeah.

I'm hearing that from almost everybody: they were 56, 57, 58 when they left. And a lot of them left about the same time; they thought there was going to be a change in ownership.

Well, there was, yes. See, like I say, when this Goldsmith came in. And they sold off the mills, and that's when I think, uh, Simpson took over first?

Well, there was James River first, I think.

Well, yeah, okay. James River, but he bought out, and then he sold to Simpson, so therefore, yeah, he did not run any of the mill, but he just sold them. And that was sort of a sad day because actually, it was our own fault. We had people—we called it the White House, Ivory Tower, down in San Francisco—and that's where they stayed. And our stock was about $40-$45 a share, but actually it was worth about $90 a share. So it didn't take too many smart people to recognize that. Goldsmith came in here with these accountants and said, hey wait a minute! All this property—we had river property, we had timber, and we were the first company to replant the forest. We'd done a lot of cutting and selling the timber to the Japanese people, but also we started planting. Right after we cut the timber we started planting.

We also were one of the first companies to start fertilizing the timber, the trees, so they would grow faster. It takes about 50 years for a tree to be ready to harvest again. So therefore, we had done everything right, and like I said, this Mr. Goldsmith came in there, and he knew that, and that's what he done—bought 51 percent of the stock and said, "So long, boys." And of course, that's our system in this country, and nothing wrong with it, but it shouldn't have went that way, because—our people did not realize that, I guess. And of course we had people that were working in Washington, D.C. We had two reps, because, you know, when you think they have money, everybody wants to sue you, and so we had two attorneys back there, doing the lobbying and so on, but I guess maybe they didn't work hard enough. So that was really the story. He wouldn't have made $400 million, because we had see, [for]100 years, we had the people that we'd tie our log rafts up, we'd probably give them $100, you know, to use their property. And therefore they could not use, or they could not have any boats on their property because we had rented that space from them. And so therefore—but they had that $100 a year that was making them money, so that's the way it went. You know, that was the story.

I like to ask people what their experience was during the Christmas week flood? Were you working in Lake Oswego then?

Let's see…yes. Well, it just flooded. Actually, it was about two days, and then the water started going down, and then it was all right. But it did a lot of damage in the mill. A lot of the electrical switches were at about six foot, but the water raised up to about eight feet in the basement, so therefore it just knocked out the electricity. And, yeah, we were down at the mill with lanterns and all kinds of portable equipment just in order to see where we were going, because down in the basement, those places were just absolutely dark. They relied on electricity 100 percent.

The river was rising up all around these rooms, up above your head.

Well, we had big logs down in the basement. These logs were probably three feet in diameter, and how they got through the doors nobody knows. That's right; they just happened to be there. One log, anyhow. They had a lot of debris, you know, and then of course a lot of silt. After the water went down, everything was just mud; it was slippery; it was silt, from the farmlands where it washed down, washed into the river.

The mill got phone calls saying take the motors up another two feet, you know, as the reports of the flood came in.

From Eugene, yes.

I mean, it's a wonder people didn't drown in there.

No, yeah, but the millwrights, they had all kinds of jobs that they had to do. They had to go up on the roof sometimes, maybe 20, 30 feet above the roof, to change the signs or something like that, or the roof was leaking. It was very dangerous; all the jobs were very, very dangerous. Yes. That's why they got paid a lot; it was very dangerous. And that's why some of them stayed on, and some of the people that worked for a while, they said, "No, I don't like this." And away they go. At least they had a choice.

I remember somebody talking how the log got down the basement.

That's very true. Or a door. The doors were knocked off the hinges, yeah.

So did you ever run into fish or nutria or beaver or rats down in the basement? Or did you meet the ghost?

No. I've seen a lot of fish down in the basement. But you could really tell if you wanted to—a lot of the people had the metal lunch pails, you know, but a lot of times during salmon season, they would always carry their thermos bottle on the side. You know, normally there's a place for the thermos bottle inside; they had the fish inside and carried the thermos bottle on the side. Yeah, you had to have humor too, you know, throughout the whole day, because a lot of things did go haywire, especially in the sawmill, you know. I was there when they started the new sawmill, but it must have been in the '50s or something, and you know, as time went on we got smaller logs. We were put together at times when the logs were huge, big logs The mill was built to handle large logs. Our mill was made to handle a 72-inch log, which is six foot in diameter. We had a lot of these logs, especially the spruce logs. And then as the time went on, the logs got smaller and smaller because we were selling so many logs to the Japanese people. And so therefore our logs got smaller and smaller, and having this big equipment handling small logs was quite difficult. We had to change a little bit, even to move these small logs through our mill.

So it became mostly Douglas Fir rather than having some of the other kinds mixed in?

There was all types of species of wood. We tried alder, which was unheard of that that time, but we had an experiment on alder. Cottonwood was used a lot, but the cottonwood, it was a very heavy grain, so therefore it would not hold up in the water. It would sink; it would sink down to the bottom. So therefore they did not use cottonwood much. And hemlock and spruce were the main species that we used in our West Linn mill, and that was ideal. It made ideal paper for us in our plant. But I realized at one time that we had a lot of logs that would sink in the raft. You know, you'd walk the raft, and pretty soon there would be a space, there would be nothing there, and so therefore some of the logs would sink. And I suggested that we start dredging or salvaging these logs. And we had enough logs one summer to run all summer long, run our wood mill, with these salvaged logs, which was a real plus, because we saved a lot of money with these logs that we used that were sunk. And actually, the Japanese people, they sink a lot of their logs because they don't have enough storage on top of the water. So they sink them down below the water. And then when they need them, they bring them up from underneath. Spruce is normally a high floater.

Okay, what sank? Cedar?

No, it's probably the cottonwood, maybe alder, some of the Doug Fir, some of the hemlock would also sink, but not too often. Hemlock is the one that was heavy-bodied, yeah.

So because of your idea you were able to pull enough one summer to really run it?

Yes, we hired Bernert tow company to bring their dredge in and dredge the river. And we found a lot of good logs, yes. Yes, it was really something.

Any adventures at the mill, in such a gigantic-scale industrial operation?

The big scale that I really enjoyed, and I thought it was really good, we were always learning things every day, but our boss wanted us to go visit other mills. Every month he'd want us to go and visit the mill in Estacada, and actually I took a trip to Canada and some of our mills in northern Washington, and the Wauna mill, so see how they were running their operations to be able to fix some of the areas in our mill for improvement. We have a job known as the off-bearer in the sawmill, and the off-bearer stands about four or five feet from the main saw, and so therefore it's very scary, and it's a very dangerous job. And we put a hydraulic piece of equipment in there that took the place of that person. The union was very upset about that at one time, but even our union people realized that it was a real comer to bring in something like that, a hydraulic piece of equipment, to do the same job as this person was doing.

But take the liability out of it.

Yes, yes. It was very, very scary. And we had a lot of material that we cut through. We had an ax that we cut through, and it was probably left by a hunter, probably in the crotch of the tree, and it growed over. And anyhow, we cut into it, and of course it just ruined our saws. But we also had armor-piercing bullets that the fellows from the war would bring home, and then they would shoot into a tree for target practice. And that would rip our saws up also, because the armored points would rip our saws right off. Our saws would cost about $2,000 for that one—it's a 60-foot band saw, is what our main saw was. And it was a lot of money. The biggest thing was the danger of flying shrapnel hitting people.

And this is just in the sawmill, at one point by the band saw?

Yes, known as the off-bearer; that was the name of the position.

I saw a big contraption shooting the bark off big logs, so that must have replaced people taking the bark off the logs. That ring a bell?

It does. But these pumps were run by 800-horsepower motors, extreme pressure. If you'd leave the jets on one area, it would cut a log right in half. This log would have to be spinning at all times, and it would take off bark. But there was so much maintenance that we went on to the other barker. We just couldn't keep up the maintenance on the high-pressure washer, and so therefore we went to the drums known as the drum barkers. It would take about an hour and a half for one block of wood to go through the whole system of being debarked. And then sometimes with the fresh bark, it would not debark. And we had people on conveyors that would pull them off, and we'd run them through again, because there cannot be any bark on them at all to make top-quality paper.

So first it was manual, then they tried the water jets, and then there were problems, so then they went to the drum debarker.

To the drum. Yeah, it was probably about 20 feet in diameter, and possibly about 40 feet long. And it was sloped in such a way that the blocks would run—after running about an hour, it would move from one end to the other and then onto the conveyors.

And then the bark would go...

To the steam plant to be burned, to use the steam to dry the paper on the paper machines.

The sawmill just sort of grew around old, existing buildings. I think it's all still there, basically.

That's true. They made these buildings so the weather would not bother them. They were actually made out of Masonite, and so therefore they stood the weather well, and they're still standing.

What the best thing was about being a Crown Zellerbach employee for all those years?

Well, the best thing is that you worked for a large company, and you seen the finished product, with paper going out, or reading the "LIFE Magazine," "Sunset Magazine." And a lot of times you met people. It's a lifetime membership where it's associated with people and their families, so therefore it was family-oriented. And like I indicated before, there was never a check that ever bounced from Crown Zellerbach. So there was several things—and it was close by. And Crown Zellerbach was very good to students. I was given a job in the summer time. We had doctors that came in—or they became doctors and attorneys after working the summer time for our mill.

So what was the worst thing about working at the mill, if there was one?

Well, there's bound to be, you know, the good and the bad, but I think the good overshadowed the bad. So that was the story.

That's a very diplomatic answer. And do you remember your last day? Did you have a party or anything?

Actually, I retired in July of '82, but I worked until December, and then retired in '83. And we had a big party in the Portland area. That's where we got our picture, the big picture on the wall. That's what we got, and also that little—I don't know; I'm not sure what you call it.

It's a trivet. Tell me about that.

Well, when we retired, I think one of our vice presidents was an art person, or loved art, and he became acquainted with this company that made these figurines or these little plaques, and he presented one of these to all the people that retired in 1983.

I'm not sure that you could read it. It's "designed exclusively for Crown Zellerbach Corporation, copyright by Ruth Asawa, 1979, cast by Berkeley Art Foundation."

Let's see. We had about 35,000 employees worldwide. Did we talk about that too?

No. You still have the "we." Twenty-three years later you're still promoting Crown Z …

[Referring to his prepared notes] We talked about the paper, the clay, 10 paper machines, yellow pages. Let's see, we had coffee filters. You know, we invented the coffee filter and there was started in West Linn, and there was coffee coming out of our ears, because everybody had a cup of coffee because it was run through that filter. And we also made all different types of toweling—the blue paper that they used on windshields on cars. That was made at West Linn. Oh and also, making paper starts with trees. You know, I think we mentioned the trees; they had to be about 50 years old before they cut them down. Oh, and another thing that was important to the industry: after you cut a tree down, they had a logging landing where they would segregate the trees out. And the number one log was known as a peeler, so that peeler went to the plywood plant to make plywood. Actually, the log is peeled about a quarter of an inch. It's peeled until it's down to about a 10-inch core. And then the number two logs were sold to sawmills for building materials. And the number three logs were sent to us to ground up to make paper. So there was three different logs. And they also you see, they got top dollar for the peeler log, and then the number two log was very expensive. And then the thing that they couldn't do anything with, but we could, we could grind it up into pulp to make paper.

So you get logs with big knots and…?

That's very true, you know. When they cut our logs, sometimes on a hill, or a downhill, and sometimes they split. They can't make any lumber out of it for building materials, so we get it.

Anyhow, you know, paper reaches every person, and [pause] yeah, that was the end of 100 years of paper manufacturing by Crown Zellerbach when Mr. Jim Goldsmith retired. And paper reaches every person. What would you do without paper, you know? It's very important.

[Post interview conversation]

I had one woman that came into our mill. Well, if they had to lift up a log or a piece of wood or something, you know, something, I had to run up there and help her.

You have some of the best hearing of any of the men that I've talked with.

You know what I used to do? I used to work with the nurse at the mill down there, and the instrument she had [audiometer?] Okay, I would go in there at least once a week to set that thing for her. But I always wore ear protection. Absolutely. And I still do when I run the tractors and stuff.

Rosie's so proud of that trip he made to Holland—where he worked on the Dutch machines.

Yeah, very true. And he had another assignment down in South America the same way. Actually, it was a floating paper machine. And we made a lot of money, but we spent a lot of money. They spent around $400 million or something like that floating the paper machine on the Amazon River. And they would be taking the timber off both sides of the river, you know, and feeding it to the mill, and they'd keep going down. Of course, they didn't care about the pollution or anything, you know. …in the river, but it didn't finally work out very well. But it was an experiment in making paper and stuff. You know, they experimented with all different types and grades of paper, the colors, and all that stuff.

Anyhow, we shifted gears, and I joined the master gardeners 23 years ago, because that's when I retired. I seen the little blurb in "The Oregonian" about master gardeners at the Clackamas Community College, Claremont Hall. And I applied for it, and I've been there with them ever since. I've got some nice plaques that I've received also. I don't teach; I coordinate the new classes. We've got classes that we start first of the year. We have it at the VFW hall in Oregon City. And I'm the coordinator. We had well, with the better master gardeners, our similar classes were 144 people, but we had 82 new people this year. This is Multnomah, Washington and Clackamas County. And in our county we have over 300 Gardeners.

Milton Wiese

Interview April 8, 2006

Would you please tell me your name and how old you are?
Oh, I'm 80 years old.

And how old were you when you were first hired by Crown Z?
I don't know; I'd have to subtract. It was in '43 I first started working there, and then I went into the service, and then I come back in '46.

Was that your very first job, or did you work somewhere before that?
Oh, yeah. I worked out on the dairy. Yeah. I worked at Publishers, or Hawley's, or whatever you want to call them; there, too.

So why did you move to Crown Z?
I liked it better— The way they did things and everything was better, a lot better.

So what was your first job?
At Crown Z? Bucking rolls on No. 9. About got killed.

That's heavy work, huh?
Yeah, yeah. They—at that time they only had sawhorses. You'd take them off of a dolly and put them on a sawhorse, you know, and then roll them down. And that's the way they did it then. They didn't have the cranes and everything then. Then they changed it all over.

How much did they weigh, do you think?
Those rolls? Oh, they were probably 1,500 pounds. I'm just guessing. It's been so long ago I forgot. Yeah, better not quote me on that.

So you started in '43, and then you got drafted, or—
Yeah, I got drafted. Oh, yeah, I was in Germany; France and Germany. I was in the infantry—And the Air Force saved our—saved me. You know, they had all those B-17s flying over and everything like that. They really saved me, you know. And I got a lot of respect for those guys. And they lost a lot of men, too. Were there nine men on an airplane or something like that?

But you started working at the mill in '43.

Yeah.

You got drafted in '44?

No, I think it was later than that. It was '44, late '44, real late.

So you were over there for a while, not too long, in Germany, and then the war ended. And when you came back…?

I thought I would have to go to—they had us all packed up, ready to go to Japan. But the atomic bomb saved us; didn't have to go.

So then you came back to Oregon City?

Oh, yeah. Yeah.

You were a single guy.

Oh, yeah, yeah. Yeah. Never got married until 1950.

Roy said that when he got out of the service, he and a friend decided to travel around.

Yeah. I didn't want to travel no more. My folks would take off for the beach or something, and I says, you go; I'll stay home. I didn't want to go; I didn't want to leave the house any more.

So when you went back the second time to the mill, what job did you get?

Grinding room.

Do you remember your first day there?

Yeah, it was graveyard, and everything, and you know I damn near went to sleep standing up, and everybody was watching me, thought I was going to go to sleep and fall over, but—that graveyard got to me; I didn't like it. So I said, I want to get out of there, you know, and so they put me in the sawmill. Sawmill worked days and swing, so I worked on the chain, you know—the chain would deliver wood to the grinding room, you know—

The chain delivered wood that was already cut up?

Yeah. It was already cut up into two-foot by no more than 11 inches. Yeah, but they were heavy. From there I worked up—in the sawmill—worked up to the carriage, where the carriage ran back and forward, sawing off the cants.

I don't know what that is.

Well, you know how they make lumber? Where they saw off a slab? Well, that's what we were doing. No, the carriage was on a track. It went back and forwards. And they had two big circular saws come down like that, you know. And the carriage would take the log and go right through, saw off a cant—you know, a slab, all the way from three inches to probably up to 11 inches, you know. It was up to the sawyer. He judged how much he was going to cut off, or how to cut it off, you know. And then they had a hook on it that turned the log over, and they'd cut off a cant, and then they'd turn over and everything. Well, I was running the carriage, and every time they'd turn over, well then I'd have to dog it back in to hold it on the carriage like that. Oh, yeah, we sawed off lumber about two inches, three inches, you know—big slab.

Did you ever make any actual lumber?

No, no, no; we didn't make no lumber. The only lumber we made was—say somebody wanted a walk, like up there at the logging shack. We want a walk, you know, so we'd cut off about a three-foot wide cant and maybe four inches thick, or maybe more, you know, and then we'd have a long cant. But that's what we'd use. And then we'd use it like a diving board to walk out to the boom. Planks, yeah. You got the idea. It's a big plank.

They must have been big logs.

Oh, yeah, yeah, yeah. We had some big five-footers through, you know.

I learned from somebody that there were three grades of wood.

Well, I don't know. They had hemlock and spruce and sometimes cottonwood, and sometimes cherry—very little. Maybe a little alder too, but—but mostly hemlock. And the wood was not for lumber. It was simply for making paper, you know. They'd sort the logs out at Canby, and we'd sort it out. If we see a peeler log or something, we'd shove it aside, because that would be for lumber. We'd sort them all out.

Some logs were sinkers?

Oh, yes, yes; we had a lot of sinkers that would sink down and everything. I bet there's still logs up there in the river that sunk. Oh. Yeah, they would sink, you know, left and right.

What sank?

Uh, hemlock would sink, mostly, and cottonwood. But spruce was nice floaters and everything. Like they made that airplane.

Oh, Spruce Goose?

Yeah.

Were you there when it went through the locks?

I went up there on the bluff and watched it through, her and I.

Photo Courtesy of Jane Richardson. Spruce Goose fuselage in Chamber 2, Willamette Falls Navigation Canal and Locks, 1993.

Did you take pictures?

No, nah. I wonder how they got the dumb thing off the barge to truck it up the road to the spot?

They must have had a crane, I imagine.

Must have had. That would have been interesting to watch.

I think the locks are interesting— everything had to go through there…

They even went through with gas.

You mean propane?

No. They said it was gas. In a barge and everything—and when that went through, Jack McFall was the fire marshal, and he'd go out and walk along with the barge along the dock. And if anybody was smoking, he'd tell them to put it out, you know. Safety, you know. So that barge went through…

What barge line?

It could have been Bernert, you know.

So you worked on the carriage for the saw?

Oh, yeah, we're going back to the sawmill again. Yeah.

Somebody told me this was a 60-foot saw, really long?

The carriage? It must have been the carriage, yeah. It was on a track; it went back and forth. See, they had cables on each end and pulleys down there to pull the thing back and forwards.

Did it ever break down?

Oh, yeah. You'd always shut down and clean up the sawdust and everything like that.

Did you have your favorite millwrights that came and worked on the saws?

No, it didn't matter. No, no. They were all nice guys, you know. You know, we'd kid around. If I needed any advice or something, like how to mix cement or something like that, I'd go talk to them. And they'd sit down with me in the lunchroom or something, and write me out a piece of paper and say, this is how many shovels of sand, and how many—this and that. Yeah, and then come home and get the cement. Had the wife and kids mix it. They helped me a lot.

So what happened after that?

I went from there to the millpond out there. And then all you had to do is take the cables off—well, no. I got ahead of myself. I had to punch slip. There's a chain that drug the logs into the mill. They had big dogs on them, cleats on them, you know. And you'd put the log to the end of that, and it would pick it up and take it right into the sawmill, you know. That's the way it would go. And then from there I worked further out to the millpond. And I took the cables off the logs—pulled the cables off. And when they'd get them off, well then they had kind of a—oh, I don't know—a cradle there. But anyway, they had a wall on one side and wall on the other side, and you'd shove the logs down there by theirself, you know. And then the locks would draw water, and they'd pull the logs down to the slip, you know. They would be running a barge through or something. They'd open up those guard locks—you see, it was above the guard locks.

Where was what you're calling the millpond?

The millpond was where the guard locks ends. Right now, they got a crane out there. But right on past that crane, the millpond is still there.

There's a big thing called the log haul. Is that what you were feeding the logs into?

No. No, no. This log haul come in later.

So you were out standing on the rafts when you were taking these cables off?

Oh, yeah, you were standing on the logs. Oh, yeah. Yeah, you'd walk right out on them and then like that. Fall in once in a while. If you didn't fall in once in a while, you weren't doing anything. You'd lose an ax too, sometimes, you know. Some of the guys had a wire on the end of the pole, pick the ax up off the bottom, because the ax would—the handle would stick up. I never was any good at it. Some of the guys would—I couldn't get the dumb thing up.

So when you went from one job to another, did you have to bid on them?

In the sawmill, no. You know, they just—I just said, yeah, I want that job or something like that. Or they'd say, hey you, go do that. That was the way I got on the carriage. I was down there, and he says, we need you up there. And so they put me on the carriage, and that's the way I learned how to run the carriage back and forth. Sh-sh-sh-sh; back and forth like that.

All day long?

Yeah.

What shift were you on?

Well, they switched around every week.

How did you like changing shifts?

It was fine. Swing shift, I had to come home here and work all the way around the house, and then I'd go work, and I'd be dog tired. So I had to watch that.

I've heard about people that had jobs down in the grinding room, they said it's just really wet. Was the sawmill drier?

Yeah, sawmill was a lot drier. No, no, no, there was nothing like that. No, that was the steam and everything down in the grinding room. Some of those people, they liked it. I didn't like it, no.

It looked like really hard work—just tossing those blocks all day long.

Yeah, you had to handle those blocks all the time, throw them in the pocket, things like that.

So wasn't the saw really noisy? Did you wear ear protection?

No, they didn't have that stuff then. No, no, no. They didn't have anything like that. I don't recall even wearing a helmet or anything like that.

But you sound like you hear pretty well.

Well, I've got one bad ear, and she's got the TV on pretty loud.

Well, we're going to go see Art Dorrance, and he's really deaf.

Oh, you tell Art hi for me. We were pretty good friends.

What part of the mill did he work in?

He was a plumber.

Oh, he was a plumber. I thought maybe millwrights did the plumbing too.

Oh, they probably do. But he was a plumber; he could plumb anything, you know. He's a smart cookie. Yeah, he'd build houses; he was pretty good. And not only that. Remember PACC [Physicians Association of Clackamas County], you know? He was an officer in that.

So when did you start on the river? I mean, you went to the millpond…

They had to put a bid up, so I bid on it. And it went by seniority and stuff like that, so I—that's when I got up there to the loggers' shack. We called it a loggers' shack—up there, and that's when I started working on the logs and doing with the boats and like that.

Then you liked that?

Yeah, yeah, yeah, I stayed to that until, I don't know what year. They cut back, and I got laid off on there, then I went back to the yard gang, and I worked in the yard gang there. But they were pretty good about it. When they needed a man to move the barges and everything, they'd call my supervisor on the yard, and they'd call me, and then I'd go down and start working on moving the barges in, and all by hand, moving them in by cable. See, the one barge, the first barge, would go down with the boat. And then, let's see, there'd be maybe, there could be up to five barges going down. Most generally there was just two. But they'd just put the one barge—it would go down first with the boat, with the tugboat.

And then the next one would go along while we'd pull in with the air winch, and pull it into—I don't know what you'd call it. But I mean in the middle there: you'd put it in there—right to the edge, where the gates were. And as soon as they opened up the gates, then the bridge would go up, and then they'd either wash it in with the water—you know, the current—that way, or if there was a— wind was coming from the north or something like that, well, they'd have to put the winch on it. They—the government—would put their winch on it and pull it in that way. Then there could be two of us on the end of it. Have you ever looked in the locks?—you see a hook, a cable down with a hook on it. Well, anyhow, it had a hook on it, and then you'd have a rope, and you'd hook the hook in the— them together, and that would stop the barge.

So it wouldn't hit the gate.

It wouldn't hit the next gate. And then you'd give the other end of the rope to the guy that was standing on the dock, and—there's posts there; I could show you pictures of them—and he'd throw it around that one. And that one would keep the barge from surging back and forwards and everything. Then as the barge went down, well, he'd slack off just let it go down. Then when they went to the next lock, they'd put the winch on it again and do the same thing over again, all the way down. Then he'd come up and—so just opposite.

It's a delicate maneuver, all that weight going, trying to get it stopped at the right place.

Mm-hmm. And that one picture there, where they're up against the wall there—that barge had 600 tons paper. That's what I was told. Now, I don't know; I never weighed it. [grinning]

So it was the mill's paper being barged down.

Yeah. And they'd barge up pulp too. Pulp, and sulfur. Whole barge of sulfur, you know. And then hog fuel, you know.

So the mill used the locks a lot.

Oh, yeah, they used a lot of chemicals. They had sacks of starch, sacks and sacks of starch that would come in that way. And they did some trucking too, trucked it in, chemicals and stuff like that.

As the only way on and off—any time the barge is going through, the bridge had to go up...

The guys getting off a shift, it was always a bad deal to take a barge through when a shift was changing. They could still walk across the locks, that one locks there. The government set up pretty decent for you could walk across. Where, you know, you can come on home; you ain't going to wait, wait for a dumb barge. So you'd get going, you know.

So when you were a log handler, you worked with a pilot and two log handlers on a tug?

Oh, they just put you—I call it west side; that's across from Canemah there, a little bit up from Canemah. And they'd take us down there, and they'd dump you off, and there was a shack there where we could eat and get—maybe eat out of the rain or something, and then they'd bring in logs from Canby, you know, and down, and we'd have to sort them. So we'd take the cable off, and they would have a gap there, with the timbers on both sides. Then they'd have a big log, two or three logs across, and then—these here boards I told you they sawed in the sawmill?—they'd put one of them on top, you know. And we'd shove the logs underneath, underneath that, and we'd have logs sticking out certain ways and everything. And a peeler would go over here, and one that was going to the sawmill would go there, and then there might be a spruce, and you'd shove that other this way. It was a sorting deal. You just sorted the logs. And that's what you did all day long.

There'd be tugs there, waiting for you to give them enough to make a load?

They'd shove them in to these other cradles, you know, and then there'd be guys down there putting cables across. Cables and dogs and everything, and wrapped them up, you know, like that.

So you'd sort them, they'd tie them?

Well, I could tie them, and you know, do whatever you feel like doing, you know. They always had an "HB," they'd call him, a head brailer. And he would say, you go do this, you go do that, you know. He got a nickel or two more an hour there. But that was just sorting, you know; that's where you started. Then after that you worked up to the barges. Then you'd get to work on the boats a little bit.

First you worked the millpond, then you worked the sorting?

Yeah, I worked up there at the logger shack, you know.

So did you work on Paul's tug?

No, no. Paul never run a tug. Paul went from the logger's shack down to the sawmill to be a lead man…

Supervisor?

Yeah, yeah, that's the word I wanted. Yeah, you know, he's salaried.

So after the sorting, what happened?

Well, then, you just—you went from sorting up to moving barges, you know.

Moving barges through the locks?

You had to work on the logs first, you know, tie them up, put breast lines on; they had to be tied into the bank, you know. And they had cables, wire rope, and to get them tied up, you know. And after that, then—if they needed you and you had the seniority, then you went down onto the barges. Much softer job. Instead of handling a five-pound ax, you just handled a 65-foot rope, like that.

How long did you work on the barges?

I don't know; a long time.

Was that your favorite?

No, running the boat was my favorite.

So you went from the barges to the boats?

Yeah, from the barges, you know—you become deck hand, you know. You learn how to, you know, put the barges in there like that. Tie them up to the dock—

When you were on the boat, you ran between Canby and the mill or…?

No, from Wilsonville to, oh, down to the Clackamas River, the mouth of the Clackamas. Once in a while, you'd have to take a boat into Portland to get it fixed, but that was rare.

How many years did you have when you retired, total?

Well, they give me 43 years, but well, my retirement deal and everything they only said 38.

Only 38 years. So did you have any impressions from your very first day at the mill?

I don't know. It was scary, you know. It's just like a new job; you don't know what's going to happen. You don't even know where the bathroom is, you know. And you don't know this guy or that guy and everything like that.

And you were on a bad shift.

Yeah, it was a bad shift. They had those big paper rolls going around and around and around, and the guy on the other end, he had experience, and he'd stick a—I don't know—a stick in it, and then he'd shove it out onto the dolly, you know. One time it came out when it hit me—it hit me up here and it knocked me against the deal and everything, and he kind of laughed. Then you get it on the dolly, then you have push the roll out.

Okay, so you went down there. It's late at night; you're falling asleep. You get hit in the chest—that wasn't the first night, when you got hit?

No, no, no. That was down in the grinding room when I went to sleep. I stood there and just—going around and around and around. I'd come home, you know, and my mom would say, "Go to bed!" And then I wouldn't sleep. Any little noise, you know. Because I was in the service, and I was prone to listen and see what was going on.

Oh.

So I—so that was still with me.

I'll bet it was with you for a long time.

Yeah, yeah.

When I run the tugboat, you know—I pulled 22 people out of the river, and one dog. The dog was the most fun. Yeah.

Was this all at one time?

Oh, no; over the years, you know. Oh yeah—boat overturned, three people; we pulled them out, you know, and one dog. And so the deckhand got the dog out, and he went and got the paper towels and dried the dog off. He didn't give a damn about the people, you know. And then that wasn't good enough, so he went down in the engine room and got some old—some towels, you know, and then he dried the dog off. And we couldn't get the people in to the dock because of the logs, you know. And so guess what? We had to park outside the logs. And the logs wasn't bundles then; they were just flat, you know, and everything. And so he picked up that dog, and he went running across the logs, up the bank, and put the dog down. And the dog run back over to him—they fell in love right now, you know, and like that. It was good, you know—and I had to get out there and I had to let the people—of course, they didn't have no cork boots on or nothing, you know—and I says, you step on this log, and you step on that log; don't step on the end of that; you'll sink, you know. And we crossed the logs there. Oh yeah.

So you were doing some rescue work when you were out on the river.

Oh yeah, above the falls—you know, anything can go wrong. I missed one. We had to move a barge, and I was under a lot of pressure. And they had to have this barge in the mill because the machine was going to go down, and everything. And I had come out of the engine room, and I looked over there, and I seen a 16-foot boat going down the river, you know, and everything. And I says, oh boy, oh well. So I made the decision to let it go, because just above the falls it's shallow, and the Crown Z draws five feet of water and everything. And they said, let's go. So I says, okay, let's go. And that went on down, and there was a man and woman in there—I think they were doing hanky panky—and a fisherman picked them up down below.

Company tug at lower entrance to locks. Photo courtesy of Milton and Ted Wiese

They didn't go over?

Oh, they went over. They survived. The thing was, I couldn't see nobody, you know, or else I probably would have went out—

Looked like an empty boat.

—risked wrecking the boat and everything to get them. But that's the way it goes.

What happened when they closed the locks for two-week maintenance, and everything started to pile up on the river?

Well, we had a real good boss. He said, "Go out there and put that raft up." You know, we had to put cables on it, you know, and then they had plank on it, and we'd go out there and pound the darn logs in, for one week. You know, put the raft up and everything. And the next second we'd come along, he says, "Go out and take it off and put it all back." And they said, "What the hell is going on?" you know, and everything. And he would say, "Well, you're working, ain't you?" But we couldn't say nothing about that. We were working.

So that was in these two-week periods when you were stuck up there.

So we got smart enough to know—they would say, well, this boom had to be fixed, you know, when this would come up, or something and everything. Or we'd build a new shack, you know, because they had to have shacks on the river, you know, where we'd go and eat our lunch, put our tools in there and everything, where they wouldn't get stolen.

Did you guys fish from those shacks?

Well, Mootry, you know—I told you about Mootry and the pitcher there? He was good. He had a little stick with a string on the end and a hook on it, and he'd go along this way, and he'd catch a crappie, you know, then he'd catch another crappie. He says, "Well, that's enough for supper. Here, Milton, you want to try it?" Well, I did and he says, "Hell, you'll never catch a fish." And I didn't either. That was the end of my fishing.

I think Olaf said that sometimes in salmon season, salmon would actually get up into the mill, in the grinder room.

My brother worked there too, and he got some, but I never did.

Did you ever interact with the fish, when they were by Sullivan? Or is that away from what you were doing?

That's away from us. You go out the guard locks, and you see the millpond on the left here. The end of it. And then you got the Warehouse 3. Well, it's between the end of the guard locks and Warehouse 3.

I'm not a swimmer.

Well, I'll give you a death preventer. Put on a death preventer and then you'd be all right.

What? A Mae West?

Yeah.

That's good. Did your dad or any other people in your family have jobs at the mill?

Oh, yeah. My dad worked at Hawley's, and I used to go down and help him. That was before we got married. I'd go down there, and they had sulfur, you know, and he had to put it in the wheelbarrow? I'd go down there—I'd take her to church, then I would go down there for an hour and help him load up, work and everything. And then he would catch the work all up. So he'd do all that work and he'd get everything done. Then after an hour, they'd go home and pick her up and take her home.

If you have kids, did your kids work at the mill?

Oh, no. I had two brothers work at the mill, though.

What kind of jobs did they do there?

Well, one worked in the sawmill, you know. He sorted logs—I mean blocks of wood like that. And the other one, he worked in the yard gang. He covered everything. He worked on the lime rock, and I think he ended up being a janitor.

Did they both stay for a long time—

Oh yeah, yeah.

So it was a good job.

Oh yeah—good enough for me, you know. The deal is, I didn't know what else to do, you know. I marvel at these people like she and you. You've got something to fall back on. See? You've got something to fall back—I didn't have nothing to fall back on, you know, so that was really important for me to keep that job. I darn near went to Wauna.

What's that?

Well, it's halfway to the beach on 26.

That's another mill?

Yeah, it's a Crown mill.

Ah. They almost sent you there?

No, I damn near went there on my own, because they were cutting way back, you know.

But you didn't.

No, no, I didn't. I went down there, and I asked the mill manager, of all people, what should I do, you know. And he told me. Said he'd steer me straight. He was really a nice guy. He was one of the white hats. We had them all. The white hats, good guys. I think he was in charge of the maintenance, and when he retired, the maintenance guys all got together and gave him a party. And that's pretty darn good, to give a high mukimuck manager a party, you know. Showed what they thought of you.

What was his name?

I think it's Clarence Enghouse. Yeah, Clarence Enghouse. Anyway, I worked myself up to run the job.

Run the job. On the Crown Z?

Well, the whole shebang? I'd get there in the morning, and I would answer the phone, and they say, we want this barge moved, that barge moved and everything, and we want these logs in the mill. They had to be in the mill or something like that. Or sometimes I'd just go down to the mill and say, oh, well, they need some more spruce, so I'd go get some more spruce, tow it in there.

So you were your own boss.

Well, no. They cut back—this here Whitey, when he retired, then I took over there. And he—and the yard gang boss—supervisor—he was in charge of the logs. Didn't know a damn thing about it. So we got along fine, you know. So in a way, I was kind of in charge, you know. But he was a nice guy, you know.

You've said a lot of the guys were nice. So is that the good thing about the mill—that you made friends there?

Oh, yeah. Of course, then, maybe we're forgetting something. Up there at the logs and everything, there were only five of us. That was all that was left. They had one time in the heyday, you know, and everything, when I first started up there, I think it was 30 or 40 guys up there, you know. But you had guys splicing cable, and you had—we had about 50, 60 railroad cars come in, you know, and they'd dump the logs in the river, and they would all be sorted, like that.

So by the time you got to the end of your job there, it was down to five men?

Yeah. Once in a while I would maybe have nine men, but most generally it was just five.

Then they got rid of some of the tugs?

Yeah, and they got rid of the Viola, and then we only had three. That little Billy K—the other two boats would break down, and that little Billy K would take over. It was surprising what that little old bathtub would do.

Did you guys have to fix the boat motors too, or who mechanicked on the boats?

Minor details, you know. If the hose broke, you'd fix it, or the impeller that pumped water through, you repaired that, you know. But really when it come to the diesel engine, no, no. They had some guys come out and do that. I don't still understand the diesel engine. Do you?

No.

I don't know about the locks. The railing along used to be cable....you know, cable, with an eye hook on it, you know, and like that. And now they've got this railing, you know, and pipe. And you've seen those big rocks for steps? Well, they would hold those walls in with rods.

You didn't have to bid when you were in the sawmill- but you had to bid to get the river job?

Yeah, up there, yeah.

Did lots of people want the river jobs?

I don't really know. I think there was a few, you know. I don't know. That was office stuff.

You didn't feel you were competing with people for that job?

No, no. I figured I had seniority, you know. Mill seniority. That's what they went by. And they went by, too, on your qualifications, how healthy you were, and so on and so on. If you went up there to get a job, they'd probably let you try it out.

So do you remember how much you got paid for your first job at the mill?

I think it was 47 cents.

An hour?

Yeah.

And that was for day shift, or—

I imagine, yeah.

Were you there when strikes happened? Any strike that you really remember?

No, not much. The only thing that I remember is that we went to line a barge up against the dock, and one of the mill yard supervisors came out and watched us, and he says, "You know, when you guys were on strike, it took me two hours to do that." And we just, you know, put it right in there. Nothing to it.

So the supervisor that had to do your jobs was pretty impressed.

Yeah, he came out to watch us, to see how we did it, you know, and everything. He says, "Man, that took us two hours." I remember—that's the only thing I really remember.

Well, do you remember how long you were off?

I think it was six months. Long time. Kind of wrecked my Social Security.

What did you do to support your family?

Oh, the union paid me $200 a week, and she was working, and I did the housework. I didn't like that job.

Okay, can you talk about what it was like during the big flood, in '64?

Oh, yeah, yeah.

What happened when the water started to come up, and what did you have to do?

We did a little sandbagging up at the guard locks, and sandbagging around different places, you know, fill up sandbags. I was in the yard then. I was laid off from the upper logs, you know. They started cutting back, you know, and like that. So like I said, we had 30 men up there or 40 men or something like that, and then down to five.

But you were laid off? You didn't retire?

Oh, no, no. I wasn't laid off, but I got laid off the boats and everything, so I went down in the yard. So that's where I was. And then afterwards, well I guess they would kind of talk about whether they were going to shut the mill down permanently or something like that, but they decided to keep it going, and so I got to work help cleaning it up.

What did you have to do when you were doing that clean-up.

I ran a bucket. It had a motor on it, and it had a bucket on the front. And they'd throw the stuff in the bucket, and I'd run it off to the—outside the mill, you know, and dump it. Then I'd run back and get another—pick up another load and run it like that.

Dump it on the ground or in the river?

Well, it probably would have ended up in the river, but I wasn't dumping it in the river.

Like on the riverbank, and if the river came up—

Yeah. Someplace alongside there.

So where was this coming from? Like in the mill, or—

Oh, it all come down the river.

Right. But the stuff that you were scooping up, was it from in the building?

It was in the building. Oh yeah, it was like that.

Like in the basement?

Yeah. it was deep down there. Oh boy—you know, they were always having—afraid of a war—and they had all kinds of deals down there to—food—I forgot what the heck you call that. [pantry?]

Like a bomb shelter?

Yeah, you know, they had all this stuff down there. And it all got washed away.

Oh, like a pantry.

Yeah, you know. Oh, it had anything you can think of. The only thing I can say about this here is that down underneath the paper machine, the paper machines used to go down the basement with the felts and everything. And here was a cow pie like that—perfectly, the straw and everything, right there. Inside the mill down there.

I didn't know they floated.

I didn't either, but I guess the straw and everything…[laughing]

Okay. So how about fires? Did you ever see any fires down there?

No, no, no. I told you about the Viola, you know, put out fire and they had a motor and everything like that. They had a fire there on the bank there and everything, but the deal is, we had the Viola up in slings and had the shaft out. And the guy says, "Bring that thing down here and put the fire out." And he says, "It'll sink." He says, "Well, tow it down then." It would have sunk. Couldn't do nothing. These other fire things, I never got involved with them at all.

So when the Viola was built, was it built as a fireboat?

It was a tugboat. They went and sold it to somebody. Got rid of it. Cutback.

What is this wonderful magazine that you're looking at?

I think the dumb union put it out. It's '47. Yeah. That's Publisher's, or Hawley—Beater Room four. I worked there. That was underneath No. 4 paper machine.

What's that?

Oh, that was my raft book.

Like your log when you're out working? Your hours?

Yeah, you know. That was the Canby number, you know, and then down here—Canby, you know, and—East Rock Island. Here's the piers. And had to name all these places.

But you couldn't find them now by the name, I'm sure.

No wise. Then I scribbled down here, you know—say clay had to go down, 53 to Warehouse 3 at noon, and Barge 3 hit a spot on the rock, and so on and so forth, you know. You'd just scribble things on there.

That's your notes to yourself while you're working.

Yeah. And you get on the radio.

This is your anti-theft for your motorcycle?

Yeah, you know. I used to ride it up the siding. And I'd put the cable around it and then tie it to a tree.

So you'd ride your cycle to work instead of a car?

Yeah. And that's a dog.

Oh, they stick in—

In the logs, and they drove it in. That's what held them together. But they'd splice two of them in there, and then they could have 30 of these dogs, you know, that many on the end of it. And the cable was 5/16—no, 9/16. And you'd take it across the logs, you know—wide as the locks—and then you'd have what we called a boomer log, a long log on the side. And then you'd take the cable, and you'd go down around it, and you'd throw a half-hitch in that, and then you'd put a dog on each side, one of these, on each side, like that. Then you'd go to the other boomer; the logs come together like that. You know, you'd take the end of it, and you'd go around that log and put a half-hitch on, and then you'd have two dogs on the end of this, and you'd drive them into the log, and then you'd have a dog in between, and you'd put the dog in the lower log over the side, and you'd put the other dog on the other side, on the other boomer, on the other side. And then when they'd tow it down, this here would throw this log out, and they'd give it a shear where you could go through if you'd get in a close spot or something, so it wouldn't hang up, or go through the locks.

And they call it a shear?

Well, that's what I call it, yeah. It's just the way you dog it.

So you spliced this yourself, one day when you were sitting on the boat, just nothing to do?

No, I did this at home. Yeah, this is mine.

Arthur Dorrance

Interview April 8, 2006

So, would you tell me your name and your age?

Well, I'm Arthur Dorrance. I'm 78 years old, and except for a few minor ailments, I feel like I'm 40 again, but the body don't go with the head.

How old were you when you went to work at the mill?

I was 19, I think, or 20, one or the other. I put 42 years in there, a little more than 42. And actually, if you're going to work at someplace—I went to sea before that—but probably the nicest bunch of people I ever worked with was down at that paper mill. It was all about half-crazy, and you've got to be. I'm worse than that. You've got to be that way to work there, so you fit in pretty good.

So what was your very first job at the mill?

Well, the first job I was there, I was tying up and untying a barge that was unloading for ships, and they had to move it because the crane didn't move. So I done that job, and then I went into the finishing room for a while. I've got a story to tell you about that. I was going to save it until last, but—

You can tell it any time.

Frank Hammerle—they've got Hammerle Park there in West Linn—well, that man worked there 50 years, see? And he was my boss, and Rudy Perrin, he was running what they call the rewinder in Mill C. And I had worked there a month or two, and Hammerle was the superintendent of that whole department they called the finishing room, see? Well, he came down and told us one morning after the 10 o'clock meeting that he was going to have to lay off some people—and Christmas was the next week. And Rudy Perrin said, "Well, God, you can't lay them off now, because," he said, "Christmas is next week." "Well," he said, "I'll tell you what: I'm going to come around in the afternoon around three." He said, "If I can't find you, I can't lay you off." So man, we made ourselves scarce. There was three basements, a triple basement, there in Mill C, and we went down in the sub-basement and sat on the pallet board and looked out the window and watched the steelhead fishermen until Frank had went through, and I never have been laid off, but I never got laid off over Christmas. So that old gentleman was about the finest superintendent you could ever have.

So do you remember what it was like the first day you went to the mill?

Well, that was really funny, because I had been down there about three times, and I walked into the personnel office, and Steve Coney was the assistant to George—I mean Reams; I can't remember his first name, the real personnel manager. Steve Coney was assistant, see? And so he said, "Well, if you can be back by quarter to four," he says, "you're hired." Well, God, this was about a quarter after three, see? And I had to go over to the doctor in Oregon City, up above one of the pool halls or something, he had an office up there. And he bumped my knee and looked at my throat and felt my pulse and asked me, said "You got anything wrong with you?" I said, "Not that I know of." He signed the paper, and I went over there and went to work on swing shift, see? And I made it a quarter to four. Well, unbelievable, I never was back in that personnel office for about 10 years. And I walked in the door, and Steve Coney says, "Well, Art, how are you doing?" I about fell to the floor. I couldn't believe anybody could remember my name, and I'd only seen him that one time, you know. And that was really something, because I think I went in there to ask him to see if he'd hire my wife to work up in the converting, see? And she worked up there; Ed Witherspoon's wife worked up in converting, and like I say, the women, once they got there they never left. They stayed there. But that was really something that Steve Coney remembered me; I thought it was, anyway.

How long did your wife work in converting?

I can't remember, but she worked up there quite a few years. She worked up there till she went to college full-time. I can't remember; like I say, I've got a bad memory; I can remember what happened 40 years ago, but I can't remember what happened yesterday. So it's…

Were you on the same shift?

I think they worked a week of days and a week of swing. They didn't work no graveyard, see. I worked on the paper machines for a while. When I quit the paper machines and went to Mill B, I can't remember the boss; his name was—he really gave me hell. He told me, he said, "The money's on the machines". And I went to work for a guy name Jan Haugerod. He run Mill B, the digesters where they cooked the wood, see? And I'd knowed somebody over there, so I went there. But yeah, I think his name was Harris. He was the boss of all the paper machines, and God, he really gave me hell for going up there, but I went anyway. I probably should have stayed on the machine, because like he said, that's where the big money was, see? And I spent quite a bit of time in Mill B, and they shut it down. And then when they shut it down, they got the pulp from Wauna, see? And Wauna had a big continuous digester like the ones we had that run steady—automatic one—and they got pulp from Wauna to run the paper machines. Well, then when they done that—unbelievable—I was pretty shook up because I'd just got to the top to be a cook, and I was up there—it took me 12, 14 years to get to the top and one day to get to the bottom, see? But when I started going around working in other departments, I knew all these people, but never really close, you know; just to say hi. Well, I went to work in different departments, and hell, after I went to work in one, they said, "Don't let Dorrance come in your department; they'll shut it down." Well, I went to work in—I forget which the hell department, and then they shut it down. And then I went over and went to work in the environment, where they had all the sludge and all the stuff that took care of the environmental. Well, then I bid out of there and got in the pipe shop. And I stayed in the pipe shop long enough to get to be a B man, see, and a journeyman is next. And then they laid people off there, and that's when I bid and got on the river, see? And I spent the rest of the time on the river. I could have went back in the pipe shop, but I'll tell you, working outside every day you feel so good that you could go to work with a hangover, and by noon you felt good, see? If you went to work with a hangover in the pipe shop, by noon you thought you were going to die, see? You'd be in on the paper machines, where they were hotter than hell and the steam blowing in your face, and you had a fire inside anyway, so I mean—but like I say—

What was working in the pipe shop like?

Well, that was a good place to work. All the millwrights like Olaf—he was a millwright and I was a pipe fitter—they'll do a lot of tricks on you, but if you need help, they're the first guys to come in to help bail you out, see, because if they're waiting to start a paper machine up, and you're the guy that's holding them up, well, man, your name is high on the list, see? The first thing the boss in there in the pipe shop told me, he said, "Don't ever hold a paper machine up, no matter what you have to do. If you have to throw a wrench, or a monkey wrench, to get them to go someplace else, but we're not holding them up," see? So that's the first thing that I learned. But like I say, they were good, and I think the second day on the job we were down underneath No. 9, and a friend of mine was in the welding shop. Well, they told me I was on fire watch, so they handed me a hose and stuff. And there was a whole bunch of millwrights working down in a pit on a pump, see, and Glen Jubb, he said, "Squirt them guys." I said, "Hell, they'll get madder than hell." He says, "Well, give me the hose." Well, the first thing he squirted was me, and then he turned it around on them millwrights. See, them guys are crazy, you know. They'd squirt anybody. In fact, this same guy—we got a new manager one time, and he was going through the mill; they were giving him the first time to go look at it. And they're up in the rafters…working on some pipe up there, and they had a water hose up there, and somebody told Lugg Gerkman and this Glen Jubb, said, there's our new mill manager. Somebody said, "You don't dare squirt him." He said, "Well, you mean that guy?" And he turned the hose on, and God, old Gianotti, he got about half mad. He said, "Well, who are them guys?" He said, "You're lucky," he said, "they didn't squirt you more," he said, "they'll squirt anybody." He says, "Well, I believe that." And that was the first time he'd ever been in the West Linn mill, and he got squirted with water with his new suit on.

You were going to tell me a story about Rosie.

Well, I was picked to work in the dandy shop, see? They've got a roll that goes on top of the paper machine on the wet end that rolls on it and stuff, and they call it a dandy roll. Well, sometimes they've got to re-cover them. But this guy that had that job, he done a lot more than just work on the dandies; he worked on all the guides that guided all the wires, and the felts kept going straight, see? And he put in all the insecticides in the tanks and stuff to kill all the bacteria and stuff he could do, and he had quite a few jobs. Well, they picked me to work with him. He was going on a six-week vacation, but we were the same age, so I finally told them, I said, "I don't want to stay here because," I said, "he'll retire and I'll get the job, and then the next day I'll retire. So there's no use being a relief man to him." So I turned it down. But anyway, when [he was] working there, when they shut paper machines down, I had to go in and check the dandies—not dandy rolls—rolls inside the head box that kept the papers from going into little balls… called them dandy rolls. Well, I had to check them, see? Well, I checked them on Number 4, and these guys that worked there, I'll tell you, talk about squirt water on you—they washed up the paper machine, and oh, they got me about four or five times.

Finally their boss told me, he said, "If they don't squirt you, they don't like you." He said, "If they didn't squirt you, then you'd be worried, see?" So I was inside the head box on Number 4 and I checked them, and some of these rolls, if they touched each other that would create a ball, see? It would rub, see, and build up. So I was telling old Rosie how close they were, so he was going to go in there and check, see? And when he went in there, the guys were washing up the pit and stuff, and they seen me and Rosie go in there, and sure as hell we got here, they stuck a hose up one of the pipes that had what they call a go-devil on the end. It spun around and threw water all over hell, see? Well, they got us—well, they got Rosie wet from about the knees down. We come out of there, and he went over there and told them guys, he said, "I'm going to fire every damn one of you; fire the guy that got us wet." And another guy said, he said, "They all done that; you can't fire the whole damn crew." And Rosie, he was just laughing; he had a big grin, because he was a hell of a good boss, you know? But he said, "I told them, if they don't get you wet," he says, "they don't like you." He said, "They don't have to like me that damn good." That was quite a place—

It was quite a place to work. No matter what you did, something was going on. And on the river—unbelievable—we used to think up things to get the boss mad, see? Oh, criminy.

Were you on the Crown Z or one of the other tugs?

Well, I was on the Crown Z most of the time. The only time you'd run the Mary Jane is if the boss and the lead man were gone. Then I got elevated to lead man, see? So then I had to go over and work on the Mary Jane. That was the one that hauled all the logs into the sawmill and all that, see? Otherwise I'd have been on the Crown Z. In fact, I like the Crown Z better than the Mary Jane.

Wasn't it the biggest?

Yeah, it was big. It was a better boat to run. I liked the logging stuff, but like I say, it was kind of a challenge, you know, when you're running a crew. I'd tell them that, "Hell, you guys know what to do; you've been here long enough; just go do it, and when we get done, we'll go home," see? And the guy says, "Is that all the orders you're going to give?" I said, "Hell, I don't have to give any more. If I've got to give you that, I don't need you." But no, we had a good time. But, oh God, we used to have crawdad traps, see? And we'd set them out every evening, and then every morning we'd take the Crown Z and we'd run around and pull the traps up, and then we had a big holding tank, about yay big, four by four or so, and we'd dump all the crawdads in there, and dump lettuce and all that stuff in there to hold them. And then if you were having a party, all you had to do is go out and pull up the holding tank and get a half a pail, a five-gallon pail full, or pail full of crawdads to have for your party, and then put the rest of them back. So we always had crawdads, see? Well, one morning the boss come down, old Whitey Van Domelen, and he gave a big pretty good speech. "Now," he said, "the mill just called." And he said, "They're about to shut down. There's a barge at Warehouse 3, and it has to be in right now," he said. "Otherwise, they'll be out of pulp, and they'll shut the paper machines down." I says, "Okay, we'll get it," see? So I pulled off of the dock, and the mill is this way, and I started going the other way, and he's up in the office looking. And he said, "Where in the hell are you going now?" And I said, "Well, we didn't put the crawdads in the holding tank. I've got to go up to the holding tank, dump the crawdads in, and then we'll go down to the mill." Man, he was still hollering over the radio, and I hung up on him, see? Well, we got down to Warehouse 3 to get that barge, and this friend of mine, we worked in Warehouse 3, he hasn't got the ramp pulled. And I said, "Well, God, this was supposed to be ready to go. Otherwise, you've got to pull the ramp." He said, "No, they just called and claimed we got the wrong pulp on this barge," so he says, "you can't take this one," he said. "You've got to go get another one in the mill and change it and then come back out and hope we've got this one going." So my radio kept going off and going off, and I didn't go answer; I knew it was the boss trying to tell me. So finally here he come with his car, and he drove down there, see? And he says, "I've been looking for you." I says, "Whitey, I'm waiting for this barge; as soon as it's ready, I'm gone." He said, "You know what's going on," he said. "Get in the goddam mill and get the other one." So all that big hurry-up bullshit, and then we didn't have to do nothing. So that's kind of the way it went.

You were up at the Pulp Siding—

Yeah, that's where we tied the boats up. See, we'd tie up there every night, and then we had all our supplies and stuff up there. After they shut the sawmill down and I retired, they finally moved the tugs down to the sawmill. But all they did then was barge; it was no more logging or anything, see? So—but no, we used to have a good time. Old Whitey treated us good, but we liked to get him irritated a little bit.

You must have a million stories—?

Well, the one time that we had a good time, Marshall Leisman was running the Mary Jane, and I was decking for him, see, and seeing this picnic table come floating down. There was nothing you couldn't find on that river, I'm telling you. They talk about the kitchen sink and everything. I've seen kitchen sinks come floating down the river. So we seen this damn picnic table. Well, that afternoon, after we got done logging, it's hung up on the falls. So I said, "Let's go get it; I need a picnic table." So he said, "Okay." Well, we got out there, and about this much water is going over the falls, see? Well, with all that pressure on the side, we got the boat out there and we got the thing loaded, and they had another deck hand damn near broke his ankle getting it on the boat. And then we couldn't get the boat away from the thing; it was holding us there. We'd go back and forth, and we'd back up wide open. And the water kept pulling us against the concrete. Well, finally it started almost lifting us. Being so much pressure, it was damn near going roll the tug over the falls, see? So we backed up, and he says, "What the hell am I going to tell them," he says. I said, "You tell them you were stealing a picnic table for me." And we finally got out of there like that, and then he told me, he says, "God damn it," he said, "don't ever ask me to go get a picnic table again." But he was a good guy too. Nothing he wouldn't try. He was quite a guy. See, and he was our lead man, and he says, "Now what'll I tell them?" I said, "Hell, that don't make any difference, you know."

You were working there in 1964—what was that like for you?

Well, I was in Mill B then, see? And the water come up so high that it run right in—I was the acid maker then—we had a big burner—it got in there full of water and all full of brick. Well, I could have come back down to the mill and help shovel mud and stuff, and I think I was building the back bedrooms on this [house]. Anyway, I told them no, I didn't want to. I stayed—and I was off just about long enough to where I was going to draw unemployment. And I went down to Berry's Pool Hall right across from the courthouse, and in the afternoon—it was about, oh, three o'clock—and pretty soon Tom Berry hollered.

He said, "You're wanted on the phone." I said, "Who in the hell's calling me? Nobody knows I'm here, you know."

I got on the phone, and here it was Merril Cashman. He was the assistant personnel manager and safety man. He says, "Art, you've got to come to work at four o'clock," he said. "We're going to start a big gas thing, and we're going to dry that burner out, get the water out from behind the brick." I said, "No," I said, "a couple more days and I'll get my unemployment check, and I'll be in on Monday." He said, "You've got to come now, or I'm calling the unemployment and telling them you refuse to work. You won't get no money anyway." So—in fact, his daughter worked for PACC—Marilyn, see? And I told him, I says, "God damn," I says, "you're the worst damn friend I ever had." I said, "I come within one day; I never had an unemployment check in my whole life, and I still didn't get one," see? So—but I went to work that afternoon, and—

So you weren't there when the water was high and coming in?

Well, the water came up—well, the funniest part of that flood, you know, all the millwrights and everybody, they moved the motors from the sub-basement and the lower basement up to the main floor. Well, guess what? They should have went one higher, because the water come up on the main floor and got them all wet anyway. They could have left them down where they were, see? And the same way with Mill B. We got it going. They damn near didn't start that mill up. They had about a three-day meeting in Frisco over it, and we're just lucky we had the old-timers that had been managers at West Linn and whatever, and they finally decided to spend the money to start it up. Otherwise, they were going to junk it, see? And so, like I say, that was a long time ago…. Well, what saved us then is, everybody had worked there was up in the top, you know. I mean, the guys that were down in Frisco in the main office, they'd been managers and stuff of mills and stuff. And see, and then when it started going to hell, well, they hired a CEO from Carnation Milk, which he was a good guy, you know, but he didn't know anything more about making paper than I do, you know.

Things went downhill from then and got worse and then they finally raided the outfit, you know, and James River took over, and—before that, you know, like I say, all the bosses worked their way up from the bottom. And man, if you went in the office and told them you had trouble, they knew what you were talking about. And if you go in—the guy that had just come out of college and the only thing he'd seen is walking through it, well, he didn't know, you know. In fact, you'd break them people in, and then a couple months later they'd start telling you how to run the job, and you're the one who showed him how to run it, see? But like I say, it was the best place I ever worked, or whatever. I had a lot of fun down there.

What was the best thing about working at the mill?

Oh boy. I mean—well, it's kind of hard to say because you've got so many friends you work with, I never did mind going to work, you know. In fact, I had a lot of fun down there, so I would say that the people you work with down there, and all the fun you had on the job. And somebody doing something, you know, you had to be on your toes. And like I say, if you did run into trouble, they'd help you get out of trouble. Then they'd turn right around and make a big story up and tell the whole world about it, you know, and make it twice as bad as it was. But like I say, that was probably the best thing, is the good friends you work with. Like on the boats, we'd bring something to eat, you know, and Christ, the younger crew, they couldn't wait for Thelma Louise to make soup. They wanted me to make sure that on any cold day I had at least two quarts of soup with me, see, because they wanted to start eating. And hell, they'd be eating lunch at 10 o'clock in the morning, when we had a break going through the mill. When you started going through the mill, that's when you had a break because you're done logging, and then you'd go through the mill and you'd have 10 or 15 minutes, and if the barges weren't ready, you had to wait for them, see? Well, that's the time when you could grab something to eat, because we ate on the fly, see?

So people brought big lunch boxes?

Well, like I say, they asked one guy down there what he made down there and everything. And he says, "When I went to work there I made a dollar and nine cents an hour." He said he made a dollar and nine cents an hour, "plus salmon," see, because he was down in Mill A, where they were snagging salmon in the spring. And he got a lunch bucket full of salmon on the way home, and a dollar and nine cents an hour.

Do you remember what you were paid when you first went to work there? What year was it?

A dollar nine, see? It was in 1947, I think.

A lot of guys went down in '43 or '46, and they were making 40 cents, 42 cents an hour…

Well, we had a good union, you know. Like I say, when I went to work there, I think J. Lavier was the president of the Local 68. And then after Lavier quit, Bill Perrin took over. And he was president for a good many years. Lavier was the one that put me on the welfare board, see? We called it the welfare committee that handled the health and sickness and the off-the-job insurance and everything. And up at the union hall, see, when I went to sea, we had a real good contract, you know, and medical coverage and stuff. So up at the union hall I was bitching all the time about how lousy coverage we had. Finally he called me over, Lavier did, and he said, "Now," he said, "I'm tired of listening to you giving me a bad time about the medical coverage." He said, "I done everything I could to get it better," but he said, "I'm going to put you on the committee, and I don't want to hear no more complaints, because," he says, "you can bitch to the people that are going to count," see? So I was on that committee, I think, about 35 years.

Were you ever a union rep or officer?

The only position I ever had was being a shop steward and being chairman of the health and welfare committee. And that kept me pretty busy, because you handled all the grievances, you know, of people that figured they didn't get the coverage they should have or whatever, and made sure they did. But I never run for office. I kept my job on the welfare committee.

What years were the big strikes? Do you remember?

Had lots of fun on the strikes. Oh, man! I mean, standing picket duty—we called it picnic duty, see—twice.

What was the first big strike when you were there?

The first one we were on, I think, that was our big, long one. I ain't sure. We had one that lasted damn near nine months, see? And we just, we didn't quite get what we should because we went back too quick, see? But they talked them into coming back and getting the guys' vote on it. Well, after guys have been off so long, they start getting chicken-livered, see, and the company was smart, so they'd get them to re-vote again, then they'd vote to go back to work, see? And they did. And like I say, we had good contracts. Part of the wages, they weren't quite as high as other industries, but we had real good benefits, see? We had five floating holidays. After being there so long, you could get the day off. Had 24 hours' notice, then they had to give you a day off, see? Because that was really a sticker.

And little tiny pipsqueak bosses, they didn't want to have to bring in somebody to take your place. So we had it in the contract. Then they had to, see? Plus you could work your way up to six weeks' vacation, see? So it was six weeks' vacation and five floating holidays. Well, there you had seven weeks. And they'd call up and want to know where you were. "He's on vacation." They said, "Hell, he's always on vacation!" But, no, we had a good contract, and like I say, we had good representatives, you know. You can always argue between them in the back room, but when we met with the company, well, you know, we were all together.

Did you hear anything about how the supervisors did when you were on strike?

Yeah, see, because on the second strike, then I was on the river, see? And we were out about five, six months then. And like I say, I heard a lot of problems that they had putting in barges and running this, and when we'd only have two men, they had four, see, and they still couldn't get the job done, see? But they weren't really working as hard as they could either, because they wanted to get us back to go to work while they could walk around and do nothing all day, just keep the time, see? And like we used to kid them, all you had to do was keep the time and make out the schedule, and you couldn't even do that without goofing it up, see?

Roy Paradis was a supervisor then?

Yeah, he was on the machines, see? Well, Mike and he was running the sawmill. I never did work for him, but he was a good supervisor. And then we had George Droz; he was the boss on maintenance. You couldn't ask for a nicer guy than George Droz. And like I say, most of them guys, you know, if you wanted to borrow something, you could borrow anything. They'd just sign their name, and you could take it home and use it over the weekend and bring it back or whatever, you know. Crown was a good place to work. They took good care of their employees.

Some workers were really sad about the James River takeover.

Yeah. Well, you see, old man Zellerbach—J. D. Zellerbach—the times that I had met where he was in the audience or there, you know, well, the first thing, you know, he'd say, "Well, me and my employees, we built this corporation." Him and his employees, see? And he'd say that more than once. Well, if there was anybody there from like Weyerhaeuser or somebody else, you could see them sit there, and they'd just cringe. They hated that. See, but Crown then had 75 percent of the paper market on the west coast. So I wasn't there one time, and I heard he told some of them—either Weyerhaeuser or Boise Cascade or something—they were butting heads on what their contract was going to get, and he said, "Well, if you don't vote for this," he said, "I'll ship paper into your district for half the cost." He says, "You won't be around next time," see? So old man Zellerbach, I mean, he knew what was fair and what wasn't, see? But he didn't take any crap off the other manufacturers either. But like I say, at that time Crown had 75 percent of the business on the west coast, see?

Paul was saying they had a floating paper mill in the Amazon River, and after they'd stripped the banks they'd go down the river a little bit farther.

Well, I've heard about it, but I never seen any stories about that, see? They've got problems down there of the timber companies cutting all the wetlands, you know, the rainforest. And once it's cut, then it kind of dies, see? And it's about like around here, where they want to cut the last old-growth trees.

Well, I believe in, you know, a company making money and going along, but when it gets down to where there's only a few left, and them trees are four or five hundred or six hundred years old, and your grandkids and their grandkids and somebody else's grandkids is never going to see a tree like that again—and once they cut them down, they're going to have to replace the lumber with something else anyway, so why not start a little bit early? So I'm, I guess, an environmentalist. Just like the oil in Alaska. They say that when they went into

Prudhoe Bay, that was only going last 20 years. Well, they're running out of oil up there, see, so they're about done pumping big, big amounts, because it's running low. Well, they claim that wildlife refuge would only give about another 10 or 12 years, and it would be empty. So why go in there and goof it all up for just 10 years of oil when you're short of oil anyway? I mean, it ain't going to solve the problem, see?

When you were on the river did you do any sorting on logs or working on the millpond?

We had to sort them out. A lot of times they'd want small ones—they had two different rigs in the sawmill. They had two head rigs. One run small logs and one run large logs, see? So it depends on what they were using and what they were grinding. Miken would call up, and say he wanted 20 bundles, say, of real small stuff. Well, a bundle was a log truck load, see, so you'd have to sort them out and put them together and then tow them in, see? In fact, I kind of like logging. I mean, barging didn't take much brains to do that. I mean, anybody could run a barge into a rock and put a hole in it and get a name for yourself, you know. But logging you had to kind of—well, like I say, you didn't only go very fast when you were towing them, but the current—and you had to pay attention. Sometimes you'd get to BS-ing and eating and doing, and then say, "What the hell am I doing here?" Well, there was the falls, and you had to really do something in a hurry to keep from losing some of the logs over the falls, see?

Well, one guy told me that they had to have a tug on each end of some of these big rafts or they'd get away.

Yeah, yeah, in the wintertime we did, see? We'd have a tail boat they called it, see? And the guy up front, he'd be towing, but then the guy in the back, he kept us from swinging around. I showed that picture where the point is here, and then where you look across to the falls it's here, and then you go a little ways and you run into the guard locks, see? Right in here was a sawmill pond. Well, when you go by that point, the logs wanted to go around that point, and the boat was here, so you kept another boat to keep them from going around and going over the falls, see? And sometimes we had an eddy on the side that would stop them, but we'd get them halfway around there and we'd be pushing, then we'd start breaking them up, you know, the logs coming loose and stuff. But we'd get them in to where they got on the bank, and then we could go get them one at a time and stuff. But most of the time it worked out pretty good. They'd be hollering at each other, these boat operators, "I got my end; you go get yours."

Milt showed me a 'dog'—a little spike. He still has one.

Well, I got one out there too; I saved it. It's just like an arrowhead that an Indian would use with a hole in the end, and that's where the cables went through, see? And then they'd be free on there, and you them into either loose logs or whatever, see? And then when you'd bust up a bundle—this one guy we had named Beverly Salsbery, well, he was kind of chunky and stuff, but he was real light on his feet, you know, and God, I'm running over logs about this big, you see. And they'd start sinking on you. He says, "Man, you've got to think light and move fast." Because if you didn't, you'd be sinking clear down to your knees. The logs wouldn't hold you up, see, so you think light and move fast, see?

So what happened if somebody didn't take a dog out of the log, and it went up into the mill?

Well, if it hit the head rig, it would probably bust the damn head rig, and man, I don't know what them big saws cost. But it really had people flying, you know, because like I say, when it come into the head, the saw was about this wide, and probably a quarter of an inch thick, but the wheel was about two stories high or three stories. And then the wheel was as big around as this room, and one on the bottom, and then that saw run on them wheels, like a bandsaw, but it was that big, see? And man, it would really raise hell with stuff. I was never up there when they run a piece of steel through it, but it would do a job on a saw, see? In fact, my neighbor across the pasture, her husband was a saw filer down there for, oh, not too long, because they hired him after somebody retired. His name was Otto Kudrna. He worked there about 10, 12 years. And I learned more from him doing the small saws over here than what I already knew about sharpening saws. But that was a real art in itself.

I'm sure. So the saw that they're sharpening is as big as a room?

Oh yeah. See, the wheel was as big as this room, and then it was here, and then two stories high or three stories high was the other wheel. And then that blade was that big, continuous, see? And some of them had teeth on both sides, so they'd cut through this way, and then when they got that cut, they'd move the head rig over and bring it back and cut again, and they made two cuts, just one going one way and one going the other, see? They didn't have to stop and turn. All they had to do was—the carriage would move in whatever thickness they were cutting, see? Paul didn't tell you much about the saw? Because he was superintendent of the sawmill.

He did—he said that was one of the more dangerous jobs, standing near the saw, because if somebody left an ax in a tree and it grew around it, then people could get really hurt.

Oh, you bet. Yeah, the guy running the chipper plant, one time he brought us this thing—the sawmill didn't hit it, but there was a double-bitted ax that was inside this big spruce tree. And the spruce tree was probably seven, eight foot through, diameter, and here this—we called him a Swede— must have put his ax in that tree, and then it grew around it. And man, the handle was all gone, but that double-bitted ax looked like it was brand new. And here it had been inside that tree I don't know, probably 40, 50 years, see? See, at the chipper they'd make two-foot blocks. They were about, oh, must have been close to two foot long. But they were about a foot square. And then they'd run them across into the mill next to where I worked, and they had big—oh, it was about as big as this room going around—it had knife blades stuck on it, and sticking out so far. And as this log come down the chute, these blades would make chips out of it. Well, when they run it in there, it raised hell with the blades on that. But after it clicked a couple of times, well, it threw the block out of there, see? Because when they hit the seal, the thing was spinning around.

And like I say, Walter Lee was the head of the chipping department, and he brought the thing up to us and showed us the double-bitted ax that he got out of one of the chippers. They'd had three of them. But no, we used to go out, and they had a flume that come across from the sawmill and brought the wood over there. Then it went out of the flume into the chain that brought it in. Well, we'd go out there and snag fish and throw them up on the other belt, and they'd come in there, and guys would get them, and God, the place smelled like a cannery down there. The wood would come down the flume, but then they had another belt—I think they took the flume out and put in a belt. And we'd throw them on the belt, and they'd come right in on the chippers. Them guys up on the chippers, when they had one guy that would relieve them, well, he was out there snagging, see? So they'd take turns. All the guys that was around here gave everybody a fish. We all got fish, because you're probably sometimes doing two jobs. So the guys that were fishing, they'd split their fish with you, see?

A bunch of outlaws if I may say so.
Yeah. Yeah, oh God.

Did you see very many injuries before safety got to be a big thing?
I never seen many bad injuries. The only thing, like Milton, he kept me from getting killed a couple of times, I think. See, when I first went to work there, he said, "If you hear the throttle crack on that Crown Z," he said, "turn around and look; see what's happening." Because the boom sticks were 64 feet long, and then they had a hole in each end, and they'd chain them together with a chain, see? And then you'd have, oh, I forget how many sticks it took; 20-some sticks or something to make a log raft, or 30; 29 sticks, I think. Well, if they were pulling on these sticks, you might be oh, from here down to the road away from where you're working, but these sticks were all hooked together, and if they're hooked to that Crown Z, it had a lot of power. It'd yank, and if you were standing on one, man, you'd go flying, see? So if you heard that throttle crack, the best thing you could do is turn around and look and see what they're hooked to, to make sure that they ain't hooked to something you're standing on, see? And even if they were going to start moving logs or tightening up cables, you wanted to make sure you weren't in the bight of the line; it was going to go that way and not towards you when it got tight, see? So Milton was right there. But I never seen any real bad accidents, which I was lucky.

The one guy that got his leg all smashed flatter than a pancake, well, the gravel barges would go through and come down and go through the locks. Well, they was in the Number 4 lock, and—I forget; the guy's line got tangled up or something, and the gravel was real steep. And he stood on it, and he slid down to where it come to the edge of the barge, and then it slid over with one foot, and he got it in between the barge and the lock. And then the barge probably weighed, oh, like say we used to have a couple hundred ton of paper on. I imagine the gravel barge weighed probably 200, 300 tons, see? And when that came against his leg—but he worked for Bernert, you know; they had the dredging outfit up there. But that was the only one I heard where a guy really got hurt. But we came in later, and they'd already taken him away. But it really smashed his leg like a piece of paper, damn near, see?

So were you working on the river— any connection with the clay barges at all?

We run the clay barge twice a week and sometimes three times. Depends on how much clay. They'd have to run it on the weekend. And then when I was decking on the Crown Z, we run the clay barge. That was quite an experience, especially in high water like you've got now. You come around there; that was quite a deal. But everybody knew their job real good, you know, and you'd have to catch a line. And boy, you caught it. It didn't make any difference where you caught the thing; you'd get a hold of something to tie it. Just get a hold of one end; then you could work on it. But if you didn't have a hold of either end, you were going downstream.

Somebody said that some of the clay barges just came down by gravity; they weren't necessarily towed...

Yeah. See, what they did—same way on paper barges—you'd set a barge in, and then government employees had a cable and a winch, and they'd hook onto it. You'd unhook the boat, and they'd bring it down, just pulling it down. And then the boat was down the other end of the river; he'd taken the barge down already. And then you'd wait for the barges to get down there. Then you'd go in and take them out of Number 1 [lock chamber]at the bottom hole, and they came down on a wire, see? And when you're barging, the biggest thing was—I don't know; did they talk about the locks, the chambers, where the water came in on the end instead of the center? See, like almost all other locks, they fill the chambers up from the side fill. And the locks there, the wickets were in the gates, and the water came in the ends. Well, the minute they opened up a gate, that hit the barge or logs or whatever you had in the chamber and shove it back. Well, the funniest part of that was you'd have to go back for about maybe a fourth of the filling of the chamber, and then the water would hit the back end of the lock, and it would start coming forward. Then it'd push you the other way. And you had your big lines—two-and-a-half-inch lines and stuff—and man, you'd have to take the slack up on them because when you had your paper barge with a couple hundred ton of paper on it or whatever, it'd hit that slack, and you could break the rope. And then when that line broke, it sounded just like an ought-six rifle. If you'd've had your hand there, it would blow your whole hand off, see? And that was probably the most dangerous part, you had to really watch what you were doing, see? Because when you're going down, it ain't no problem. But coming up with a full barge when they've got to fill the chamber up, that's when you've got the problem, see?

So you know, like I say, you paid attention…when you were coming up, because they'd turn that water on you, and some of the employees in the locks sometimes, just for the hell of it, would really give you a pile of water. And man, I'll tell you, you're just waiting for your line to snap, but they wouldn't. Once in a while they'd snap on you, and then if you did, the barge would go ahead and hit the gate, and then it would bounce back and hit the one behind you, and back about three or four times before you could get it stopped, see? And then everybody's mad, see, especially the government, see? If you broke one plank, you'd swear to Christ that you took a shot at the President, you know.

I heard that if the barge hit the gates it would break them.

Well, it didn't really break the gates. They all had planks on them, see? And they were four by 12s. And you might bend the wicket a little bit so that it would let the water in and out, so they'd have to straighten it out. But they could shut the locks off, and one ahead and one below to get it down low enough where they could work on it, see? But the government thought if you cracked a plate, boy, I mean, we had some new planks given to them, because Crown had plenty of them, and we had to work together. Without us, they wouldn't have had a job either, but—most of the guys on the locks were good, but boy, they really thought if you broke up some of their planks, it was—you'd swear to God you'd dented their fender on their car.

So were you barging when the duplexes were in the park…when the Corps had three duplexes.

Oh yeah, yeah. They had their houses right next to the walkway. I think there was three houses. You see, before when they opened up the wickets, you had turnstiles, and they'd open them up by hand. Well, then that's when they use a lot of the water to help open and close the gates, see, because when they were ready to close the gate, well, if they'd start drawing water, then the pull on that would shove the gate shut, and the guy didn't have to work as hard. So they used a lot of the water to help them, you know. But like I say, I was working inside the mill then, but I'd walk across the top of the locks when they had the bridge up, and them guys would be walking around their turnstile, tightening up to open the wickets. And then they had to turn around, go the other way, to close them, see? Then they went to a hydraulic system, and it made it a lot easier. I think they cut down on about half their force in the government then on running the locks, because they had one guy on the upper chamber run Number 4 and 3, and the guy in the bottom one run 1 and 2, see? So then you would have had a maintenance man, and he could fill in if they needed him, see?

When you were there, the lockmaster still worked in the building that's the museum now?

Yeah, that—well, they had that tall, skinny building at the end, see? They built that new one. Hell, when they were building it, it was about ready to shut the place down anyway. I forget—they spent a million or a million and a half bucks building that new—we called it a restroom, you know, their coffee shop, see? But the government's got lots of money, so I guess what the hell? But no, they really built a nice building for them, and I guess now it's only open on—you've got to call ahead and they'll run you through or something. Bernert would probably like to have it open, because he'd like to take some of his equipment up and down. He's got cranes and barges and—down there in the lagoon they've got a place where they can take it out. But that's where they used to haul it in, and then Knappton, you know, they had a big log dump just underneath the new bridge on the downstream side.

Well, if you drive down to The Dunes—whatever they call that motel, hotel, that's right on the river [Now Rivershore Hotel]—well, you turn and go down there, and then instead of taking a right, going to Clackamette Park, take a left, and that'll take you clear to the end; it'll take you down to Sport Craft Landing, see? Well, just before you get to Sport Craft, you drive under the bridge. Well, just back from there there's a little shack. They used to have a crane in there and then they unloaded log trucks there. And then they had a bunch of dolphins out in the river where—I think they could haul three or four rafts at a time, fill in there, see?

And in fact, Julius Bogalov saved my life down there one day. They'd built a float for us, contracted and built a float for our clay plant, see? Anyway, in between that they built a float to put in there so the guys could stand and walk, and then when we tied the barge up between the two black, big cylinders full of gravel, they had that float in the center. Well, we went down to get it, see? And we hooked onto it. I was decking, so I put a safety in there; that's got a big dog on each end and a cable between them—then we put the hook in the center for the tow line. Well, they didn't wait till I got back on the boat. The minute I hooked it up, he took off and left me standing on the float. Well, that's okay when you're pulling this float; it's a little bit bigger than this room. But we got out in the river, and it'll dive, see? So Bogalov pulled up with the boom boat, because he's fixing the raft and they're dumping the logs, see? He said, "You'd better get off that thing unless you're going swimming."

So I hopped on the boom boat, and I wasn't on there two minutes, and that thing went about six foot under the water, and here come the big boat with them barges. I'd have drifted right down in front of the big boat, and hell, he couldn't stop with all the water. I'd have went right underneath there. So man, if it hadn't been for him, I'd have been a dead Jose, see? So that's the only time I come close to getting hurt, and I didn't even get wet, see? No, I fell in my own share of times. In the wintertime, these big spruce logs, the damn bark would slide off. So you'd be walking along, and with the snow on them, you couldn't tell. You'd jump on the dam log, but it wasn't a log; it was just a big piece of bark, and away you went. And if it was at the end of the raft, hell, you jumped right in the river. And the water was pretty damn cold. The guy says, "Can you float or swim?" I said, "Well," I said, "I float real good till it gets up to about here, and then I swim like hell." Well, you could walk on water till it gets up to your ears; then you'd better start doing something.

So did they wear life jackets in those days?

Well, yeah, you wore a life jacket, but like I say, a lot of times you didn't have it on. I caught hell for that quite a few times. There at the end OSHA would sue the company if they didn't make it be safe, see? So then you had to wear your life jackets all the time. But, God, it was easy, like I say—you're out there, and sometimes a small log sticking out of the end of the bundle about three feet above the water, but if you didn't look back in the bundle, the thing had been shattered; it had been busted. So you'd jump on it, and hell, the log would go right with you, and in you'd go, you see. I never fell in late in the day; I always fell in real early in the morning. And on nice days, well, people would, you know, they'd go swimming. Hell, about 3:30 or so, if we was up logging and stuff, they'd fall in deliberately, see, if it was hotter than hell.

Was the river clean enough to swim in, or…

Well, the Tualatin was a drain field for Oswego, see? It even smelt like a sewer, but, you know, if you was above that, no problem; it was good and clear. But a lot of time we were below that, and then they built them sewer plants, Oregon City and West Linn, down there in the hole, and man, that was terrible. I'm glad they cleaned that up, because in the summertime—I think Tom McCall started it. Well, like I say, he was probably the best Republican governor the state ever had. His party didn't even like him, see? And they made the mills clean it up, and made the cities clean their sewer systems up and everything. And actually, basically, when I left the river it was getting in pretty good shape, you know. I mean, man, it was nice and clear, and the color was good, you know.

Art on a company tug. Courtesy Arthur Dorrance.

What it was like towards the end? You know, the last couple, few years before you retired?

Well, at the end of the Crown Z time, one of the things that really got me, you know, when they had the new management, Crown was still running, but they had new management where they hired an efficiency expert, and he'd sit with a time clock watching you and they'd lay so many people off and stuff. Well, you know, that rarely works; it all depends on if you're really good on the job, well, maybe you had a few minutes where you weren't working. But then if you put a stranger on it, he couldn't even keep up, see? So he'd have to have two guys to take your place because he didn't understand the work and stuff. Well, that's about the way it got at the end. And it was really funny when Goldsmith and—I forget the name of the other guy—but they had this big deal where they—what they called the poison pill—where they get all the stock they can, you know, to keep from getting bought out, see? Well, here come the bosses out there, and a couple of years before that they finally let us start buying stock. For every share we bought, I think they matched one or something, see? So we were getting stock at a real good price, see? But we had—I didn't have many shares; I forget. I think when I got done I had $1,000, $1,500 worth of shares. But they came out during this time they were trying to take over Crown, wanting us to vote a proxy so that we'd vote our share and vote for them, see? And we'd just got done with that seven or eight-month strike, and I said, "Good God," I said, "you guys were out there trying to starve us out and was going to get rid of us and hire scabs, and now you're begging me to save your damn job." I said, "Hell, we ought to tell you guys to go plumb to hell because," I said, "I think you're going to lose the company anyway," see? Well, they did. See, James River, they got it, and then they sold out—I forget who in the hell got it the first time—They sold all the mills, anyway, and James River—well, James River took over. But Sir Goldsmith, they kept all the tree farms, see? And that was the profit off of it, see? They didn't sell the tree farms; they sold the mill, and that paid for all the stock that they needed to take out the money out of James River. And James River tried to run it, but they were back east, you know, and they didn't know the west coast operation or whatever. The mill was—well, like I say, it wasn't getting old or nothing, but they were having trouble with their competition with their coated paper. And the big money was in coated, but it was awful expensive to make, you know, with the clay and everything like that. I wasn't that far in on what was going on in the mill, but that was the biggest thing. They were losing customers on the thing, and they weren't holding up their grade like they should have, and they were shipping paper that they'd have to ship it back, you know, because of too many holes in it or whatever it is. So they ended up, like I say, then they sold it.

Simpson, then from Simpson to Georgia-Pacific, huh? Because my pension has changed. I hope it keeps coming. Every so often I get a letter saying it's from a new company; they got taken over, Crown. Because Crown had a good pension system, see? And they put the money away so the money was there, see? A lot of companies, they talk about great pension, but they don't fund it, see? And hell, it's good on paper, but if there ain't no money behind it, it ain't worth five cents, see? And Crown had funded the pension real good, so these guys, when they took over the mill, they took over the pension fund too. And with the federal law against it, they couldn't dip into it unless they did—so they put it in like, The Bankers Trust, I think, had it for a while out of Boston, see? And then we got our checks through them, and they just managed the money, see? So...I'm still getting my pension, so I guess they're doing a good enough job. I'll hope they are anyway.

Okay, so working at the mill was not your first job?

Yeah, well, I went to sea for two years, see, during the war. And after the war was over, I made a trip or two hauling troops. And then I quit going to sea and in fact I worked at Hawley's for four or five months. And that's when I had a chance so I took it to go to Alaska commercial fishing. Well, like I say, I went up there, and right after the war was over, which was a poor time to go. The fishermen hadn't had a raise all during the war. They fished for the same price. So they went on strike, and then the minute they went off strike, I fished for about four or five days, and I made big money, then the cannery workers went on strike because they hadn't had a raise. Well, by the time they finished that strike, the season was over, see? So I went to Anchorage and I worked around there for a couple months, then I came back home. And I went back to North Dakota. The funniest part of it was—and when I was a kid I used to work for the phone company, Northwest Bell, see? Well, I went back to work there. I got a job right there, and I think I worked until I got my first paycheck. And Christ, I forget what they were paying me, but we'd stop every day when I got off work, and stop in the pool hall and stuff and have a couple beers and go home. And I told the boss after I got my first paycheck, the next day I went back to work, and I told him, I said, "I'm quitting." He said, "Well, what's the matter," he said, "don't you like it?" I said, "Yeah, the job's good," but I said, "I'm not even making beer money." I said, "Hell, I drank more beer than what you're paying me." So I said, "I've got to go back to the west coast, where I can make a living." So my first wife was with me, and she was back there, and she hated North Dakota anyway, so we hopped on a bus and came back, and that's when I went to work for Crown. And like I said, I've been there ever since.

Was it obvious to you when the change happened, because new people came in?

Well…we kept the same supervisors, see, so actually the only big change came was down in the main office in San Francisco, see? And eventually they made a few changes on the mills and stuff, but there really wasn't a big enough change at my level to really notice it, because we had the same bosses doing the same work and everything. So it didn't make that much difference, except, oh, they probably had a deal where they were trying to eliminate even bosses. All they were going to do is have the lead men that was running it, and like I say, you knew what you had to do, so the little pipsqueak boss, all he did was kept the time and made out the schedule, see? And then if you needed something, he'd call somebody. Well, if you needed somebody, you could call them just as good, see? So they were talking about that, but they didn't do much of that either until I'd already left, see? And then I heard that Georgia-Pacific or one of them was more on that too, that they weren't going to have any supervisors. All the workingmen knew what they were supposed to do, so you'd just go ahead and do it, see? So I don't know how that came out or whatever, you know.

What made you decide that it was time to retire?

Well, my knees had went to hell on me from jumping around logs and stuff, and I'd had an operation on one, and then I had the operation—was I still working, Thelma, when I had the second one? Yeah, I was still working when I had the second knee operated on. And then after I'd retired, I broke the bearing in one of them and I had to have it reoperated on. So I got three new knees, and they done a good job on them. You don't stand like on top of this chair and jump off, because your knee will bend the wrong direction. And if it does that, it kind of swells up for a while. And I called the doc after I'd done that once, and I said, "Boy, it really hurt, and what am I going to do?" Well, he said what I was doing; I was hoeing in the garden. I didn't pick my foot up; I turned, and I turned my damn knee out of the socket. He says, "Well, quit hoeing in the garden, and pick your feet up." He said, "You'll be okay in about two weeks." He said, "Take a few aspirin." I said, "Boy, you're a lot of help. I might as well not even have called you." Well, he said, "Quit turning your knee out." He says, "There ain't nothing I can do for you."

So what year did you retire?

Let's see—I was 61, and by the time I got done with my knee I was 62. And I'm 78 now, so eight and eight is 16 years ago.

Thelma: It's February '89.

It was February of '89, see? And you know—you probably won't believe this—I felt real funny after I retired. I didn't miss going to work; hell, I'd been off over a year with my damn knee, see? But I'd had a job ever since I was a kid. I've peddled papers and always had a job, see? Well, my dad died when I was five, see, and my mother, she took good care of us and stuff, but we had to have money. Better go to work, see? So I always had a job. Well, after I retired, I just felt funny, just like I'd been canned. I ain't got a job, you know. Well, hell, here I got two checks coming in; I wasn't short of money. But it felt funny not having a job that you knew you could go to, you know. But it took about three months, and I got rid of that spell, see? It didn't bother me any more there. I was damn happy I—I wasn't out looking for work, but I mean I just felt funny without it. So I mean people can—I've heard that they couldn't stand retirement; they went back to work and stuff. Well, I wasn't about to go back to work unless I worked out here for nothing myself, but it really felt strange not having a job.

But had you been raising the steers for a long time?

Well, yeah, ever since I moved out here. You know, we had two cows when I was a kid. I'd milked cows all of my life, see? Well, after I went from North Dakota to the coast, I swore to God I'd never milk another cow as long as I lived. When I bought this place, I hadn't had it three months and the first thing I did was buy damn cows.

But did you buy this place before or after you retired?

Oh no, way before.

So you were living here when you were working at the mill. That's a long commute.

Oh, God, the thing when I bought this was kind of hard, I thought—nowadays it's nothing—but I drove back and forth from here to the mill—must have been

10 times, thinking that well, maybe it's too far to drive to work, you know. God, now they drive from Silverton to Portland to go to work, see? Here I only had about eight miles or so to go, and I was thinking maybe that was too far, see? So I think about people now, how far they drive to go to work, and I thought, man, when I was first going I thought boy, eight miles was way too damn far!

People at the mill who lived in West Linn and in Sunset walked down the hill every day.

Oh yeah. See, because Willamette—Crown used to own that town, see? That was a mill town. Yeah, well, like I say, most of them that worked there at West Linn lived in Willamette or West Linn. Yeah, they walked to work. In fact, friends of mine that lived in Oregon City walked to work.

This here woman, she got madder than hell at her husband. His name was Al Lytle. He was a little short guy, and he run the filter plant up there, see? Well, we were down at the coast, staying at his place, and this big Henry Dietrich—he was a journeyman in the grinder room—they were real good friends. They had a place at the coast right by each other down in Netarts, and we'd stay with them and dig mud clams and horse around and stuff. Well, after we got done digging clams, we were up there in the yard, and he was talking about some damn thing. Anyway, in 1935, I think it was, he and Henry were going to work. They only worked six-hour shifts then or something. And they were walking across the bridge, and I think Henry said something to Al about "Hell with going to work," he said. "We ought to catch the Toonerville Trolley and go to Portland and go cat-housing." And whoosh! He said, "What was that?" He said, "That was my lunch; ain't we going?" And he threw his sack up. And he got on the tear.

And his wife said, "Why, you son of a gun; I knew you were horsing around on me." And he kept saying, "Honey, that was in 1935," trying to calm her down. She was madder than holy hell. God, I had to laugh, because man alive, here this was, oh, probably in the '60s, you know, and it was 1935 when he told him, he said, "Let's go to Portland." And he says, "Whoosh! What was that?" "Well, I just threw my lunch in the river." We were sitting right in his kitchen when they were talking, and he was telling me about it, and they were laughing. And Henry was telling it on Al, see, because Al's wife was there, and Henry's wife was down in the next cabin. "We ought to go to Portland," and he threw his lunch off the damn bridge. "What was that?" "That was my lunch." [laughing]

Did you work with some really old-timers that told you stories about before you were there?

Yeah, see, it was Fred Bietschek. He'd been there about as long as me, but he'd been on the river all the time most of the time, see? And they used to unload logs up there at the Pulp Siding. They had a big crane up there, and they'd make the rafts right there and stuff too. See, that was before I went up there. And that was—well, they had big crew, 25 or 30 men up there, because all the logs would come in single, and they'd dump them in the river and then they'd raft them right there and then take them and tie them up. Well then when I got there, the rafts come already made up. All we had to do was tie the rafts up, see? And we took them and tied them up for in the wintertime and stuff. And during that '64 flood they lost, oh, 10 or 12 rafts, I think. There was logs stacked up on the Oregon City bridge, on the West Linn side. Oh man. You know where the hole is in the side of the bridge? Well, the logs were higher than that because the river was up that high, and the rafts, there was enough wood in there to—man, you could have had firewood for 10 years. But see, that's where the raft, the logs, hung up in there, see? They got some of them back, but man, they had logs from Pulp Siding to Portland, see?

Wow!

You know, and I had to laugh. I came down there, me and Thelma, and we were going to her brother's. Her brother lives in Oswego. Well, that night we came down there, and the cop was at the Oregon City side, and he said, "I don't want to let you go across." He says, "All them logs are there; we're thinking the bridge might get washed out." I said, "Well, hell, I've got to drive through to Sellwood and then come back." I said, "Good God." "Well," he said, "I'll let one go across. If you'll make it, maybe we'll let some more." I said, "Boy, that's nice." He's going to let me, and if I make it, then he'll let somebody else cross it. I said, "Man alive, here I'm a guinea pig," but I said, "well, the bridge wasn't moving or anything," …" so we went sailing across Oregon City/West Linn bridge. But man, there was really—well, the water was up on Main Street in front of the Elks, you know. So it was up pretty high.

Were you working on the river when it was really icy?

Oh yeah. Yeah, up the Pulp Siding—what they call Rock Island, just above Pulp Siding—well, we had to take a raft out of there, and the ice was about four inches thick, see? Well, then the river dropped, but the ice is still there, and here's about four or five feet above the damn water, see? Well, you go out there and start chopping, and you didn't know when it was going to fall in. If you were above the raft, it wasn't too bad, but if you was over water, you'd go swimming, see? And so—but they, you know, banged around with the tug enough to break some up and stuff. In fact, Milton was running boat one day, and he turned around and asked me. He says, "Art, you want a deer?" And I said, "Well, yeah," I said, "no problem." But I said, "What the hell do you mean out here?" "Well," he said, "right behind the boat. No kidding." Here was this big four-point swimming right behind the boat, see?

Well, I had a tag that I hadn't used yet, so I got the boathook and I grabbed his horns, and I hit him behind the head with a dogging ax, and we drug him over to the bank, and I gutted him out and stuff. And then after work I went down and got him. And like I say, I made sausage and stuff, and I split it with Milton and the other deckhand, the three of us. I made sausage and give them some steaks and stuff. But I got a four-point with a dogging ax. But that was really funny because we first tried to catch it. Milt would back up, and the deer, they can swim pretty speedy, but it was trying to get on the raft. If it had got on the raft, we never would have got it. It could run, Christ, better than any logger, you know. It didn't have corks on either. But we kept it from getting on the raft, and finally I got hold of a horn and hit it behind the head. And then I cut its throat and we took it over to the bank.

Then me and Thelma's dad went down and got it that night, you know, and hung it up. But no, that was the only deer I ever got. I never shot one with a bow and arrow, but I got one with an ax. Yeah, it was in November, I think. Hunting season was still open, so I got a legal deer; I just didn't shoot it. It was just swimming across the river. They swim that back and forth there steady—where the power lines go up the other side? Well, that's quite a deer crossing right in there. Yeah, it always was, and that was nice big black tail. God, it was a nice big deer.

What happened at the sawmill when it was frozen?

Well, they didn't have too much trouble. The only trouble that the mill had when it was colder than hell, a lot of the pipes were outside and they weren't insulated enough. And if they didn't keep water running through them, they'd freeze up and bust. In fact, when I was in the pipe shop, we had a big deal in Oswego; they had a chip thing there where they stored chips and loaded barges for Wauna and Camas and stuff. Well, they had a lot of their pipes freeze up, and when I was in the pipe shop I had to go up there and help the other journeymen and stuff. There'd be two or three of us working, replacing pipes because they'd froze up and busted, see? But they kept pretty good track. They knew which ones they had to take care of and run water through them steady so they wouldn't freeze up.

Milton called you a plumber, and I'm seeing you as a river captain.

Yeah, I worked in quite a few different departments, and I never worked in a bad one, you know. And like I say, when I worked with Rosie, and—like I say, them guys told me that if they didn't squirt you they didn't like you. Well, oh boy, they've got to quit liking me so good, see? But that was a lot of fun in there then. You always had two or three changes of clothes anyway, so you'd go up and change clothes. If it was late in the day, to hell, you never even bothered. You'd just wait till you go take a shower and take them off anyway.

Did the men socialize?

No, they never had any parties. Not that there wasn't some times you could get a drink in the mill. I mean, but you weren't supposed to bring it in there. But like I say, the working people, they never had no Christmas parties or nothing. And the firemen used to have a party, but then the firemen were all made up. I never got to be a fireman because in Mill B, that's where the fire was. Running the acid plant, I had a thing—have you ever seen a cement plant where they've got the big tube turning, where they burn the lime rock, they get it real hot and they burn it? Well this here, you fill this big tube—it's, oh, damn near as big as this room here to the end of the swimming pool—and you put liquid sulfur in there, and you got it burning, and then you burn it, try to burn all the oxygen out of it, and then you take—the gas runs into a big combustion chamber, your sulfur dioxide, and then that goes up through water and lime rock, and that's how they made the acid, see? And see, I couldn't be a fireman because I was sitting there by a big fire, and I couldn't leave that fire to go fight another one, so I never did get in the fire department. But they used to have parties outside the mill. They'd have a party someplace, but I don't think they had any in. And then they'd— whatever they had was girls and booze or whatever, you know. But I never got to go to them.

The firemen had other jobs in the mill, didn't they?

Yeah, like most of the millwrights and welders and a lot of the paper makers, they were firemen. So if the fire alarm went off, they'd drop everything. They had good firemen at Crown; they had a good fire department. And then after I got on the tugs, they made the tug a fireboat, see? So there was one fire in Canemah down there where a houseboat burnt. And we went down there and helped them, and they sent us hats from Oregon City—they hopped on our boat, and we took them up so they could run their lines, and then they had a pump they just threw in the water, and then they had a rope on it, and it had the hoses and some, and we helped them put that fire out. That was the only one we ever worked on. And like I say, we got a hat for doing that, see?

Milton was talking about one time when the shaft was out of the tugboat, and it couldn't move to fight a fire.

Yeah, if you got no shaft, you got a hole in the bottom of the boat about that big. If you dropped it in the water, like I say, it wouldn't go 10 feet and you didn't have it, see?

When you retired, had most of your friends already retired?

Yeah, see, when I retired, Milton had already retired, and Fred Bietschek, he was old; he was a skipper; he'd already retired. And Marshall Leisman—he was the lead man before that—he'd retired. So Gene Kleinsmith was still—he was a little older on the river than me, so he was the lead man then. And then Whitey'd retired, and the boss in the yard gang that had taken over running the river crew. So they had just about all gone except for me. I had more time than most of them, but I hadn't had that time on the river when I'd come to work, see, as far as seniority was concerned.

You seem like you were a natural to go on the river because of your ocean background.

Oh well, we had to take a test. See, they grandfathered all the other guys to get their skipper's license. Well, I had to go and take the test. Me and Ernie Colley had to. And it helped a lot by going to sea, but I sailed in the engine room, see; I didn't sail on deck. But you still had to know the rules of the road and everything else. I liked going to sea; that was a good life, but if you were going to get married it wasn't, you know. But it was a good life for a single person. Good contract, good wages, and man, on the boats they fed good, you know. Hell, you really had a home away from home, see?

So what was the worst thing about working for the mill?

Well, about the worst thing was the split shift where you worked three shifts—day, swing and graveyard. It seemed like every time I was on swing shift, then Sunday afternoon or Saturday afternoon you'd get ready to start barbecuing steaks, and when I lived in town or after I moved out here, everybody would holler, "Hey, Art, come on over." You know, "We got a cold case of beer, and we got some steaks." Well I told them no, I said, "I can't come; it's three o'clock and I've got to go to work," see? So that was about the worst in the summertime, was working swing shift, see, because like I say, by the time that people had mowed their lawn and got done, they was ready for happy hour. Then guess what? You had to pick up your nose bag and head down the road, see?

So that was about the worst time at the mill was—then at other times them shifts come in handy. If you wanted something to do, you were off in the morning. I mean, it worked out good, but—so I didn't really mind it that bad. And graveyard, when I moved out here, well, I was buying houses and putting foundations under them and doing the plumbing and wiring and stuff, and I was working swing shift. So I didn't mind it at all. And that was before; I wasn't on the river then. I still was doing it, but then I had to wait 'til I got home on days and work. In summertime you could work till dark, but on spring shift I'd work swing shift, I'd get up at six in the morning, and I could work till one or so in the afternoon and take a shower and eat lunch, and maybe a half an hour nap or something, and then go to work, see? And take it easy on the job and rest up. [grinning]

You said in the summertime when it was really hot, sometimes people fell in on purpose.
They didn't really jump in with all their clothes on, did they?

Oh yeah. Man, you know, you might kick off your corks. You didn't want them on, but most of the time they bailed in clothes and all. Bietschek, he lived right on the river, see, and a lot of times his boys would be down playing along the river, so we'd stop and pick them up and take them over where we tied the tug up, and then they'd ride home with their dad, see? Well, we always picked up all the cans that was floating down the river and stuff and put them in a sack, and then we'd throw them on the bank and give them to Fred's boys, see? And a lot of times the boss, when he'd be looking out the window and we're supposed to be going someplace in a hurry, he'd turn around and go, "What the hell are you doing now?" Well, that beer can was floating, and we had to stop and pick it up for Fred's boys, see? So they had to stop the operation for something important, see? Like I say, I doubt it would have made any difference how much it cost. I had to stop for a nickel beer can; you'd have to do that.

Did you have kids that went to work at the mill?

Oh, my boy went to work at the mill. That was the funniest damn thing too. See, he lived in Seattle, but he was coming down and staying with us, so I called up, and yeah, they'd hire him; said no problem. They hired him, and so he was going to work on swing shift. A friend of mine worked in the supercalenders, and I no more got home on the first day he went to work, and he called up and he said, "You better go up to the hospital, Art. Your boy got hurt." I said, "That's the first thing I told him: don't you ever get hurt down there." I says, "God damn, you get hurt, I'll break your neck", see? And like I say, I'd just got home, and Bob David called up and said that he worked in the supercalender. He said they hauled him out of there and took him to the hospital.

Well, what happened to him, they had carts down there, like electric ones, that—oh, you'd see them; they'd go along. Well, they had a big long one like from here to the TV. And it was only about this high and had electric and you pulled it down, and then the platform on it raised up and down with the hydraulics so when you lowered it down to put these big long rolls of paper on it, see? And God, you'd put three or four ton of paper on it or more. Well, what they did was, he was having a guy break him in, and he walked up too close, and this leg only come off the ground about this high when they raised it up. Well, he walked up too close to it and put his foot under, and his little toe was underneath the edge of that. When they let it down, it come down on his little toe and damn near split his little toe open at that first joint and broke it and stuff. When they set it down, it damn near cut it off, see?

Well, they took him up to the hospital, and Christ, I went up there to see what was going on because I didn't know what happened. And God, the personnel manager was there, and Bob Carter was there, the head of five and six and the supercalender. They had all the big ones—I thought he'd got killed, you know. And they'd caught his little toe like that, and he got put on what they call extra duty or soft duty. A friend of mine's wife was running this place, making out the stencils and stuff, and they put him in there with her. Hell, I think all they did was play cribbage and horse around. And he didn't really want to go back down to the supercalender. He finally had to go to work, but man alive, he worked there two summers, I think. But like I say, he got hurt a half an hour after he went to work; I couldn't believe it.

So he was working there between college years?

Yeah, see, he was—he was still going to high school, because at 16, you could work at 16, I think.

Thelma: No, he was done with school, I think.

He was done with school? But anyway, like I say, God damn, that was really something. Only been there a half an hour and already hauled him to the hospital. I never had any other relatives work there; just people that I knew, you know, and stuff. Good friends. Well, I'll tell you, there were so many people from North Dakota and Minnesota, that if they'd all went back to North Dakota and Minnesota, there wouldn't have been no paper mill. In fact, this one friend of mine who I knew well, the Holsworths, when they left North Dakota and come here, they shut the post office down in the town they were in. But man, there were so many people from the Midwest that was working there. They had a good name, that if you was a farm kid you knew how to work. So man, they'd just about give you an occupation with Crown or hire you, because in the grinder room and stuff where they had hard work, well, them guys, they stayed there like it was nothing to it, see? So the people from the Midwest had a good reputation that they knew how to do a day's work, see?

Paul Miken said that Lithuanian immigrants had a reputation for being hard workers, too.

Well see, them Lithuanians, I worked with them there in Mill B. One was named Frank Sekne and Joe Tross. And the Gerkmans, they were Lithuanians; they were millwrights, you know, all big people. God, they had hands on them like, you know, whatever. But they'd always say "I figure up you're going to do that," see? I forget what the hell I'd done, and he said, "Yeah, I figured up you were going to do that," see? But they were good people to work with. Man alive, in fact, Frank Sekne's boy, he worked for Portland Cement in Oswego. And he was a hell of a guy. He was a pitcher for the teams they had around there. Guys would get together and get a ball club going, and Frank was a pitcher. They said that he was kind of a character anyway. And if he had a guy up to bat and there wasn't anybody on base, and say there was two strikes and two balls or whatever, he'd wind up and let one fly clear up in the stands, like a wild pitch, you know. Then he had one ball left, you know. But God, he said that they really had a time. And they had some pretty good ballplayers, but he said they had some real characters playing ball.

You know, the Corps always said they would never shut longer than two weeks because it tied up the whole river. So how did that affect what you were doing on the boats, when the rafts started to pile up?

Well, actually, it affected us quite a bit in some ways. We done a lot of work that we wouldn't have done otherwise because we didn't have no barge, you see, but we had to move barges down while they unloaded them to put them on the truck. Sometimes they'd use the barges for storage, see, and they'd pull the barge down, oh, about in Number 4. They'd hold water in it—where the bridge went across going into the mill?—they'd hold water in that and they'd shut the other three locks down below that, see? Well, we'd pull the barge in there, and then they'd unload it on the other side of the canal so the trucks could haul it. And then we'd pull it back out, so we had enough in-and-out work, little tiny damn moves and stuff, but enough to keep a crew going so you had to move the barge, and then you'd have to move it in the morning. And I think we had to come in at eight o'clock at night move a barge, see? So it didn't—then we'd do a lot of maintenance work—so it really didn't affect us that much as far as getting laid off or anything.

When they had to work in Chamber 4, then you'd have to go?

Yeah. Well, usually they'd get that done pretty speedy to where they'd only be down maybe a day or two. See, they'd get right on that and get enough people there to get it going again, see? And I was never there when they shut Number 4 down; that's the top lock, right where the bridge goes across into the mill. We used to shut down and drain the locks, drain the whole darn thing, and that would usually go from, oh, three or four in the afternoon till just about daybreak the next morning. They we'd come back in, and they'd fill the canal up and fill the locks up, and then we had all the barges out at Warehouse 3. We'd have to bring them in and [re-spot] them again.

Kenny Bietschek told me a story that his uncle or his dad had lost a Timex in the locks.

That was probably his brother Ed. See, his brother Ed worked for Knappton. In fact, he could probably tow logs better than any guy I ever seen. I mean, man, he was really a boat operator when it come to towing logs. Because he went to sea like I did, see, and as soon as he quit going to sea he went to work for Knappton. And that was Fred's younger brother, but there wasn't too much age difference in them, I guess. And then his dad—I think Fred's dad worked in the sawmill. He retired out of the sawmill, Fred's dad did. And I didn't know him there, but I knew Fred real good because I'd worked with him. And then Fred—when they shut the upper river down, the logging end of it where they didn't raft any more, then Fred got laid off from the upriver, and he went back in, and he was working in the [oiling] department for a couple of years. Then when it opened back up again, he went back on the river when enough guys retired, see? And Leisman was doing the same thing too.

You have the greatest memory for names.

Well, it's getting so I'd be outside, I come into the house to get something; when I get there I can't remember what I'm going at. Then I get so damn mad, I get halfway outside, I remember what I'm coming after, see?

I think that we should let you take a rest, unless you've got another story.

Well, like I say, I'm not too tired. This is pretty easy work.

Do you remember how many years you worked on the river?

Well, God, it's hard to tell. I mean, when I left the pipe shop—well, when did I go on the river, mother?

Thelma: It must have been darn near 25 years almost.

No, I don't think I was on there that long. After I got my license, I had to renew them every five years.

How much were you affected by things like safety committees?

Well, we had a safety meeting, but we weren't affected. I mean, all we had was, make sure we took care of each other. The barges had safety lines, because those older barges only had about a walkway about this, and then when you're carrying a big line, a hook line to throw, and you had it over your shoulder, you were leaning over the side and you had hold of this cable—well, one day coming in with the clay barge into the—tying it up, well, I went around the corner to catch the cable to get it, and I never come back. And Leisman was running the boat, and pretty soon the boat went on by where I was supposed to be, and I'm standing on the—they had some short piling they drove alongside the other ones because they'd rotted off and tied them. And when that cable broke on me, I was going to get crushed in between, and we were coming like a bat out of hell, so I made a dive for this piling, and I got both arms around it. And I hit my head so hard on the side, it knocked my glasses. In fact, the lenses went flying out of my glasses.

And that's the only thing that I really got hurt, was hitting the side of my head and my knees where I grabbed that thing so I didn't slide down the damn—and I got a bang out of Leisman: he said, "Dorrance, what the hell are you doing over there?" he said. "You're supposed to tie the barge up." And I told him, I said, "Well, back up so I can get on the barge," see? So I got on, we tied it up, and then—see, this barge had a wooden hull. Well, the damn cable didn't break, but the line that was holding it that would put it in, it just pulled out of the rotten boards, and then the line was so long that it, you know, about six foot of slack. Well, the line wasn't six foot long, so I just bailed off the thing because I couldn't hold onto it no more. But that was the closest I ever come to getting hurt—and then that clay barge, when I bailed off that and jumped over on that piling and got hold of it. Then I went down and—I think we even found my lens was laying down there, so I even recovered my lens back. But the damn thing never wanted to stay in the frame again after that.

It sounds like a wonderful life.

Oh yeah. Well, see, I could have went back in the pipe shop, and I turned it down because, like I say, you never done the same—you done the same thing over and over, but never quite like it. Well, you know, you always had a different job every day, and it makes life more interesting. If you do exactly the same thing every day, pretty soon it'd drive you to drinking, you know, or something. Yeah. But the river was a good job. And you felt good all the time, like I say. And something's going on. We used to dream up ways of getting Whitey mad at us. But he worked on the river same as we did, so he knew what was going on. But he was kind of—well, he really wasn't short-fused or anything, but he would get irritated at us if we were rubbing him too hard, see?

You guys had this playful thing going?

Oh yeah. I mean, hell, sometimes, Christ, guys 60 years old are worse than a 20-year-old. Yeah, especially if you get, oh, like Glen Jubb guys, or God damn, or Lugg Gerkman and them, Christ, they couldn't wait for you either to fall in or squirt you with water or do something, you know.

Well, it sounds like a good life.

Yeah. Well, God, you know I didn't mind going to work, even when it was colder than hell and stuff, because then you had to laugh. You know, you'd put your rain gear on, and when you'd listen to the weather report, if it was supposed to be sunshiny all day, you made damn sure you had your rain gear with you. But if it was supposed to be rainy that day, a lot of times, hell, it wouldn't rain a drop all day long, you know, because it was supposed to. But it didn't do it, but—then in the real cold weather your rain gear kept you warmer than anything because it cut the wind, and that wind—that east wind blowing off the river at about 20 above, or 15, well, hell, no matter how many clothes you put on, it didn't do you any good. But your rain gear would keep the wind from blowing right straight through you, see?

When you guys were making pulp it had to be pretty stinky.

Camas, you know, smelled like rotten eggs, like Albany. And see, when the wind blows up the river in the Willamette, it's actually blowing north, you know, from north going south. But the wind comes down the gorge and then it turns the corner and goes back up the river, and then that's—it's colder than hell up in eastern Oregon. That's where we get the cold weather, see? And man, we used to pray for rain, because when it rained it warmed up, you know. Well, we didn't notice the smell—once in a while when Publishers or Hawley or whatever was still running. They kept their digesters going longer than Crown. But they didn't smell like rotten egg; you'd smell the sulfur dioxide. You could smell the gas from the sulfur, see, but it was never that bad up there.

Did you see a lot of other traffic coming from upriver, when you first started doing this?

Well, we had the gravel barges we had to watch out for, or they watched out for us. In fact, down on the lower river when we was waiting to go home in the evening, well, they'd come up with an empty, and if we didn't get a barge in right quick they'd sneak one in in front of us, and then that would put you about 20, 30 minutes later, see? And if that put us on overtime, then poor old Whitey, he'd get all shook up we were getting a half hour overtime, and here come a gravel barge. And we'd tell him, well, we don't own the damn locks. Other people can use it too, see?

No, there'd be a lot of pleasure boats. You had to really watch out for them. God damn, I'll tell you, if they'd have knew that we couldn't stop and that how—like I say, we was good enough operators, but when you get a barge loaded, and them damn water skiers would get in front of us with them there—I forget what they're—jet sleds, you know?—And you'd see one going across this way to make sure he came out the other side. If he didn't, you wonder where in the hell is he at, you know, because boy, if you run over them, no matter how careful you are, it would still be your fault, you know. The Coast Guard would blame you. And there goes your license. And then if you lost your license, there goes your job, see?

Was the Coast Guard up there checking on you?

Well, no, they don't check, but if you have an accident, they've got to come and investigate it. And the minute they walk on the boat they want to see your logbook, see, to see how you've been keeping track of it. Man, they go through that with a fine-tooth comb, see? So you want to keep your log book up, because if anything goes wrong, well, either you solved your problem, or it wasn't your fault, or you cleared it off, so it was in there, you know. Because a friend of mine—Bernert, he ran over a fishing boat—and this friend of mine, Randy Shaw, he was fishing down there, he pulled one of them out of the water. And the other guy I think made it to the bank, see? Well, they had a big hearing on it, and Randy went to the hearing, and when they did, well, Tommy Bernert was there. I wasn't there, but he told them he had his lawyer with him and stuff, and the lawyer wanted to tell Tommy's side of the story, because they were taking a gravel barge, and they'd turned from the Willamette to go into the Clackamas.

Well, the river was running so hard they didn't have enough power, and it was sweeping them downriver, see? And they hit this fishing boat. Well, the biggest problem was, I guess, whatever they done, was he wasn't blowing five or more whistles, which is a danger signal. And if he'd have been laying on that, then the fisherman should have heard it and got out of the road. But I guess he only blew it once or whatever.

But the Coast Guard, according to what Randy told him, he told his lawyer, he says, "You sit down," he says. "When I want you to talk, I'll tell you." So it was just like going to a kangaroo court, you know. I mean, you don't tell the Coast Guard nothing. They got all the rules, and they got all the discipline, and they know what's going on, and they read what's happened, and that's about the way it's going to be, see?

Mm-hmm.

So actually, they treated me okay because I never had a problem with them. You know, that's the way it is down south. When I was in New Orleans, we took a ride on the riverboat, see? The Old Natchez, and they'd fixed it up. And I'd run down and looked at the engine room and everything, and then I went back up and I was in the wheelhouse. And here the skipper, he blew three whistles, and we keep going along, and pretty soon I ask him, I says, "Well, what does that"— three whistles usually means that you're in full reverse, see? And hell, we're going. "Oh." he says, "Down here that means 'have a good day,'" he says. He says, "These crazy dumb Frenchmen," he said, "hell, they'll blow a whistle and you don't know what they're blowing for."

Robert Kunz

Interview April 9, 2006

Would you like to give me your name and your age?

I'm Robert Kunz. People call me Bob. I'm 70 years old, and I started work at the paper mill in 1956. I had worked for Portland Cement in Oswego just prior to coming to the paper mill. They weren't hiring before, or I would have been at the paper mill earlier. But until they had a hiring siege, I couldn't get a job down there. So I just turned around and waited until they had an opening, and I got the job at the paper mill. And I stayed there for about—over 38 years. Actually, I quit working in '94, October 31st of '94, but my official retirement date was January 1st of '95. So now you know as much about it as I do.

It sounds like you had some paid vacation stacked up on the end of your work there.

Yeah, yeah, I did, I did. But no, I started down there in the grinder room, which was at that time probably the worst job in the whole mill, piling wood. And I stayed there for several years, and then I went over to which is the extinct Mill B. That's where they made the sulfite. That was something that has went out. Oh, I think it went out in 1968, if I'm not mistaken. And I'd seen a lot of changes down there since I went to the mill.

Do you remember your first day of work there?

Yes, the first day I started there, I went right down to the grinder room, and it didn't take much instruction to learn how to pile wood. And I piled lots and lots of wood. And it was damp down there, and noisy, because at that time the sawmill was running three shifts, which is something that they had gotten away from later on. They cut down to two shifts, and then they finally went down to just one shift; they only run a day shift in the sawmill. The sawmill was located above the grinder room, so all the drum barkers and all the conveyor chains, when they ran it was real noisy. And then the wood clunking down on to the conveyors, because they dropped it about a floor to hit the conveyors. And then it would run over on conveyors. One of them went to what they called the Number 3 grinder room; that's the one that's located out by the falls. Number 2 grinder room was located in the bottom of the sawmill, basically.

Wasn't it Number 3 where they had the collapse? Did you hear about that?

Well, that was above Number 2 grinder room. It actually collapsed down into the grinder room where people were working. And fortunately, nobody was killed, but there was people, I think a few people, hurt, but that was before I came there, and not very long either.

Did they ever figure out? I mean, just too much weight up there, or...

Well, I'm supposing the reason that that floor collapsed was, yeah, too much weight, because they piled wood up there, and they had the grinders and stuff up on top. I don't really know; I wasn't there at the time. There was a lot of people in that grinder room when I first went there. But, you know, it's—slowly they eliminated it, so they kept shutting things down. Due to the Department of Ecology or whatever, they couldn't use the water like they normally done, because when I went in there, everything was water-powered. And we had electric motors, but we ran them only when the water was so high that we didn't get any power out of the river. We had—there was usually they classified about a 40-foot head between the top and the bottom river. And if you had 40 foot, then you had plenty of power to run the grinders. But when the DEQ and all those came in and decided that it was affecting the fish runs and everything, of course then they eventually eliminated all that stuff.

Where in the mill did they have the turbines?

Well, the turbines that we generated electricity with were located between Number 2 and Number 3 grinder room. And there was—let's see, there was 17 and 18, and I believe there was—I think there was 20 and 21. You know, how they number that I don't know exactly any more. But there were several of them in there—turbines. And when we could use the water power, we could generate quite a bit of electricity. And then of course, the electric motors on the grinders, well, we could generate electricity with them if we weren't grinding wood on them, if we had the water power. And of course we ran some of these generators—oh, shoot, I'm going to say right up into the '80s, I think. You know, we was allowed a certain time of year that we could run some of that water power, yeah. But then things got to be old and obsolete. We done a lot of rebuilding down there, you know. And that was a big problem, is it was so expensive to have to maintain those old generators. And then fighting the fact that we couldn't get water to run them. So it didn't make it feasible to keep them running.

Then they got the power from just the regular grid?

Yeah, yeah, I think they took it off the grid. They was going to go into cogeneration there with PGE, and somehow that fell through. Because we have a good steam plant over there, and we used to use all the waste from the sawmills and—well, the grinder rooms, the sawmills and anything else went into our boiler system over there. And so they could generate steam with it, and most of the paper machines—well, in fact, all of them—were turbine-powered, steam turbine-powered, to start with. And I think when I went in, there was around 10 paper machines. And they gradually got down to where I think there are just three left now. We used to call them Number 9, Number 5 and Number 6. Now I think they're designated 1, 2 and 3; I have no idea. But I worked on all the paper machines, too, when I was down there.

So when you first started, you went to the grinding room. That was really hard work—

Yeah. Well, you just had to keep up with the machines. They had tattletales on the grinder. And when you threw the wood into the cylinders and closed the door, well, the tattletale would come down when you pulled the lever to push it down against the wood. And yeah, it kept you busy. It could keep you busy. And other times we had good days.

When you were first working there, did they still have flumes bringing the wood?

No, we used carts. They were about a third of a cord apiece, and we used to pile—oh, I think we used to pile between 25 and 35 carts of wood a day. So we would run between, oh, eight to 15 or 12 cord a day. Usually, if you piled 12 cord a day, that was a hard day's work. But you got used to it. We was in good health. We were young and in good health. It's nothing to do for an old man; that's why I got out of there.

What kind of wood was it then?

Well, at the time that I first started there, they was running primarily white fir and spruce. Well, those were the two main ones, white fir and spruce, and once in a while we'd get a little cottonwood or something like that in there, which stunk like bad fish. But primarily it was spruce and hemlock, lots and lots of hemlock too. And hemlock, that made a man out of you because it was so stinking heavy. Now, spruce, that was a real snap. If you had enough seniority to get on the spruce line, you loved that because they was like lifting a block of lead and then going to a pillow. There was about that much difference in them.

Between hemlock and spruce?

Oh yeah, that was the worst wood we had, that hemlock. And white fir was in between. But they used those for different grades of pulp. If they wanted to brighten the pulp, they ran more spruce, because it was a whiter wood, and it put out a whiter pulp. And if it needed it darkened, of course you could run more hemlock or white fir. But they had a combination that they figured out. Well, they mixed it. It all went to one screen. There were screens—after they ran through the grinder, it went into a trough and ran down to a central point. It was screened there, where they took all the heavies out, and then just the pulp that they wanted to keep went through. And that was pumped up to the screen room. But they controlled it. When we first came to work, why, the jiggerman, which was kind of like a working foreman, I guess you'd call him—but anyway, why, what he would do is, he'd get an order in that said put so many stones on spruce—because there was three stones to a line in the No. 2 grinder room—and they would say put so many stones on spruce and so many on white fir and hemlock. And that's the way they run it. Normally we ran a No. 10 grinder primarily on spruce. The rest of them, which were when I started, 9 through 16, well most all the other ones ran on hemlock and white fir.

So what are the numbers? I know the paper machines were numbered 1 through 10, and the grinding rooms were numbered, you know, 1, 2 and 3—but what are these other numbers?

Well, 9 was the first grinder that they had on the far wall, which would be south, in the grinder room, because you're about, oh, 40, 50 feet below the water level there. The water's all backed up against the wall. They were old brick walls and stuff, so that backs up a lot of water against that. And then of course it goes down and goes through the turbines. But No. 9 was on the very, very south end, and No. 16 was on the north end. And you had No. 9 and No. 10. No. 11 and 12 and 13 had electric motors. And then 14, 15 and 16 were just straight turbine power, water power. So if you were lucky, other than the heavy wood, you got down on the water power, and you can kind of control the speed on those; it didn't work you so hard. But on the electrics, they could turn that wherever they wanted to, you know. If you happened to make the foreman a little mad, I'm sure that they kind of boosted the power on it to keep me busy.

Then you didn't have any control over the grinding speed? You just had to watch the tattletale and keep that wood going in there?

Right. It was a constant thing. Those grinders ran at a constant speed. I think it was around 240 revolutions per minute. Now, I may be off 30—low or something—but I think that's about what it was, close to about 240 RPMs. And it would eat a lot of wood. And they had what they called a jigging burr; did anybody mention that to you?

Not yet.

Well, a jigging burr was a round tool—I suppose it was diamond-impregnated or something. You know, it was real hard, and it cut grooves in those stones. And those stones were made of carborundum. And they weighed—oh, I think the whole setup with the shaft and all were between 20,000 pounds. So they were a big piece of machinery. And anyway, why, they would run that across and put a—actually put cutting edges on those stones. So if they wanted to speed it up, they—and that controlled the fiber too, because there was one jig that's, say, 80 per inch or whatever, and it controlled the size of the fiber that that stone cut. And so they would have some smaller ones, and they would have some that cut larger fibers. And that's what the whole thing was about on those. But they sure could keep you busy down there on those.

We've heard that the drainpipes didn't have screens in the old days, and fish would come up into the room.

Oh yeah. Well, we used to do a little fishing down there. Now, we weren't supposed to, but we done a little fishing down there. And I think the biggest salmon I ever caught out of there was about 65 pounds. And I put it in a wheelbarrow and I run it around to the back room. But we never got hoggish with it. We split that up amongst a bunch of lunch pails, and we packed it out in our lunch pails. Yeah, we had fresh salmon every once in a while.

But a 65-pound salmon couldn't have actually gotten into the room.

No, no, no. We snagged that right down in the river itself. I mean, we had a hole through the floor between No. 9 and No. 10 grinder, and they used to catch people fishing there, so they put bars over it and set them in concrete. Well, before the concrete was dry, they jerked the bars out. And so we always had a hole there. It was probably in the vicinity of three-foot square. And we'd crawl underneath the floor; we'd go down behind the motors and drop down underneath the floor. In fact, we had a trapdoor in the floor, because they used to put trash and stuff in there. And anyway, why, we'd go down there, and we'd take a big snag hook and fish. And of course, it's illegal as heck, but we sure had a good time doing it.

Somebody yesterday said sometimes it smelled like a cannery in there.

It did. Oh yeah, if you'd come in, say, on the night shift, you knew that the shift ahead of you was fishing because you could smell fish where they dressed them out down in there. The company frowned on it. I mean, if they caught you—I got caught one time, and they decided they'd just put me on the electric line to keep me busy. Well, the guy that was running the electric line liked salmon so well that he would do a lot of my work so I could go down and fish in the hole between the grinders. So we never did quit fishing 'til we got out of there. But you could walk across the river on the fish, almost. Well, see, that was a dead end up there; it was a cul de sac. And the salmon used to come up underneath there—that was before they put the fish ladder in—and you could catch salmon off of the conveyor belts; or you could catch them down there underneath the grinders, or off the seawall—wherever.

We used to work real hard and get ahead, and then sneak out there and do a little fishing. Somebody would say they wanted some fish, and we'd go out and do it. But most people didn't get greedy; they just took enough to eat. But like I say, it was illegal.

But there was a lot of salmon in there. But then, that's due to that bottleneck up there at the end, because we'd have to shut down—don't quote me on this, but I think it's right close to the first of July—we used to shut down and shut all the grinders, all the generators off, and then there would be no water coming out. And then the fish would drop out of there, go down and hit the current coming from the falls, and then they would go on up and take the fish ladder and go on up over the falls. But otherwise you couldn't get them to back out of there. You know, as long as there was current in there, for some reason they want to stay up there and fight them.

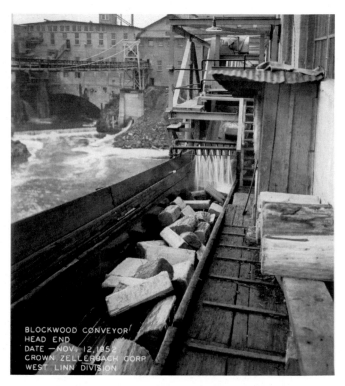

BLOCKWOOD CONVEYOR
HEAD END
DATE —NOV. 12, 1952
CROWN ZELLERBACH CORP
WEST LINN DIVISION

1952 photo courtesy of West Linn Paper Company.

Yeah, yeah. You know where the arches are? See, that's the old No. 3 grinder room. That's the outfall from No. 3 grinder room. There was four grinders over there, and those had four stones to a line. On normal operation there was two grindermen plus a wood piler, because a wood piler used to pile for what they called four stones, meaning a whole line over there. That is, at normal water height. If the water got too high and dropped the head—what they call the difference between the upper and the lower river—then we would have to go to five-stone operation, and that means you had to pile for five stones. So it made it kind of bad. But yeah, there was two electric motors on those grinders out of the four, and that was on No. 23 and No. 24 grinder. And boy, those could work you to death over there. And the others were turbine-powered.

But when I first went in there, and for a lot of years, well, they even run the electrics on water if they could. But if they were not using that many—because they ground a lot of spruce over there, and they didn't need a lot of that light wood sometimes—well, they would just generate with them. And they put out lots of electricity, so—but it's kind of interesting, you know, to see all that gone.

We used to go up there with our boats and drive up in to the arches and throw our anchor and back out. And we could fish—I think it was about 15 or 20 feet from the wall. And of course, them salmon are coming up there, trying to fight that current, and wherever there's lots of current, they fight it. So what did we do? Why, we anchored there and fished. I don't know if you're old enough to remember the hog lines up there they used to have. Well, they had a deadline then. Yeah, the actual deadline was off of Black Point on the Oregon City side, ran to the lower corner, which would be the north—I think it's the northeast corner of the grinder room. There was a deadline there. In fact, they had a cable strung out over the top so that you knew. But the building itself, you had to be—I think it was between 15 and 20 feet down, or 25 feet down, from the building. So they kind of watched them pretty close.

So there were 24 stones?

No, there was four grinders in No. 3 room, and there was four stones to each one, so that'd be—what, 16—sixteen stones.

What's No. 24?

Twenty-four was a big electric grinder. Twenty-four and 23 both had electric motors on them, and that's the only two in that room that had them on them. And the other two were just strictly turbine-powered, water-powered. And then as you come out the door to go to the elevator, to go back upstairs into what we called the screen room, why there was a generator off to the right-hand side there that generated; that used to be used a lot. And I think that was either No. 20 or 21; I don't know for sure.

Then everything that was waste became hog fuel?

Everything, everything that—yeah. I worked in the sawmill too, later on. I was a lead millwright over the sawmill, so yeah, the old hog, that's where all the waste went through, and they chipped it up and run it to hog fuel. That's what they classified as hog fuel. It went on a conveyor belt from the grinder—or from the sawmill it went on a conveyor belt over to the steam plant. And then they had storage bins in there, and they put it in there. And then later on, why then they put it up on the hill there in West Linn; you see a lot of storage up on top where they hauled it in; by truckloads they hauled a lot of stuff in there. But primarily, our hog fuel is what created our steam to start with.

Now then they're fired mostly with gas, I think. And they still have oil there, but I think—the last I heard they maybe had eliminated the oil altogether—those green tanks down by the locks? Well, those used to be just strictly big oil tanks. Yeah, they held thousands of gallons of crude oil in there. And they would pump that up to the steam plant, and that's what they used to fire a lot of the boilers with, too. Whenever we got low on hog fuel or something like that, well then they used the oil.

How long did you stay in the grinder room?

Well, I started in the grinder room in '56, and then within about three years I graduated over to what they called the chippers, which was Mill B. That was across the canal, by the steam plant. And so I worked on the chippers for a while, then I bid into Mill B itself, the sulfite department. That's where we made sulfite, which we made real good sulfite over there. But that's what used to stink things up. Remember the old sulfur smell? Well, that's when we blowed those cooks down there—we called them cooks—when we blowed those digesters, that's where that smell comes from, because that was all exhausted into the air. And the liquor at that time was pumped, or drained, directly into the river, which I don't—it had a lot of sugar and stuff in it. And I'm sure it didn't hurt to get it out of the river, but actually there was a lot of nutrition in there for the fish and stuff. That's what they always said, anyway; they know more about it than I do.

Do you remember how much you got paid for your first job in the grinding room?

Oh, I think when I first started in the grinder room, the pay was, oh, less than two dollars an hour. And that's why I kept changing jobs. Every time I changed a job down there, I got a raise. And then I stayed at Mill B until that shut down in about 1968, when they completely eliminated it. And so there's no sulfite or anything made like that any more down there. And I think the grinder room is eliminated. We buy most all our pulp on the open market. We had machines in there to grind chips up. We used those for a long time. And after the market got so high—that's what kind of controlled that mill down there. Pulp got so high. It used to be like $40 a ton, and I think when I got out of there, it was like $400 a ton. And see, that's the reason they quit making their own was it was cheap. They had lots of it they could get, and it was cheap. But it didn't turn into that. It turned into be very expensive. And you need lots of that to make your paper.

What shifts did you have to work? Did you ever change a job so you could get on days?

Well, I worked shift work from the time I started there until I got into maintenance. Now, I worked a short time on the upper river loggers. Yeah, that was quite interesting—in fact, Paul Miken had something to do with that at one time. He was working up there, and then of course he worked in the sawmill too. Then he became the operator, or the boss, at the sawmill. Very nice fellow. But no, that's primarily why I changed, is the wage level. And the job; I liked the jobs, you know. And after sulfite shut down, which broke my heart, because I was working up to, hoped to be, a digester cooker at that time, and I went to the paper machines. And I worked on the paper machines from 1968 till about 1974, I believe. And then I luckily got into maintenance.

I'd been trying to get into maintenance, and when I was younger they didn't need anybody because we had a full crew up, and then when it finally opened up for maintenance, if they hadn't passed a law they couldn't discriminate against you for your age, I wouldn't have gotten that job. But because they couldn't discriminate me, they had to give me a chance, and I done real good. And I loved it. I mean, I loved it; I really liked that. It's something different, and it's better pay, and you didn't have to put up with the crap on the paper machines.

So you've just about done every department in the mill, well, except maybe converting...

No, I worked in finishing a bit. And I worked in the beater department. I worked in the clay plant. I mean, I—there's not many jobs in that mill I didn't do. Yeah, yeah. At some time there I had worked in them. You know, between jobs, and like during that flood, while I was off work for a couple of weeks, I either lost work or went down and cleaned up. Well, it just happened to be I got to work with a maintenance crew while the flood had us all flooded out and we was cleaning up the mud and everything. So it was—I've had a good, interesting career down there.

Were you down there in '64 when the flood hit? You must have been working then.

In '62 we had the wind. I was there working the night that the Columbus Day storm struck, and I seen parts of the buildings blow off. They looked like feathers floating across, and there'd be big chunks of corrugated metal roofing. And then they'd hit them power lines to the PGE plant, and they'd sock into that, and then they would arc and just weld themselves right to it. It just cut lines and every other darn thing. It was something. It was kind of frightening there. Of course, I wasn't worried; I was inside a great big concrete building most of the time, so the worst part about it was getting out on top of the digesters—or, well, the liquor tanks—and shutting off the valves, you know, because when things went down, they just went down, and it stayed down for quite a while. We had emergency power, but you didn't [have] enough to operate everything.

I've heard people say the dangerous job was working on the big Crown Zellerbach sign.

There was a big Crown Zellerbach sign up on the roof. And yes, I've been up on the roof and worked up there quite a bit, because there was vent fans and stuff up there that had to be taken care of, repaired, and so we went up there and worked quite a bit. It wasn't so bad; you just had to be careful when you get out on the slope down there towards the canal. But if you're going to fall off of the roof, you want to fall in the canal; you don't want to fall down on the concrete.

I was down there when we had the barge explode. I'm trying to think when that was. I'm going to say somewhere around '57 or something. They had chemicals come in in a barge, and they were the type of chemical—it's like a powdered, bleach and stuff like that—which is basically Purex. Purex is diluted, and this is the raw powder; this is the pure thing. And evidently, why, it got water into it, and it started to burn, and then of course it started to burn some more. And then they put water in there, trying to put it out. It was a heck of a fire when I first seen it; it was burning clear across the dock; out of the barge, clear across the dock, and up into the roof of the mill. And yeah, it was quite an experience, and you could hear the blast up on the top of the hill in West Linn. My wife heard it, and then she looked out and could see the smoke coming up from the fire down there at the mill. It was quite a fire.

Yeah, they used to barge all their chemicals up, you know, at that time. Of course now all that situation's so much different, and probably much safer, different type of chemicals and stuff. But I think that was bleaching, paper bleaching, material, and which is probably, like I say, basically I think it's the same derivative they make Purex and Clorox and all that out of.

Was that barge in Lock Chamber 4, by the dock?

It was up by the docks. The barge was above the guard lock, and—oh, let's see; I don't remember the gate numbers down there now—but it was a long pool in there between the guard shack and the next one up. Yeah. It was pretty exciting at the time.

Did the Crown Z come down with hoses, or how did people fight the fire?

No, we fought it—we had our own fire crew, but you know, that was about—I'm just going to guess—about six or seven in the evening, so it was on the night crew. I was on swing shift when that broke out. And so the men that were on the fire department at the mill, they're the first ones to start throwing water on it, and then of course they brought the city fire department down, and they finally got everything put out. And like I say, it burned part of the roof of the mill, burnt the barge pretty good, and luckily no one was injured. Yeah, because it blew; when it blew, that was a pretty good explosion; she really boomed. But I think most of it went up instead of out.

Good thing it wasn't at shift change or something.

Yeah. It could have been worse; could have been much worse. But that was one of the fairly exciting things I seen happen down there. And the [1964] flood was a doozy. That flood was a doozy. I mean, the water did come up good. And in fact, I think it was flooding over the lock gates, if I remember right, the upper lock gates. Flooded into the basements of the mill. In fact, way in underneath Mill C, or Mill D, there was lots of water down there in the basement. There's marks on the wall there in some places where I suppose it's eight or 10 feet high in there. So that—yeah, that was quite a time; that was quite a mess. And it darn near shut us down. I think at that time—which doesn't sound like much now—but then I think it was about a million dollars, they said, damage. Yeah, they lost a lot of stuff.

Well, you know, the falls was level; the falls was level; it just looked like a big rapids up there—but it came in the side of the mill there. If you look on the West Linn side of the mill, which would be facing Oregon City, but right there at the falls, those windows, those were great big metal-framed windows. And the water come through there, broke out all of that. See, they've got—glass brick in there now, I think, is what they used finally to repair it. But the water was coming in there and running out the doors below. So I mean it was high; it was really high.

When we used to get prepared for the flood there, why we would go in and put seals on all the vents on those big motors down there. We'd have to pull them off, pull the vents off of those big electric motors for the grinders, and then we would put a flat plate on with a rubber seal underneath it, and bolt it down good. And even on the top, you pulled things off—the vents on top—and sealed them off solid. And we pulled a lot of pumps. Some of that caught them off guard so fast that they lost a lot of stuff because they really didn't realize that was going to hit like that. I remember when they closed the bridge there, the West Linn bridge.

It was Christmas; I think it was Christmas Day, if I remember right, if I'm not mistaken. And we had a family function up there, and we had rented the Sunset fire hall because we had a large family to start with and the whole bunch together. And when they went to go back to Oregon City, they couldn't get across the bridge because they had closed the West Linn/Oregon City bridge due to the fact—I think they said about almost a million feet of logs and stuff that was back up there against the side of the bridge, back in that kind of cove there that swirls around. And they was a little concerned with the bridge, because it kind of shook it, you know, when those things banged against it, because the water was so stinking fast down through there. And if I'm not mistaken, it come up and crossed over [Highway] 99 there, right at the bend where it comes up in Oregon City there. Even back in the 40s, I think it flooded down there around Tony's [Fish Market], because Tony's, he was the low spot there. Oh yeah. Oh, you bet. It was really a doozy at that time. It was kind of spooky, because there was so much water there. I mean, unless you had seen it, it's just hard to believe how much water was there. And of course I've seen it flood there several times, you know, but nothing of that magnitude. I think that's probably the worst one I've ever seen. And '96 it flooded again, but not nearly as bad, but it wasn't far from the '64 flood, actually. But it didn't catch them off guard so much, so there wasn't nearly as much damage.

That's right. I mean, a lot of stuff got ruined down there, because like when you're getting in there to the guard shack, you know, you're getting into the nurse's office and all that and the offices down below. I mean, everything was just flooded to beat heck in there. In '64 I was working still up at Mill B. So I got to go in one week out of two for cleanup. That's the only time I filed for unemployment, which you didn't get anyway, but in over 38 years I never drew one unemployment check from the mill.

So they kept me—they kept me busy. And that's why I went to work there, because I knew that I had a paycheck coming in I think it was once every two weeks to start with, and then of course later on it went to once a week. But I knew you'd have a steady income. And that's why I didn't take logging or something like that, because at the time they didn't log in the wintertime much. Now they log more in the wintertime than they did, but—so I was real happy, you know, to get there and have a steady income, because that was pretty important.

We was raising a family and everything, and you knew how much money you had coming in. And of course, like I say, I changed jobs; if I liked the job, I stayed longer, but most of the time I changed jobs for the financial gain, and the work. But then I loved maintenance so much that I couldn't think of doing anything else. I loved maintenance.

But you never made that jump to supervisor like some of the guys.

I was a relief foreman. Yeah, they mentioned to me a couple of times that why don't you put in for foreman, but I didn't. I just stayed a union man. I liked the union man. I didn't have to worry about somebody getting mad at you and firing you. You know, I kind of liked that.

Well, that brings us to the strikes when you were there.

I think our first strike was in 1964. And I think that's probably the first strike in that paper industry, especially, I know, in West Linn and that area. I don't think there was ever a strike there before that I know of. My dad was a charter member originally, when they started the union down in West Linn. And they were pretty nasty to them when they tried to get that union in, of course. Now then they have the Taft-Hartley law and everything. Now at that time I'm not so sure they did, but I know that they threatened a lot of people with losing their job if they got a union in there. But yeah…

How long was the strike in '64? Do you remember?

Well, now, see, you've got me by the ears there. I think six and a half months— one of them went about six and a half months. I think we've had about two or three while I was down there; two for sure. Sixty-four and then it seems to me '72 or something. Well, you don't pay any attention to that stuff. You just—you know, it's a funny thing: you dread a strike, because I always say nobody gains. Well, you don't, really, and yet you do. But after the first paycheck—you lived payday to payday anyway down there—after you got your first paycheck you was back going again. So it was no big deal, but it seemed like it at the time.

Did it really help out to have the salmon supplementing their budget?

Well, we caught those salmon down there because people like salmon, for one reason. No, I suppose you could afford to buy them if you wanted to, but it was more fun to go down there, I guess, because it was illegal to do it. And then, a lot of those people, they had big families, and they didn't fish; they couldn't afford to fish. So we supplied them with fish. Oh yeah, we enjoyed it. In fact, I love to fish anyway, even legal or illegal at that time; didn't make no difference. I was young. But it was pretty exciting.

So while you were there, the union paid you something if you were on strike?

Well, yeah, we had a strike fund that we had started to build up. And I don't know—it seems to me like it was—boy, I'm just going to say $100 a week. And that just about run out, I think, at the last. I know that we had a six-and-a-half-month strike; it seems to me we had one that went about seven months, it seems like. But that's about all I can tell you, other than we made it. Yeah, it was real tough because, you know, you've got bills to pay. Electric bills and stuff, that's what worried you more than anything else. And there was a food bank or something that we got some food from. I don't remember what organization it was, but they helped everybody. Well, that's the same one that handed out the cheese and the butter and all that; it was a federal thing. And at the time it was a good thing we had it. You know, that sure helped.

Did you have to take your job back from a supervisor?

Well, no, I don't think so, at the time. But yeah, they had been doing the work. But then there was some jobs there they weren't able to do. But most of them were filled in by the supervisors. And after you went back to work, you didn't hold it against them. You know, we knew that they were salaried people, and if they told them to go to work—they had to go to work or lose their job—well, if they would have told me that, I would have probably went to work. But yeah, yeah, there was a lot of them, though, that did change with the supervisors, because—and they were happy to see us come back, let me tell you. They were happy to see us come back. They didn't like to have to work any more. And I don't blame them, you know.

I'll bet they were happy.

Oh sure. When you go from a job where you pack a pencil and you're not allowed to do—you know, the supervisors down there weren't allowed to—like my maintenance man—we had a few of them that would get in and help you, but they weren't allowed to. I mean, if—and of course you're not going to bite the hand that feeds you. If they want to help, I'm sure not going to turn them in to the union or anybody because I thought it was really kind and nice of them. And I enjoyed it, you know, because most of the supervisors I worked for I enjoyed anyway.

Well, did you know Rosie?

Schultze? Oh yes, I knew Rosie. I worked for Rosie. I worked on the paper machines, and he was the super—head honcho on the paper machines when I was there. Yeah, Rosie Schultze, real nice fella; big old jovial fella. I understand he's still alive and doing well and in his 90s.

He's 96. He started in '27, you know…

Oh, you bet. Well, my dad went to work there right around that time. It was around '28 or the '30s, there, early '30s. And he worked there for a while. He worked there until the war broke out. And then they needed people in the shipyards, so he went to the shipyards. And at that time your job was supposed to be guaranteed. When you come back from work for the government in the war effort, you was supposed to have a job guaranteed back. Well, they was a little nasty over there at the mill about that. They resented the people leaving them to go to the shipyards. And we needed the ships; they built a lot of them in Portland. My dad—I think after a few words to tell them how he thought it was unpatriotic—he never went back to the mill. And I went there in 1956, and then my oldest son was there until the paper mill shut down. He went in there—oh, it seems to me it was right around 1972 or something like that. So he had around 23 years in the mill when it shut down.

Did he go right out of high school?

Yeah, yeah, not long out of high school. Then he took training after the mill shut down, and he became a heating and ventilation expert. He's got a real good job, makes a lot better money, and he told his mother—today on the telephone we was just talking about it, and he said, "I think the mill done me a favor shutting down." You know, when I got out of the paper mill, people asked me, well, why are you retiring? Well, I could have worked a couple more years; in fact, I could have worked five more years. But I got tired of the way things was going down there, and I said to them, "You know, since I've been here for almost 40 years people have said this paper mill was going to shut down, and I said that was bunk." But you know, I said, "I see the handwriting on the wall." I said, "I believe—and I hope I'm wrong—that this paper mill's going to go down." Well, it did shut down in just a couple years.

So that just shows you, you know. But there had been all indications. You know, every time you turned around they was after the paper mill, you know. The outfall or some silly thing, you know. And really, it wasn't as bad as people thought, you know. And the smell, but it was just a sulfur smell. You know, it was nothing going to hurt you. In fact, it would keep your nose clear. [laughing]

Olaf told me that the acid rain of the sulfur had eaten his cyclone fence within 10 years.

Oh yeah? Well, that could be, yeah. But I mean, you know, we lived with it. Well, you know, the steam plant, there was nothing but, oh, hydrocarbons or whatever come out of there, you know. Depends what they burn. Mostly hog fuel there wasn't much that come out of there, you know, that would bother you. But when they blowed one of those digesters, then of course that's when you got your sulfur smell. And then you had it in West Linn, and then you had it across the river there at Oregon City. Yeah. See, we was going to change and put in a modern system—what they call a Kamyr digester. In fact, I think if I remember right we had a small one there, a small Kamyr digester. That was an experimental one which cost probably beaucoup bucks to put it in there, and we made some like Magnaflite or whatever they called it. But we was going to originally skin the old digesters down there and make it into a different process, to eliminate the smell of sulfur and everything. But they decided it would be too cost-prohibitive. Like I told you, then I think pulp was around $40 a ton or something. So you know, they thought, well, heck, you know, it's not worth the money that they would have to spend on the original outlay.

So you worked for Crown Z, James River and then Simpson.

I started with Crown Zellerbach, and then I thought I would be there until I retired. And then, of course, it changed to James River Corporation, which was a very progressive move as far as I was concerned, because when you worked for Crown Zellerbach you had no stock options or anything. When James River come in, they offered us a lot more than Crown Zellerbach did as far as additional benefits. And they were pretty good people to work for. They were, I believe, safety conscious, real good about that. And then we went to Simpson. Well, Simpson was a family-owned outfit, and they were supposed to be progressive; they just couldn't make a living down there, evidently. You know, I don't know what the cause—they're still in business in the timber, you know, so they're a successful business. But I think of all the three, I liked James River because they offered more incentive for the employees. See, we had no stock options or anything, but when they come in, that's the first thing they did. It wasn't very long down the line that we could buy into the stock options and stuff, and IRAs and everything, and it made it real nice.

A lot of the men are very loyal. I mean they like to tell stories about Mr. Zellerbach coming down and bragging about his employees…

Well, the Zellerbach family, that's right; they knew what was going on. Old Harold used to come down there once in a while, I think, and visit. I heard nothing but good about him; everybody liked him. And you knew what was going on. That's the trouble: it used to be kind of one big family down there, and then it just spread out to where it was just strictly business, you know. The attitudes of the people changed, and the attitude of the supervision changed too. And I'm telling you, they've had some bad supervision down there, but they've had some good supervision, too, you know. I mean, a lot of their troubles was in their supervision too. And of course some of the hourly employees, you know. I mean, we're all guilty of something. But no, I was almost brokenhearted when Crown Zellerbach went out, because I had started there, and I had—well, that's all I know. And I was born in West Linn; I was born at Holly Gardens. Do you know where that's at?

Okay. Well, I was born right in a house at Holly Gardens, almost probably where the new bridge goes across, down there above the river, just above the river, is where I was born. And you know, Crown Zellerbach—it was Crown Willamette, you know, and then Crown Zellerbach. And you know, that was just an old name there, and you just depended on it. I kind of hated to see them change. But when they changed to James River, it was a very progressive thing as far as I'm concerned. The shame was that James River couldn't hang onto it, you know. But I think the worst part of the whole works was that—what was his name? Sir James Goldschmidt [said disgustedly] or whatever his name. He caused most of the trouble when they come over and bought the Crown holdings. All it was was a money thing for him. And now he's up with St. Peter explaining that. But yeah, we was real disappointed when that happened, other than it brought the stock up, if you had any stock. Why, it boosted the price of the stock. So I didn't have to worry; I couldn't afford any in them days anyway. I was like everyone else, living paycheck to paycheck. And thank God I had one coming in every couple of weeks.

Art gave an interesting perspective in the way management changed.

Well, the management did change down there. Because most of these people earned their way up originally. You know, you could depend on—if you was a good worker and done your job, you had a possible chance of getting an improvement and an advancement in pay and a better job, and up into supervision if you wanted to. But then when they started bringing those people in from the outside, boy, we got some doozies; we did. They, might have been selling popcorn for all I know, but they didn't know how to make paper. And a lot of these other jobs they didn't know how to do there. And you know, that makes a big difference. The people you could depend on, the old-timers—like, well, Rosie or any of them—they'd come up from the ranks, and if they told you that this or that had to be done to make things work, you could depend on it. But these other people caused more problems than they—you know, they should have just went to the office, sit down with their pencil and stayed there, and let the men run the job, and it would have been run fine.

But that's right; that's the change right there. And maintenance too. When I first went in there, you couldn't hire maintenance people from the outside. Your electricians, your pipe fitters and your mechanics all came up from the ranks. You started with a pair of pliers and a Crescent wrench, and you worked your way up from there. And then when you got up to where you was capable of doing things and became a journeyman, you knew what was going on around there. Well, it's the same with the paper machines. Those people that started down there as a helper and then worked their way on up to where they was a machine supervisor, or the machine room supervisor, they knew what was going on. They could go down there and make an adjustment, you know, and it would be the right adjustment.

But when you get people from the outside that think that that's the right thing— you know, books is always good, but sometimes books aren't—they're not really the proper thing to use on the job in something like that. Experience is a much better teacher, no question about it.

Olaf said it was like he spent the first 21 years waiting to be millwright –

Oh sure. Well, that's the way with me. I tried many times to get into maintenance, and until things opened up where I could get in, I was never really totally happy with my job. Now, I done it just because to make a living. In fact, I was much happier in maintenance, much happier. I didn't like the paper machine. When I went into maintenance, after I became a journeyman I had to work—oh, I don't know—between six months and a year back on shift work, because I became the shift millwright. And of course, the next guy up, just as soon as he got up, I threw him my keys. I was gone; I went back to day shift. But yeah, you know, I'm like Olaf. I just loved the job. It was something different most every day, and you gradually got better at it. You know, you learn new things. And then they threw a lot of new equipment in there at the last. You know, there's a lot of new things we didn't have. We used to do a lot of work on turbines and stuff, and the turbines—the steam plant primarily done that, but they went to electric drives on all those machines. And you know, they eliminated those turbines altogether. And that was quite an experience. That was really something.

See the steam plant used to take care of all that before, and then later on, why, we had to take care of it all, between the electricians and the maintenance men. I mean, they hired a few. But you know, like I say, they used to bring them up through the ranks. Electricians and everybody started at—they was a bulb-screwer to start with, you know, and they just screwed bulbs in there around the mill there and done a few little things. And they brought them up from the ranks, and they turned into real good electricians before it was over. But we started hiring from the outside. Well, it turned into the computer age, and a lot of these young people come in, a lot of things was controlled by computers. And they had more knowledge—I mean, most of us didn't hardly know much about a computer. And I'd not say that I'm really—what do they call it? Literate today, or illiterate? I think I'm illiterate when it comes to computers. I play games on it; that's about all I do.

Bietschek and just a whole bunch of those old-timers...real pleasant people to work with. We used to make fun out of work when we worked on the logs. You know, we'd walk a boom stick or something, and I hit the water more than once, betting them I could jump off a tug onto a boom stick and keep going. I got a little wet, but that was in the summertime, and it was kind of nice to get wet.

Pulp Siding, south of Canemah, where Georgia clay was offloaded from rail to barge. Photo shows Zellerbach tugs at dock when the Willamette River is at flood stage.

We hear river work was such an independent operation, part of the mill, but it didn't seem like it?

No, it was kind of separate altogether, you know. We had our own loggers' shack up there. That's the one located on 99. You see that old green building there? Well, we called that the loggers' shack. And then there was unloading up there for rock, lime rock. You know, we used to haul a lot of lime rock down there. If you talk to Art Dorrance, he was one of the old Mill B experts. And we made our own acid. We had towers that we put tons and tons and tons of lime rock in, and then they run that sulfur—they burnt the sulfur and run the gas and water through that lime rock, and that's the chemical they used to cook things with. It was kind of interesting.

Well, the other thing Art talked about was how about people jumping in the river when it was hot. And I said, "Well, with all their clothes on?" And he'd go, "Well, no, they'd usually take just their boots off".

Well, most of the time I didn't have the pleasure of taking my boots off. I just jumped in. Usually I just jumped off the tug with my cork boots on and get on a boom stick, and I'd get about halfway or something and get rolled off. And so I had my clothes and my boots and everything on, but it didn't hurt them anyway. They got a little bit wet, but they dried out by the next day. But we used to do some silly games. We'd get to log-rolling out there. And we had a couple of old-timers there that could really roll logs. And you really didn't want to get on with them, because they would—I remember old Marshall Leisman was one of them that was pretty doggone good, and they'd get out there and do some log-rolling. I used to really get a big kick out of them because one of them would sooner or later get dumped in the river. But they were good at it. I didn't compete with them; they were too good for me, fast on their feet. But we did a lot of things. We made a good day out of it, you know.

And I barged, and that was kind of different too. But I think probably one of the best jobs I ever had was working up there on the logs. We made the log rafts up and everything, and then we'd put them in storage. You know, we'd hook onto them and tow them up the river, and they had places they anchored the log rafts for the year. And they stayed in the river I think probably a minimum of about three years. And that helped loosen the bark and stuff on those logs. That was the reason they kept them in so long. In fact, them old sinkers; you've probably heard them talk about the sinkers, maybe—they would leave them in the water so long those hemlocks would soak up and a lot of them would sink, you know. They were just so stinking heavy. When you pulled a raft of those in, why, that was quite a lift.

I think it was a 60- or 80-ton crane that we picked them out of the river with down at the sawmill. And boy, it was a good load on them. You had to be real careful, because of them big old heavy ones hanging down, and then a lot of them would break loose and stick in the bottom of the river. And they'd be sticking straight up, and they'd have to come in—once a year old Bernert Towing usually come up, and he dug them out with his clam shovel. He'd load them all on, and I think they brought them back to the mill, but they got them washed all up and everything first.

Yeah, Paul said that he knew there was a lot of timber on the bottom of the river.

Salvage. Oh yeah. Oh yeah, you bet. You bet. There was a lot of it in that river up there. But like I say, primarily spruce floated; spruce is like a cork. It's very buoyant, and that didn't sink like that. You could leave it in there, and it would loosen the bark on the darn things and stuff, but it didn't sink down. But boy, that old hemlock—I think we weighed a block down there at the mill one day, and those were—I think they were a foot by a foot by two feet long, and it seems to me they weighed between 140 and 160 pounds. Because I know we'd get hold of one every once in a while, and it just—we was only about 145 pounds then, you know, and you was picking your own weight when you picked it up. But we didn't get too many of those. They usually tried to drag them off or something. Well, it felt like a rock. You bet. Oh, them were heavy buggers, I'm telling you that.

Did you have any Willamette Falls close calls when you were working with the logs?

Not so much. I remember getting up and dogging, having the tugboat grounded up against the edge of the falls and having to jump out and dog into some logs there right on the lip of the falls there. That was kind of fun. It was low water, you know; it wasn't so bad. But it was a little spooky, especially the first time you done it.

Yeah, I've been pretty lucky down at the paper mill. I got hurt a couple of times down there. I almost got into a couple of saws in the sawmill once. A conveyor started as I crossed it, but luckily—have you ever seen handprints in steel? I think I put handprints in steel. It started to take me into the twin-band rips, and I don't know how I kept from losing a leg. I still haven't figured that out. It didn't catch me. I got hold of the wall and hollered, and they got it shut off and reversed it. But I had hold of flat steel somehow. I hung on. But had I not gotten hold of it, it would have took me in and cut me in a couple of pieces. And it's pretty frightening when you stop and think about it.

Yeah, my cousin almost lost his leg there, Art Baisch. He was working on the logs. I think probably, maybe old Art was working on there at that time. And he got in the bight of a rope. I think they was moving the boom sticks up towards the falls there. They used to keep boom sticks along to get the riffraff off of there. And anyway, why, he got caught in that bight, and it almost tore his leg off. And he was crippled for the rest of his life, you know, as far as getting full use back. He never got full use back of his leg. He could walk and everything, but it was stiff. Oh yeah, there were some pretty good ones up there.

Somebody said, well sure, it says it was an accident-free year, but we just didn't tell them about these little things.

Well yeah, it depends on what they called an accident, see? But anything that went to the doctor, then they had to record that. But yeah, yeah, there—well, there was people got hurt there every once in a while. You know, get their hand in a nip, and especially, you know, on the winders and on the paper machines. And that's unforgiving. I used to tell the people that come in there when I worked on the—even in maintenance—we used to tell those young people, now you want to look at these gears here. They're unforgiving. I said, "You know, you can stand there and scream and holler and pray all you want, and that machine's not going to stop. It's just going to eat you. So keep your hands out of the thing. And if you're going to work on anything, lock it out. Don't take any chances." And that helped a lot to—you know, I was safety-aware, which you've got to be to stay alive in a business like that, because there's too many things, you know, that you can get hurt.

I've seen some freak accidents. I was working piling wood on one grinder one day, and there was a little hole in the wall up there. Well, our wood come down chutes. There was a turbine on the right and a turbine on the left, and there's a trough in there—well, it's just an alleyway where the wood piles up into it. It piled up clear above the door. The doors were made out of real heavy chains. They were just a curtain—we called them a curtain—that come down to keep the wood from coming out. They had, oh, like fence wire with big heavy chains so the wood couldn't come through. Well, there was a little hole in the wall up there about, oh, I don't know, about 12 inches. And it was by a pipe. And one of the guys I was working with was running the grinder, and the next thing you knew he was laying on the ground. A block had come through up above there and come out of the wall and hit him in the back of the head. It darn near killed him. But I mean, it was freak things like that. And you—well, that had been there a long time, for 40 or 50 years or whatever, and it never had come through there before, but it sure come through there that time. So—but most of the time they were pretty safety-conscious down there.

How many people in your family worked there?

You know, my family was represented real well down there. My grandfather worked there; my dad's brother worked there; my grandfather retired from there; my uncle, he retired from there; and I retired from there. And my cousins, the Baisch family—you've probably heard of the Baisches if you was ever around West Linn—in fact, Reuben Baisch, who played football, just passed away; we buried him yesterday. But anyway, I would imagine, oh shoot, around 25 people in my family worked there, at least. At least. Because with all the Baisches and all the Kunzes, and there was a lot. There was a lot of my family down there.

Ed said that if his relatives all stayed home the same day, the mill would shut down.

Yeah, he's my cousin. Yeah, that's right. At that time, that was just about the truth. You know, his uncles and everything worked there. I mean, there was a whole batch there. But like I say, I know myself, my grandfather retired from there, and the Baisch boys, I think there was—oh, shoot—four or five of them that worked down there. There was Emil and Art and Ed and Albert, you know, but probably another one I can't think of right offhand. So there was quite a few of just my uncles and cousins worked down there, besides my direct family.

Did you guys all walk to work?

My grandfather walked to work every day. He walked down the Sunset Hill there and went to work, and walked back up. At that time they had that stairway that come up over the side, and he'd climb that stairway and come on up, and he'd get up and hit Sunset Avenue and walk up the hill to his house. He lived on—was it Chestnut, maybe? Well, it used to be an old dump up there, you know. Used to be a city dump. Freeway goes right through it now. And he lived, oh, about a quarter mile this side of the dump. Well, in fact, there's some apartments built up there now. You come up Sunset Avenue, take the first road to the right, what's there now, that was my cousin lived on the corner. And then my grandparents, you went up a block and turned to the right, and then my grandparents and all lived in that area up there.

I've heard people say that when they were putting in I-205 the blasting would shake the paper machines, so they'd shut down the machines for an hour while they were blasting—

No. No, they never shut them down that I know of.

Do you know anything about the ghost at the mill?

No. No. I'm innocent on that. Yeah. Well, there's some holes in that mill that's spooky like a dungeon. I remember crawling underneath there in some of those places, and I've seen rats about as big dogs in there once in a while. Boy, there used to be some big rats around there, and you'd see them every once in a while—well, I called them river rats, you know, them big old long-tailed, slick-tailed rats—I've seen them run off like from the edge of the dock in towards the mill. And boy, they were big. It's spooky.

Yeah, worst thing I had happen, I crawled in the elevator one night. I was about half asleep. It was probably about three o'clock in the morning, in Mill B, and I was going to go down and work on one of the digesters at the bottom. So I jumped in the elevator, and all of a sudden I heard this "snap, snap, snap," and I turned around. I thought it was a big rat or something a-growling at me there. At the time I didn't know what it was, so I got hold of a wrench I had with me and popped it on the head and killed it. And it was a great big muskrat. So what I did is, I knew that at night, when the cooks would take a nap—and they had a bench there, and it took about six hours to run off a cook of chips, and so they would take a little nap in between. All you had to is watch the temperature arc, and if you'd done your work and if you had a good helper, he'd keep that arc going for you, and you wouldn't have to worry about it, and you could maybe take a few minutes' nap. So I snuck up there, and where they put their pillows and stuff, I put that dead muskrat in there. And I guess about the next day, I guess them buggers reached in there and got hold of that when they was going to take a nap, and they about crapped their pants. But it scared me when I first seen it, because heck, I was half asleep anyway, and doggone, that thing was snapping there. I think it was scared as I was, but then they're pretty good size. And I popped it anyway. I didn't know the difference. I thought it was a big rat.

Ed was mentioning Eckerts and some other people—

The Baisches, the Eckerts, the Kunzes, Witherspoons. There's a whole bunch of them are all in together.

And they all lived in West Linn?

Yeah. See, my grandparents, my grandparents—my grandmother was a Birch. And her sisters, one of them was married to an Eckert; the other was married to a Baisch. And that's how they all got together. So now you know that life history.

Did you ever work with any Sullivan people, from the power plant?

Well, I had a nephew that ran the Sullivan Plant, and so I got to go in every once in a while and visit with them. And yeah, anything we could do to help them, or they could do to help us, they would. But primarily that was, you know, PGE. But no, that's quite a place in there. I got to look it over real good. And that was quite a few years after I started there because, you know, that was kind of off-limits for us So it kind of made it nice when my nephew went to work there. So he'd take me in there and show me all around the place.

What was the best thing about working at the mill?

Well, I think the best thing about my employment there at the mill was the fact that I liked the people I worked with. You know, on my retirement, I miss the people. I truly miss the people because they were—most of them were friendly. And I think the best thing is the fact that I had steady employment all those years, you know. And it made me a decent living, and we raised our family from the money we made there. And yeah, I've seen some bad times down there, but I've seen a lot of good times. I think overall you can make your job a lot happier if you enjoy it, and especially when I got into maintenance. I'm like a lot of these other people told you. Once I got into maintenance, it made my life worthwhile, because I didn't especially like the shift work; I got tired of that. And then the fact that I just plain didn't like those machines or anything like that. But I think the best thing was the people. I think the people down there were fine people.

And what was the worst part?

Oh, I don't know. When I first started working down at the paper mill, a lot of people—you know, they didn't realize they're paid by the hour; they think they're paid for the job. And especially on the paper machines, some old-timers, you know, if you needed any help, a lot of times you didn't get it. Now, we had some real nice people on those paper machines that if you needed help, they were there and they would help you. You know, they didn't care whether you was a back tender or whatever. If you needed a little help, they'd get right in and help you. And other ones, "that's not my job," and sometimes you'd work so hard that you couldn't eat your lunch or anything. You know, you'd just be too busy, because it depends on the grade of paper you made, especially when they had the little machines in and we'd be making some oddball rolls, and you'd maybe have to put sticks through them, or maybe sticks through and wrap them and tape them and all that. And time—you'd take one roll off the machine; it would make maybe three rolls of the stuff you had to make, so you was always behind, you know. And I think the worst part was that a lot of people thought that their job ended if their machine shut down for a few minutes, they didn't have to help anybody else. I think finally, later on, they started teaching everybody different jobs down there. You always learned a job or two or three ahead of time, or ahead of you, and I think it helped when they let people know that, yeah, you can do all these jobs, but you can still go back and do some of these other ones, too; it doesn't hurt. In other words, cooperation. That's what lacked there sometimes. And other ones was great. I mean there was some people in there—in particular I think of a fellow named Ralph Keller; nicest fellow you ever met on a paper machine. But if you needed help or you needed some advice, he was there to help you. And sometimes you did. And it wasn't because you wasn't working hard; it was just because it was more work than you could handle. But that's the worst part I seen. Sometimes, especially on the paper machines, it was just too fast. Well, you didn't have the equipment and stuff to handle it.

They're better down there now. I mean, there's no question about it; it's better than it was. But thank God—it's like the grinder room, you know. That wasn't an easy job, and in some ways I'm sad to see it go, but in other ways I'm glad that there's young people that doesn't have to do that work right now. When they get to complaining on the job nowadays—I used to get a big kick out of them. I said, "You young people don't know what work was," because, you know, you don't have the jobs where you put out real physical labor like you used to at those places. And it's like those rolls: a lot of them they used to move around with hand trucks, you know. And boy oh boy, you know, it's a lot of difference now; everything is electrical carts and stuff like that. It makes a big different, a big difference.

Can you tell what kind of papers you saw being made, or helped make?
Oh yeah. We made Avalon paper. We made Catalina paper. They're just different grades of paper. We made like cash register tape paper and stuff like that. I think that probably was on Avalon. I'll bet we had about 160 different grades they could actually make, poundagewise and colorwise and everything and sizewise. They made a lot of them on those little machines. See, those little machines, you could afford to change grades, because they didn't run that much tonnage per hour. But on the bigger machines, like the coated machines, when you changed

grades, between the time that you change the grade and get a perfect product, you can run a lot of tons through. Where the little ones, they were more like a variety machine. They were small machines; I think one of the smallest was like 93 inches or something like that across. And you didn't lose so much paper when you changed grades. So that made a big difference.

They had specialty grade paper. They had so many stinking kinds of paper you couldn't believe it. I think the worst paper I ever ran was computer bond. And that's when that first come out. I guess that's computer paper. I don't remember, but it was called computer bond. But the tolerances was plus or minus zero, and you'd get a 40-inch roll or something off your machine, and that made about three sets of rolls that that computer bond had to be shipped out in. And all the slitters which cuts the paper had to be perfect cuts and everything. That was a real pain in the butt. Boy, you just sweat your tail off there because you just couldn't hardly keep up with it. That was one of the worst, but yeah, we made a lot of paper down there, a lot of different ones.

Mostly—it seems to me our goal there at one time, which is more now, but it was to get 500 ton a day out of the mill from those paper machines. And that's a lot of paper, quite a bit. Yeah, you know, they cut that in so many sizes, you know. And of course, nowadays these big machines, they run such a big roll off them. And in a lot of ways—and then they've got the winder right close where they can just pick it up and put it back on the winder and wind it off. Then they run the supercalenders; I mean, it's quite a process. But they primarily go by tonnage.

So were you there when they were making the pineapple cloth, pineapple paper?

Oh, that mulch paper? I was there, but I never worked on it. That was down on old No. 10 paper machine, if I'm not mistaken, that they made that. Yeah, I helped make the pulp that went into that from Mill B. But I think we used our screenings in that, which is the rejects. A lot of it was rejects off of our screens. I mean, we used pulp, but they used the screenings too because it was a low-grade thing.

You said that No. 10 ran a lot of little experiments.

Well, that's what I was telling you. Those small machines, see, you don't lose so much tonnage in between the time you start your experimental run till you get a good product.

Well, I think that's everything.

Well, I hope I helped you.

Unless there's something else that you want to say for posterity.

Well, I just wish the paper mill would have kept running for the young people, because that is one of the main—when you was a young man in Oregon City or

West Linn, if you could get a job in the paper mill, you knew you had a job that would last. And that is a thing of the past. There's not many places you can go nowadays and find a job that you can get out of high school and work till you either die or retire. And it's a shame in some ways. But that's the one thing I will say about it: it was a good, steady income for lots of people. And so it done a lot of good. A lot of people cussed it, but it was still a good family place to work.

So what would your wife say about the mill? Did she have any frustrations with it, or was she just glad you had a steady job?

Yeah, she just left that up to me. Yeah, she was glad that I had a good steady job, you bet. She didn't like the night shift; didn't like staying home with the kids at night and stuff. But you know, you done what you had to do at that time. And if I took and went to the yard department to work, well, it didn't pay as good. And the night shift, you got a little differential to start with; it got pretty good towards the last. I don't know what it is now, but at that time, why, it helped to work nights. But then it was part of your job anyway; you knew that you worked one week of swing, one week of graveyard, and one week of days.

Do you remember what your first paycheck was?

I think my first paycheck—gee whiz, it was—let's see—I would imagine about $250 or something. It was more than I was making when I was working at a service station, and then I went from the service station to the cement plant. And the cement plant—I took a cut of wages coming from the cement plant, but I was happy that I did. It was closer. And then they shut the cement plant down there at Oswego.

There you go. You made the right decision.

Yeah, you bet.

Henry Herwig

Interview April 9, 2006

What is your name?

I'm Henry Herwig. I'm 68 years old. I was born in Wiesbaden, Germany, and I started at the mill in 1964, right after graduation from Oregon State University. I started in the technical department, and at that time we had a big, big technical department; we had three different sections, because we had nine paper machines in operation. There was a non-coated section which dealt with newsprint, telephone directory, Crezon, and a few of the other grades. So we had a coated department for the coated paper machines, and also a pulping department. And I started in the non-coated section of the technical department. I stayed in the technical department probably about five or six years. I was both in the pulping section and the uncoated section, and I believe it was about 1971 to '72 I went down into production as an assistant to Jimmy Farrell in the supercalenders and the coating department. And then later on that developed into being full-time supervisor in the coating department.

How did you hear about the job? Why did you think it would be a good job?

Crown Zellerbach sent two people to Oregon State University for recruiting purposes. They offered me a job at the mill. I also had a couple of other job offers; one was down in the Bay Area with PG&E. I wasn't really too thrilled about going to the Bay Area. And the other one was with Alcoa in Vancouver. I had worked for Alcoa the previous summer as an intern. But for some reason, somehow this area and the mill and the people appealed to me. So I hired on at the mill. That's how I got started.

So did you have to go through the interview process?

The one person I remember from the interview was Bob Gaiser, who was also a tech supervisor at the time. They had posters in the chemical engineering building saying they would have representatives from this and this company on certain dates, and if you wanted to interview, you just signed your name onto it. And that's how the interview started. And yeah, a few questions asked, but the main interview took place at the mill, in the personnel department. And that's where the actual hiring was cemented in stone. Here, you're now part of the Crown Zellerbach family.

You had to get a physical or something, I suppose.

We had to take a physical. I graduated—I'm trying to remember—it was the first week in June. I think I started about a week later, on June 15, 1964, and stayed there until July 1, 1998.

Let me see—that's 34 years. Continuous?

Well, from 1964 to '98, yeah, 34 years. But there was about a six-month period where I was unemployed, and that was in 1996, when Simpson Paper Company shut the place down, and I was semi-retired, so to speak. I did have some retirement pay coming in. But then the mill reopened in '97, early spring '97, and hired me back, and I agreed to go and stay for a year, maybe two, enough to get the place going again. And I thought, this is it, folks. I'm going to bag it after that.

So you were one of the few that they asked to come back and help them get going again.

The startup of '97 with the new owners was quite an interesting story. We might want to get back to it later, because it's a complete change of the way things had been done. And I'm not sure you have that information right now.

Okay. The first day you went there: any memories or impressions about that?

When I first got there, one thing that really happened to me, I developed a tremendous cold about a week after that, because they gave me a tour of the mill; it was fairly cool, as I remember. And we went from a hot place, like on the paper machines, to a cool place, out to a hot place, to a cool place. I had a terrible cold for about a week. But that's one thing I remember about the first day being there, yeah.

I think Paul Miken said the first day it was the noise. He remembered how loud it was.

Oh, the other thing was that the place, to me, seemed so huge. You know, it was spread out. I was like, my gosh, this is a big place!

Can you give me your job title again?

Well, my last title was coating plant superintendent.

How many men did you supervise?

Sixteen. There were 16 people in the coating plant, four people on each shift.

Can you explain that? How is that structured?

The coating plant had four shifts. One shift was day shift, which went from eight till four in the afternoon. Then swing shift came in, which worked four till midnight. Then the graveyard shift came in, which worked from midnight till eight o'clock in the morning. And the fourth shift was off. And the shifts were staggered, so you'd work five days, you'd get two days off; you'd work five days and get two days off. But the way they were staggered throughout a month period, there was always one shift that had two days off, and the other guys were on day, swing or graveyard shift.

So if a machine needed four people or five people to run it, then there were actually 20 people that knew how to run it?

Yeah. Actually, probably a few more. Say you needed five people to operate a paper machine, or four people to operate a department, so that—say four people—that means you had to have 16 people that would train for the job. But what if somebody was sick or went on vacation? Well, you had a couple or three other people trained from other departments that could come in and fill that position for however long it took. You had to have spares. I think in the coating plant we had three spare people that we could call in from other places.

The coating plant, then—is it a place, or is it a process?

The coating plant had five different floors, and operated by four people. The coatings—actually, you might compare it to paint, okay? You put a layer of paint or coating on a piece of paper, and it does two things: it makes it nice and white, and it makes it nice and smooth. Coating mainly consisted, in the old days, of clay, which came from Georgia by rail, and it was unloaded upriver from railcars to the big barges you used to see out there. There were three hundred tons of clay on a barge. So coating in those days was probably about 80 percent clay, about 20 percent adhesive, which was starch mainly, some latex, and some titanium dioxide for brightness—little additives that did special things. That's a very basic coating, and it's mean to give a nice smooth printing surface.

And how did it move into your process?

That's one of the interesting ones. Like I say, the clay came from Georgia, and they have huge open-pit mines in Georgia. These things are mind-boggling how huge they are. I guess it's one of the very few places in the United States where you have that high-quality type of clay. So it's shipped over here in 100-ton cars, it was loaded on the barges, brought downriver. And then we had an eight-inch vacuum hose from the top floor of the clay plant, which is probably 50, 60 feet above the clay barge, we had that running down to the clay barge. And we had two guys, believe it or not with cement hoes, hoeing this stuff, keeping the stuff flowing into this eight-inch pipe. From there it either went right to the process, or we had two 1,000-ton silos that we filled up.

And the reason we got the silos, you know, the Willamette goes up and down. And when it goes up here, there are no clay barges. So then we had to take the clay back out of the silos. And we had a couple of difficulties. One summer the water was pretty low, and I got a call about two o'clock in the afternoon. The guy called me, says, I think I hear water running into this part. So I went down there, and oh my gosh, water was just pouring in. They had hit something going through the locks because of the low water, and punched a hole in it [the barge]. There were 300 tons of clay on this thing, okay?

Which is in powder form…

In powder form. So we got all the pumps that we could muster in the mill; we sent out to Portland for a couple others. An hour later, she went down, and all that was showing was about this much of the top, you know. And unfortunately—this was in the Crown Zellerbach days—we had company from San Francisco headquarters, and I never saw so many white hats down around that area. I didn't go down there until everybody cleared out, but we got some divers in; they put a soft patch on it, pumped the hull full of air, and raised it up. And we had all kinds of meetings about how we were going to recover all this clay. Three hundred tons of clay is a lot of money. Clay at that time cost about—I'm trying to think—about three to four cents a pound, something like that. So 300 tons is a lot of money. I said, don't worry about it. Oh yeah, we've got to figure out how to do that. Well, by the time we got the barge off the bottom a week later, it looked like somebody had swept the floor; there wasn't a speck of clay left there; it all washed down the river. But surprisingly, you couldn't tell; there was no white showing at all. It must have all gone underneath, because we expected the DEQ and the EPA and the FBI and whoever else to be up there, but nobody ever showed up, so—

So you did have a total loss on it.

A total loss, yeah.

I'm picturing it turning into white powder sludge and filling up the lock chamber.

It's white, yeah. Very white. They're specially treated. You know, they bleach it and they refine it, they do all kinds of stuff to it. Generally, on a scale of—you might consider titanium as 100, which is the brightest thing you could have—clay would be in the neighborhood of 88 to 90, maybe 92. So it's very white, yes.

I hope they're wearing breathing devices…

Masks, yeah.

…and it goes into this eight-inch tube. And then what happens?

When we took the clay off the barge, it went one of two ways. It either went right into the make-down of the coating, where it was mixed with the starch and the clay and the titanium and all this kind of stuff; or—anything in excess we would put into the silos for a rainy day, more or less, because we had, many, many times during the years where the water came up so high we had to pull the clay barge out because it was just too dangerous to leave it tied up; in which case we would go into the silos.

If it went into the process, which was a three-story plant, could you talk about how it got to the machines? There must be a lot of plumbing involved.

The coating plant is a nightmare of plumbing; there are pipes on pipes on pipes on pipes. We had some pretty old equipment in those days, pretty primitive. But the clay and the starch and all the ingredients were metered in on a volume basis, so many gallons or part of gallons per minute, so the blend was always the same. And it would go through a bunch of pumps down to the paper machine, where we had storage tanks, something like 1,000, 1,500-gallon storage tanks. And from there it was pumped to the paper machines.

It didn't settle out in the tank?

Agitation. We had agitation to keep it mixed up, yeah. You had to have it agitated, yeah.

So you had big mixers. That didn't heat the clay mixture?

The only thing that was heated was the starch. It was cooked, just like you cook starch for when you make gravy, you know. You have to cook the starch to develop that adhesive strength. The rest of the stuff was just normal, you know, room temperature.

Then it went from the storage tank, where it's being agitated, into pipes that went …

Down to the paper machines. The way the coating process works, the sheet of paper comes in, and you apply coating on both sides, first one side and then the other. Then you have some way of doctoring it off to make it smooth, you know, to make it even. And then it goes through a bunch of dryers so that coating will dry on the paper. And it happens pretty quick. It has to dry quick because if it doesn't dry quickly, it'll stick, and then you have all kinds of problems. So as soon as it leaves that coater and hits that first dryer, it has to set right away. And then it goes to the supercalenders, where it's polished.

There are two ways to coat paper. One is on-machine, where you make the paper, dry it, put it through the coaters on the same machine; put the coating on, dry it, and then ship it from there to whatever the end product is. That's what we had. We had on-machine coating. And the other method is, make the paper first, wind it up, transport it to a different part of the mill, and then put it through a coater. And the last quite a few years, that seems to be the way to go. There are advantages and disadvantages to both of them. If you have a break on the paper machine, the whole thing is down. If you have a break on the paper machine, but the coater still goes, that can go on, and vice versa.… until it runs out. But that's what we had: on-machine coating. The whole thing was all done all at once. And then the paper was taken from the paper machine to the supercalenders, where they put the polish on it.

I haven't asked anybody this: what's a head box?

Okay. Basically, the stock, which is the pulp and all of the additives, comes in at a very low concentration. Probably about in the neighborhood of half a percent is fiber; the rest is all water, 99.5 percent water. So it's very thin. It goes into what they call a head box, and there it's agitated. It has to be kept agitated, and uniformly agitated. And then at the bottom of the head box there's a slot—it varies—it's what they call a slice. That's where this solution of fiber and water comes out onto what they call a wire, which is just a mesh. And the fiber settles on that mesh. Some of the water drains out, and that's where the sheet forms. And then it goes through a series of presses to further take the water out, and dryers that take the rest of the water out. Paper at the beginning is 99.5 percent water.

So is that where the felts come in?

Yes, that's where the felts come in on the presses, and the felts help absorb some of that moisture. Then the felts go through dryers, and come back and pick up more moisture.

God, that must have been a hot job in the summer.

It's very hot. Very hot job on the paper machines. Nowadays they have air-conditioning areas for the operators to sit. Of course, nowadays a lot of it's automated, computerized. In those days it was all by feel, by look, by hear, by smell, whatever. You used the five senses to determine what this thing was doing. Now a lot of it is just automated.

Paul said the old-timers would walk in and say, that isn't running right, just by sound.

Oh yeah. Like I say, in those earlier days, you used your five senses. You hear something that's just a little out of whack, a little tick or something that's not rhythmic, like it should be. Or you feel something, you see something, and now those days are gone. Some of the older guys, the way they used to test moisture as the sheet was all done on the paper machine and was being wound up, they'd put their arm up and see how high the hair would go, you know.

By the static electricity?

By static electricity. Static is a huge problem in the paper machines, yeah. On the supercalenders I've seen sparks this long between the steel roll and the paper. It builds up a tremendous amount of static.

Okay, did you work with Rosie?

Rosie, I think he was still there when I was an assistant in the supercalenders.

He's very proud of them sending him to Holland. That was maybe '67?

VanGelder, yeah—I think that was the name of the town, yeah.

We learned that basically you made pulp like crazy when the pulp making was good because in the summer there wasn't enough water. And now you say they stockpiled the clay because in the winter they couldn't get the barges in.

Actually, we had two types of pulping down there. One was the ground wood, where they put these big blocks against a stone and actually grind the wood into a paste, more or less. The other one was the sulfite process. The old-timers are familiar with the sulfur dioxide smell downtown. That was to dissolve the fiber away from the wood, and we shut those down back in the early '70s because of pollution concerns, and to change it over to a different process would have been cost-prohibitive. About that same time the new mill at Wauna came on stream that we built down there, and they had enough pulping capacity to supply us with pulp. So we shut everything down and started using Wauna pulp, and had a lot of problems. One of the biggest enemies of any coated-paper machine is plastic.

What happens on a paper machine—if you can visualize this—you have a huge roll like this, and the paper comes out over here, okay? And you have coating deposited down here. Then you have this knife sticking right here, which doctors off the excess coating and leaves a very thin layer. Well, when you get a very tiny piece of plastic, it gets lodged on that blade and causes a streak. And Wauna was notorious for plastic contamination. We lost, literally lost millions of dollars because of plastic contamination. And we'd put in screens and cleaning equipment, but the problem is, you take a piece of plastic a quarter the size of a pinhead, not very big. When it hits something hot, what happens to it? All of a sudden it goes from like that to like that, gets caught on the blade, and you have a plastic streak. And you can't sell that. You cannot sell that.

The advantage of getting Wauna pulp was negated by the fact that the quality was so bad?

Almost, yeah, yeah, because we had to throw a lot of paper away, yeah.

How long did it take for them to figure that out?

Oh gosh, it took years; it took years to clear that up. It would be periodic, you know; sometimes we'd run well, and other days it would just be a total disaster. We lost a lot of paper over plastic contamination. Well, eventually, what happened is, we went to other suppliers. We went to Canadian suppliers, and that's where they're getting all their pulp from now, mostly from Canada. And some from overseas.

I'd never heard of Wauna. Which highway is it on, if you're driving?

Wauna is between Portland and Astoria, this side of Tongue Point. I'm trying to think what the nearest town—you know; it's about maybe 15, 20 miles this side of Astoria, on the Columbia there.

When you were talking about the two ways of making pulp— was that the second way?

Well, we had a wood mill. We had logs that came down from upriver, and they went to the chippers, which made the wood chips, which went into the sulfite process for cooking. And that's where the fiber separated from the rest of the big chips.

Both processes were running through that mill at the same time?

Yeah. When I first started there, like I say, we had nine paper machines, we had a converting department, wood mill, sulfite mill, sheet finishing, which made printing sheets, huge sheets. We had something like 1,500 people working there.

In 1964?

Yeah, yeah, 1,500 people.

Let's see: you started in June; there was a flood that year.

I well remember the flood of 1964. We were down for a little over three weeks. And we were talking about the silos earlier. The bottom of the silos were under about, I guess, eight to ten feet of water, and all that clay had turned into solid sludge. I was not in the coating plant at that time, thank God, but I guess it was quite a mess. They had openings, actually. There was a door on each level so you could go into that thing, because you didn't want to go into the door below the clay; you always wanted to go above it. But yeah, the bottom of those silos were totally just sludged up from water.

Wow, what a mess!

We used to have to send people into the silo. The clay barge was bad enough, but the guys could stand on a good, hard, solid floor; they had nice weather; they were protected from the weather. But the silo was a little different: these were 40 feet across, and the clay was fairly fluffy. So you

Worker hoeing and suctioning clay powder at rail siding.

can imagine what happens when you step into some fluffy clay. You go in. And we tried all kinds of different methods to keep people from having to be buried knee-deep in clay. The best thing we found was a big piece of plywood. We would lay those down, and they would step on them. But we did away with the people in the silo by putting in an automated system that would keep feeding the clay off the bottom of the silo. And that was a big improvement. That was about 1979. That was the same time when we converted one of our last big machines to a coated paper machine. That was old No. 9.

So if the days of people in the silo ended in 1979, was that when the clay started being brought in on trucks?

No, that didn't happen until probably about 1994, '95, we did away with the clay barges altogether, somewheres in there. That's when we built a bridge across the canal and started unloading the clay by blowing it across the bridge right to the silos. The one right over the locks, yeah, yeah. In fact, that's when we not only unloaded the clay; we unloaded calcium carbonate, latex, titanium dioxide—a lot of these additives that we bought in bulk are unload up by the parking lot and blown across that bridge, across the lower locks, yeah. They all have separate pipes, and they all go a separate place, yeah. I don't know how many pipes are in there, but there's a bunch of them. I would say there's probably, oh gosh, ten or so. You know, ten pipes, and they're all in a big bundle, and they're protected and they're insulated, and they have heat tape on them so they don't freeze up, except one year the heat tape failed, and the latex line froze solid all the way across the bridge. That's another thing: extremely cold weather can play havoc in that mill because of the distances between places where you have to transport stuff. Everything had better be in working condition; if it doesn't, you're going to have a freeze-up.

I never thought about cold inside the mill.

There were some places in the mill that were fairly cool. Like I say, most of the production areas, you know, did have heat in them. But the problem is, you see, you have to transport a lot of these things, like from the ground wood mill to the paper machines, there's the old bridge up by PG&E, that green thing? Those lines are completely uninsulated. They're huge lines; they're big lines, yeah, but they're uninsulated. And what happened a lot of times was like we would shut down for Christmas for a week. If something was not thought about—one year they froze solid, and they had to cut the pipes in sections and dig that stuff out of there. And like I say, we had latex lines freezing solid; we had clay lines freezing solid. And the challenge then becomes, how do you keep the mill in operation while you're doing all this? And sometimes it was quite a challenge to do this.

Would that be a job for the flying squad?

Anybody we could get our hands on. And that was the other bad part: we worked people long hours out in the cold, you know, to get things going again. And like I say, there was no particular favorites—whoever was available that we could call in to work on this with blowtorches and saws, cutting the pipes up and welding them back together. It was quite a chore.

How did the welding shop relate to maintenance and millwrights...?

Of course we had the millwrights, and each department had their own millwright gang, because we had the pipefitters, we had the instrument people, we had the welders, we had the painters. Who else? We had the electricians, and all those guys are considered maintenance; they're all part of the maintenance mechanics' group, yeah.

And you had a fire department.

We had a fire department, yeah. Joe Nixon—I'm trying to remember some of the others – Dick Buse?

The fire department must have been just as-needed, and they did other jobs.

Yeah, they were volunteer. We had one chief, and then you had people on different shifts, different crews, that were firemen, and they had a practice once in a while, and that was the fire department. They had water that would just pull right out of the Willamette, yeah. Untreated water, yeah.

The mill had a settling pond... Is that the water that was in the head box?

No, no. The mill used, like I say, well water for most general applications, like, oh, fires, hosing down floors, stuff like that. They used just plain old well water. But then they also had a water treatment plant. The water used in the coating process, in the paper machine, absolutely had to be clean. It was chlorinated to take care of any bacteria. It was filtered so they wouldn't have any sand or grit in it. It was very pure water. And then we had what they called white water. White water, just as the term implies, is white, and consisted of—I talked about the water in the paper machine, where the pulp mixture, you know 5 percent pulp, 99.5 percent water—where this water came out it took a lot of the fine material with it. And that's gold; that's something you don't want to lose. So this went back to a huge holding tank, and it was called the white water tank. And the white water was used to pulp up baled up, dried pulp; it was used in a lot of applications—number one, you wouldn't lose the heat in the white water. You wouldn't lose all those fines—a lot of the fines consisted of titanium dioxide and clay, and that's expensive stuff; you don't want to just throw that away. So you call it white water, and you reuse it as much as you can. Those were the three main water types.

When you say "fines," you mean just little fine fibers?

Very fine stuff that would go through that wire mesh, yeah.

How do you get the pulp back out of that white water?

Well, white water—actually, the amount of material in the water was so small that we really couldn't reclaim it, okay? Not by separating it, in other words. You couldn't reclaim it by separating it. But the huge volumes of water meant yeah, percentagewise there's very little, but weightwise, there's a huge bunch of solid stuff sitting out there that we can't throw away. So the thing was used as water, as makeup water.

So this could go back into the head box with new pulp?

Yeah, sometimes it was new pulp. Actually, I don't know if you've seen the big bales of pulp they have down there. I mean, they're usually huge bales of pulp, you know. And they're probably about 99.5 percent pulp and a half percent moisture. Well, those have to be broken up until there's about 3 percent consistency. You go from 99.5 percent pulp to 3 percent pulp, and that is done with white water. That's where the white water comes in. So they use that water to break that pulp up and thin it down.

Everything is used.

It has to be, yeah. Like I say, you know, you don't want to lose the heat; you don't want to lose the expensive—what they call the fillers—titanium dioxide and clay and a lot of other chemicals in there that you just can't afford to throw away.

Reconstituting a dried bale with white water, was a process that had heat in it?

Mm-hmm, yeah.

But you hadn't added the adhesives or anything yet?

Not yet.

So you wouldn't have to heat the whole thing up from scratch?

That's right, yeah, yeah. There's a tremendous amount of heat in one of those paper mills. The dryers are probably running close to 200 degrees Fahrenheit, so there you get awfully hot, and it takes steam, so a lot of heat; there's a lot of heat in a paper machine. That's why now they have nice air-conditioned cubicles with all the automated computers there.

Did you bid on jobs at the mill, or just stay in the same department and get promoted?

I didn't really bid on anything; I just got promoted. As salaried, we didn't really have a bidding process. You could ask for a job, I guess, you know, and say hey, I wouldn't mind having this job when it opens up. But most of it was just on your performance.

What did you do during the strikes, or were you there during that?

Well, unfortunately for a lot of people involved, we did have quite a few strikes. It was sad. The last two were eight and half months, very tough times, both for the people who were out of work, and also for the people working down there. We put in some huge long hours. We worked 12-hour shifts; we worked 10 days on, from 7 to 7, and then had five days off. And we had a lot of people in from other places who really didn't know anything about the job. We had to train them, so as a superintendent, there were many 16, 18-hours days, you know, unfortunately. So it was a tough go.

Roy said when they were shut down, it was spooky in the mill, it was so quiet.

Whenever—like during a maintenance shutdown like for Christmas, you shut down for a week or—you walk through the mill, and you hear all kinds of little hisses, little air leaks, that you never noticed before. In fact, that was about the only time where they could do a good job maintaining the air system and the steam system. Because now all of a sudden you can hear it, you know. Before, you wouldn't have a chance. And there were many, many small leaks, yeah. And they all cost money. Every time you throw away a pound of air, you know, you throw away so many cents.

Who would fix the steam, the hissing?

The steam department actually had their own millwright crew; they called them steamfitters. They were different than the regular millwrights. They would come in and fix them all up.

Roy tells a story about the basement, trying to replace light bulbs, and there was a lot of water, and rats and nutria swimming around there in the dark.

Oh yeah. We had a lot of nutria, and like you say, rats, especially in the coating plant, because they love starch. They love the powdered starch. Oh, they just ate that up. So we had a lot of those. Another thing we had a lot of—after the 1964 flood, I had—I don't know if you want to call it the "honor," exactly—of cleaning out the ground wood mill with five people. We didn't have any power at all; none at all. So we got a portable gas-operated pump and stuff. We washed everything down, and finally I think there was one pit left about the size of this living room. It was deeper than that. We pumped the water out of that, and the first thing we ran into was driftwood about, oh gosh, about this much, you know, just wood. So we got rid of that. We had to haul it up by tripod, if you can imagine that. So we got rid of all that, and after the driftwood came the eels: there were jillions of live eels in that pit, about this deep. And I sent my guys down there, and they were in their boots, and we had I think it was about a five-gallon bucket. They took their shovels and shoveled these eels in there, okay? By the time we had them up to the top, half of them would crawl out and fell back down into the pit. That was a mess. That 1964 flood was a mess, yeah. But

I'll always remember those eels in there; it was something of a nightmare.

So let's see: if the flood was in December, the eels must have been running in December.

I don't know; there was sure a bunch of them, yeah.

Were you ever there when unauthorized salmon fishing was going on?

I heard a lot of stories about the salmon fishing in the early days in the ground wood mill. I never did witness it, but I heard a lot of stories about the salmon fries they used to have out there on the night shift, stories about the guy walking past the guard gate, always with a raincoat, in the

NUMBER 3 grinder room looked like this on December 30 after the flood waters departed. Millwright Tony Stalick sizes up the job to be done.

summertime. And they couldn't figure out why until one day he came through and he had two fishtails hanging out from under the raincoat. The only one I ever had any experience with was—the job on the clay barge, to hoe the clay, was generally filled by what were called the extra board. These were people who were not assigned to any department. They were trained in a lot of different departments on the bottom job. And all new people who came in went to the extra board. So, this was Saturday during salmon season, and I got a call from my lead man, Bill Dominiak. And he says, "You got time to come down a minute?" I was home. I said, "Yeah, yeah, what's the matter, Bill?" He says, "Well, you need to check something out."

So he took me out there, and here this guy was fishing; sitting on the clay barge, fishing. I thought hmmm. I went down there, and nobody on the barge, no clay being moved. The guy was just sitting there, you know. "Having any luck?" "No, no, not yet." "Well, I think you better take your fishing pole and your lunch bucket and get out of here, and go someplace where the fishing is better, because you're fired." He says, "You can't do that!" I said, "Well, I just did, you know." He just figured he would take some time off and go do a little fishing, you know. And I'm sure a lot of that went on without me knowing it, especially at night. A guy could even keep working and just put their rod out with a little bell on it. I'm sure it went on. That was funny.

He was one of floaters, or whatever you call them.

The extra board, yeah, yeah. When you got a job at the mill, that's where you went until you actually bid into a new department. Some of the people were trained in the bottom job in several departments, and they filled in for vacation, sickness, what have you.

So were you in this mysterious Horseshoe Club? What is that?

We had two clubs. How do I want to say it? One was just for general salaried people, and this included people in the office, the superintendents, the supervisors, people in the technical department. Then they had the Horseshoe Club, okay? This was an elite group of people; it was the department heads. And once a year we would go out to a park out by Molalla there; I don't remember the name of it. Towards Carver. We'd go out there and have a picnic. It wasn't McIver. It was just the other side of Carver a little bit. I don't remember the name of it, but anyway, a lot of tomfoolishness went on. The initiation rites, you know, were—I don't think I want to talk about those, but they were sophomoric, a lot of fun. And a lot of beer was drunk. All I've got to say about the initiation rites: when I came home from being initiated, my wife wouldn't let me in the door. Says, you change clothes before you come in. That was fun. I was president of the Horseshoe Club for a couple years, and developed a couple of new devious methods of initiation.

Rosie talked about it.

Oh yeah, the old-timers loved it. The old-timers really loved that. Unfortunately, it kind of went by the wayside with the new philosophy, you know.

Roy said they had to let the first woman supervisor in.

Absolutely. Some of the old-timers were just, oh, were set against having—set against that. We had one fellow that refused to take part in the initiation, and boy, he was ostracized fiercely. And it was his right too. I told him okay, that's your privilege, but you cannot be a member of the club. Says I don't care; I just don't care. That's fine.

We call it networking now.

Yeah, absolutely. Well, like I say, times change, and you have to change with them, too.

Rosie had a scrapbook of the Christmas party just before he retired.

He was Santa.

Were you at that one?

I wasn't at that one. One guy I used to work with—this is just a story; if you want to use it, fine—but the guy I used to work with, Jim Farrell—he was my boss on the supercalenders and the coating—and Jim was a mustang; he came up through the ranks and became superintendent. And he was kind of the godfather of the Christmas, Santa Claus, thing. The idea was, you elect somebody Santa Claus by votes, okay? Some of the salaried people in the mill, we would get names from the welfare department, or I don't know where they got them—well, these people are really needy. And we would go interview these people. And I'd been involved in that a couple of times. It made my Christmas, I'll tell you, to talk to some of these people out there.

And then we would have a committee, we'd get together, and say okay, the Joneses over here, you know, he's out of a job; they have five kids, you know, and they need this. And we would either allot money, or we would pay their light bill for two months, or something like that. And that's what we collected the money for. And it was Santa's job to deliver those goods, okay? But it wasn't an election so much; it was a railroading, because Jim Farrell, he was the collection point for all the tickets that were sold in the mill. And two or three of us would get together and say, hmm, I think Rosie would be a good candidate this year. Of course, that's before he retired. So it was kind of an honorary thing. But that's how the Santa Claus contest worked, yeah.

So then he got to take the things out to the families?

Yes, uh-huh.

Someone said that sometimes those Christmas parties got a little out of hand—

Yeah, again, there were two Christmas parties, one for the superintendents and one for the rest of the salaried people. And they were free. Good food at the West Linn Inn. Mr.—what was his name that ran the place?—put an excellent, excellent dinner out, and it was an open bar, yeah, absolutely. Mr. Seavey. He and his wife ran the West Linn Inn, which was a duplicate of the one in Camas, by the way.

There's an Inn in Camas?

There was; I don't know if it still is. He ran that place for a long time, and put on some excellent food.

When you got there, they probably had safety committees—

Safety was a very tough thing to sell. Unfortunately, the union and the management a lot of times were this way on safety. We never did go this way, where we could reconcile some of our differences. And safety shoes, hearing protection, eyeglasses, it was a very tough fight, until we finally had to say hey, you know, you either wear it or you're out the door. It's the law; we had to do it, you know. And I think it caused some hard feelings, but unfortunately, that's the way it went. But as a kid I worked in a sawmill, and my dad also worked in a sawmill. And I worked right next to some saws, you know, huge saws about this big, turning I don't know how many RPM, but they howled something terrible.

I came home on weekends, and my golly, my ears are ringing, and what's going on? It wouldn't go away till about Sunday, Monday, and then start all over again. This was, you know, back in the '50s, '60s. And I've got to do something, so I started wearing cotton. They say cotton doesn't do anything; well, I beg your pardon, it does. It does something, because my ears quit ringing. And I told my dad, gee, dad, you should be starting to wear something. Aaah, no. Well, he's stone deaf, you know. So yeah, it was a tough, tough sell to sell that type of thing.

Roy told a story about some guy that came in and had long hair, and they couldn't make him cut it basically because his dad worked up in the office and was salaried...

I know who you're talking about, yeah.

But if you were under there and your hair got caught in the nip, my God.

Yeah. You're dead. It's that simple: you're dead, just like that. Those nips are very powerful, and things happen so fast. These machines are going 2,000 feet a minute; that's moving right along. The guy that used to live next to me here, Harold Lamb—he was in the tech department—his left arm was crushed because he worked on the paper machine, and I don't know if he was trying to feed a piece of paper in or trying to get something out, and that nip grabbed him, and before he knew it, he was up to here in it. And he could have got killed, because most of the incidents that happened like that, your head gets caught, and it snaps your neck. Well, he didn't go that far, but he had a lot of rehabilitation to go through, and he never did have use of his arm completely. But it's so fast it's just unbelievable.

Even if they had an automatic brake like a wringer washer or something, you know, you can't get it slowed down...

It's not fast enough. Like I say, the machines are anywhere between 1,000 and 2,000 feet a minute, so you can imagine how long it takes to go from here to there. Yeah, it's the blink of an eye, or less.

How much equipment was provided to people when they were hourlies?

No, I don't remember much about that. I think—gosh, I think the salaried— I'm not sure about the hourly people, but the salaried people did get a shoe allowance; we were allowed to buy one pair of safety shoes a year or something like that. Of course, earplugs were free,I think safety glasses were too, you know. Prescription glasses were free.

Oh really?

Yeah. I think it was the same for the union people; I'm not sure, though

Okay, on the West Linn Inn…

Well, the West Linn Inn would not hold 400 people, let me tell you that. I'll shoot that right now. They may have had some outsiders come in for a strike. But it definitely wasn't 400. I don't know how many rooms it had upstairs, but it wasn't that many.

So people would meet there for coffee sometimes just before shift break—

Oh yeah, we'd go have lunch. We had a lunch hour from 12 to 1. A lot of the sales people that came in, our suppliers, you know, they'd come in and make their quarterly tour, "how you doing? How's the product? By the way, we just kicked the price up a hundred bucks a ton," you know, this kind of stuff. They'd tell you it over lunch up at the West Linn Inn. And they had some excellent lunches there, yeah.

It was a very handy thing…you hear the noon whistle and you go back to work…

Mm-hmm. But unfortunately, the place was getting old and wasn't up to code, and it just would have cost too much to get it up to code, so they tore it down. Oh, we'd stop there a lot of times, some of us guys, after work especially in the summertime. They had a little veranda facing the river, and you could sit there and watch the boats going up and down and have a couple cold beers, and it was nice. Yeah, it was nice. It was part of the history of the mill, really. I mean, it belonged to the mill.

My only memory is it was really dark inside.

It was a little. It was an old, old building. The paneling was all dark, you know, yeah.

So during the strike you worked…?

The coating plant. I had to; I was the superintendent.

And it wasn't a full-bore four shifts, was it, during the strike?

During normal operation, we had four shifts, okay? We had a day shift, a swing shift and a graveyard, and one shift off. And they were staggered so there was always somebody had two days off. During the strike, we only had three shifts. We had one shift that went from 7 in the morning to 7 in the evening, 12 hours, for 10 days straight. Then you'd get five days off. Then you'd work 10 graveyards, 7 p.m. to 7 a.m. And it was always one shift that had five days off, okay? So he had three shifts, three rotating shifts, which cut down the number of people you needed and still got the job done. But we didn't shut down; we kept operating 24 hours a day. Shutting down a paper machine is—it's not good economically. You've got to keep running 24 hours a day. To shut a paper machine, like a coated paper machine—say we would shut a coated paper machine and say okay, we need to have everybody out of the mill for this holiday shutdown at 8 o'clock in the morning, okay? We would start about midnight that night, shutting things down. So by the time 8 o'clock came, everything was down, everything was clean, all the lines were flushed out, everything was ready to start up.

So on startup—okay, we're going to start up at 2 o'clock in the afternoon. I would bring people in on the midnight shift before to start cooking stuff, getting everything filled up, and it would take us from, say 4 o'clock in the morning till 2 o'clock just to get the stuff ready. And then sometimes we didn't start at 2 o'clock. We might start at 10 o'clock that evening, or 12 or 2 next morning, because of difficulties. So you're losing more—say you want to shut down for a weekend, two days. You're probably losing three to three and a half days, total, of nonproductive time. That's no way to run a business, no way to run a business. It's a waste of time and a lot of wasted material, because you can't plan—you know, you have four or five different things that have to come together; they all have to run at the same time. It doesn't happen.

That was a great story about the eels. Do you remember anything else about the flood? What was that like?

At the lower part of the locks they had a gauge, you know. And they brought us in in shifts to check that gauge every hour. I remember at one point Ray Dupuis—he was the [mill] manager at the time—had just come in. The river—believe it or not, the river had come up three feet in one hour. I can remember that vividly. It was huge. Yeah, I remember that. And we knew it was going to be bad. We knew it was going to be bad, so we started pulling electric motors, because they can be damaged. A lot of them, the smaller ones, we just undid and took them upstairs. A lot of the bigger stuff was just hoisted up to the ceiling. Well, the problem was, the water went past the ceiling. And it seems to me, numberwise, that we had to send out something like 900 electric motors to have them cleaned and dried out because we didn't get them out in time. It was just so fast that nobody had any idea that it could happen that badly.

I think the last big flood before that might have been in the '40s…

Back in the '40s, the Vanport flood, yeah. Well, if the Columbia backs up, this is going to back up.

It must be terrifying to be down below the water level.

If I remember right, the upper river came up 10 feet, which doesn't sound like much. But the falls were almost nonexistent. So you know, that's a 50-foot drop right now, so you can imagine, you know, 10 feet here, and this thing's up here, how much water there was. Just mind-boggling, actually.

Were they still making hydropower in the mill when you were in there? What do you remember about that?

Well, actually, the mill was really built because of the waterpower. When I first started, a lot of the grinders in the ground wood mill, which are huge stones—they're about, oh, this big around, this wide, and with those shafts about like this—they were driven by water. They put in motor generators so they could use either water or a big motor. Yeah, gradually, as the years went by, with mainly the fish kill, fish going through the turbines, is what caused us to shut all these things down. And I know there was experiments made with all kinds of meshes and screens to try to keep the fingerlings out, but it just never panned out. So more and more, yeah, especially whenever the fingerlings go through—this two-month period you can't use your water wheels. That's how it started. And gradually it just got expanded out 'til they shut them down altogether.

So now they use gas or something?

Well, we used to use a lot of hog fuel, which was just trash, you know, bought quite cheap at that time. Now it's all gas and oil, yeah.

Okay. I was going to ask you why you came to Oregon State and what your degree was?

Oh, okay. Well, like I say, I'm from Germany. My father was in the German army. He was killed on the western front, actually inside Germany on the western front, in 1945. And my mom met and married a G.I. from the Los Angeles area. He was in the Air Force, and they got married in 1948. And then his enlistment was up, so he came over here, and he didn't re-enlist. And for some reason I couldn't come with them. I stayed with my grandmother for three years. And in 1951 they finally settled in Brookings on the southern Oregon coast, bought a house, and saved up enough for my plane fare and ship fare and what have you, and shipped me over here. And I attended the high school down there and then went to Oregon State—married my dear wife and had a couple kids and went to Oregon State—and graduated as a chemical engineer in 1964, and then a week later started to work for Crown Zellerbach. So that's kind of the history of my short life, short long.

Did your son ever work there?

Yeah, Mike worked there—let's see, he worked in the ground wood mill on the extra board. Yeah, in between school terms, for I think one year, yeah; one term.

Paul Miken and others have mentioned that the mill gave people summer jobs.

Yeah, yeah.

…the kids would get good work experience, real work.

You know, the ground mill, was a very dangerous place to work. I mean, the wood was heavy, and they also always teach you, when you put a block of wood in, cradle it in your arm like this. Don't ever hold it on the end because there are some real sharp corners. Well, he did it; he got his finger caught. He didn't come home because he was afraid I'd get mad at him, because I kept telling him, you know, please watch yourself; that's real dangerous. But he was okay. It's hard work, though, very hard work. Yeah, the mill actually went through four different owners: Crown Zellerbach, then, I don't know, they just kind of let it slide. I don't know what their philosophy was. They didn't want to spend any money. I mean, "We're not going to spend any money on any improvements, do everything as dirt cheap as you can."

And we paid for it dearly. We lost a huge share of our market on the West Coast because our paper just wouldn't hold up to the competitors'. And part of it was because they wouldn't let us use what we had—you know, to upgrade our paper. That changed a little bit in the latter years, but it was kind of done on the sly, you know. I'm not sure they ever knew what we were doing. But then of course they folded up and were bought out by Sir James the financier; you've heard about him. And then he sold the St. Francisville mill and us and Camas to—who'd he sell us to?

James River.

James River, yes. And they lasted about five years, and they sold out to Simpson. And the story—I've heard it mentioned several times—James River bought the mill because they had to to get a bunch of other stuff, timber and whatever. Simpson bought the mill they wanted to. Simpson didn't last. They shut us down in '96, and then the West Linn Mill came into being in 1997, and oh, is it different! Yeah, a lot of difference between the way things were. The first thing, the outfit that bought the mill, they're located in Vancouver, they decided to go with non-union, which was a big step. And then we were able to do something that we had never been able to do before. In the old days, if you could drag your body into the personnel office, you were hired, unless you were half blind or something. I mean, most anybody was hired. With the new ownership, things were a little tougher.

We advertised, and the people had to take an extensive exam in spatial recognition, mathematics, just plain logic, spelling. I mean, it was something like an eight-hour test just to be able to get your application in. And we took the top percentages of those people and put them through a structured interview. And this could be anywhere between two to four hours. It involved four people from the mill sitting around a table and grilling this poor schnook on the other side. Just anything, digging into his or her background. What did you do? How did you do? What were you involved in? What will your fellows say about you? On and on and on and on. And they were graded according to that. And the way it turned out, I think for every six or seven applicants we accepted one. So we hired something like 200-some people, so you can imagine the number of interviews that we went through for about two months, a huge amount. And I must say it was quite successful. I mean, we had a few that slipped through. We did, unfortunately. But 80, 85 percent was—we got some excellent people, motivated, intelligent. It's not—you didn't have to say something twice; they would go do it on their own, taking a lot of pride in their job too. So yeah.

And then the other thing we did—in the olden days, a promotion in the hourly ranks was by the most senior qualified man. Well, unfortunately, a lot of times the qualified went out the window; it was the most senior man. And I don't have a strong argument against that. I mean, you want to take care of your senior people. But the new way we had technician levels, and to advance from one level to the other had nothing to do with seniority. You had a certain time frame where you could advance from A to B; say it was in six months. But to get there was tough. Again we designed some very stringent tests. In our department, for instance, we had a written exam. I forget how many pages it was, but I've seen people complete it in five hours, and some as long as eight or nine hours. That was only the first step. The second step was a hands-on thing, which could last all day. I would go out. Old John, he's going to be under the gun today, he thinks. So okay, John, I'm going to fix you. I'm going to go up here and flip this switch. You've got to figure out what's wrong; you've got to figure out what to do about it. And you've got to do it quickly. So as soon as I flipped the switch off, or did whatever to upset the process—I'd upset it enough so it wouldn't be just right, but it wouldn't hurt the product, okay? I mean, it might dump a bunch of stuff in the sewer for about 10 minutes. I say okay, John, you've got something haywire; you'd better start looking. And that was probably the toughest thing for the people.

And then, after that they had to sit down with the people on their shift and go through a structured interview again. Is it really worth it to us for you to advance from here to there? What's in it for us, you know. In fact, we hired back about 25 percent of our former work force. The rest of them were bankers, painters, hotel managers, people who worked for high-tech industries like Wacker in Camas, janitors. We had the whole gamut of things, okay? We hired back about 25 percent of our people, and one of the guys that we hired back had worked for me before. He was always on the low end of the totem pole. I mean, he didn't have much chance for promotion because there was too many guys ahead of him in seniority. But now he saw his chance. Oh boy! He dove in, and he failed miserably. So I called him into the office. I says, John, is this too tough for you? He says, no, no. I need to know this stuff, and I'm not going to fail again. But it was a shock to him because he'd never experienced that type of thing before. So it was a big change in philosophy, yes. In fact, I think they're running with less supervision now than they ever did before, because the people are taking over the jobs.

From your description, everybody has to be able to problem-solve…

Exactly. Yeah, there are much broader responsibilities than they used to have. And the people can handle it. Anybody can handle it if you give them a chance; if you give them a chance.

But people were competing for those jobs. They still wanted to work there.

Sure. Well, you know, the nice part of this, like I say, with the union, union days—I'll give you an example. One of my guys, he was the bottom—out of 16 people he was the one with the least seniority. The kid could never go anyplace. And he was sharp as a tack. He was sharp, he was a good worker, he was ambitious. But he was stuck, forever. We re-hired him, and I took him from the bottom job to the top job, and he performed like a champ. What an opportunity for a young person like that. His ability as a performer put him right at the top, from the bottom. It's a huge jump, you know. He would never have been able to do that otherwise.

Seniority has its disadvantages and advantages.

And advantages, yeah. Like I say, you do want to take care of your long-term employees.

The people who have invested their lives with you.

Yes, you betcha.

So you retired in '98…

Eight years, going on eight years, yeah.

Did you miss it when you quit?

No. In a way; I missed the people, because I made a lot of friends, and even with the new batch coming in. I'd only been with them a little over a year, but I made a lot of acquaintances. And I go down once in a while. I slip down and get a pass and go around and B.S. a little bit, and share a cup of coffee with some, you know, just to see what's going on. Yeah, I don't miss the job so much, but the people. I miss the people.

So what was the best thing about working for the mill?

Oh boy. I liked a lot of things about it. I've always liked working with people. And of course the location. I love the location, West Linn, and I'm an avid fisherman. I love to go salmon fishing. I love to go skiing. I love to go to the beach. So we have the best of both worlds, going either way. And the companies have been good to me—and I like to think that I've been good for them too. I think it was a mutual give and take type of thing. I don't think I'd do anything different. I had a couple other job offers. One time in my earlier years, I was a little, shall we say disgruntled because I didn't see where I could get any kind of move. So I put a couple feelers out, and I had an offer from Longview Fiber—no, Weyerhaeuser, Weyerhaeuser up in Longview. I decided no, I'd better—I'm not going to do that. Shortly thereafter I went down to production. So that was a good move, not going up there.

I also had a chance to go down to—Jackson? Jackson, Mississippi. I think it was—I'm not sure what company it was, but they built a brand new paper mill. And the guy who was the mill manager used to be my boss up here. And he called me up and said, "Why don't you come down for me, work for me and run the supercalenders and the coating? And you'll get in on the ground floor." And oh, what a challenge that would have been! But I couldn't see myself living in Mississippi; absolutely no way. And my wife neither. She says, oh God, you go without me. But careerwise it would have been an excellent move, because they were just building the thing, you know. I turned it down. I'm not glad I didn't.

It came at the wrong time.

Yeah, yeah. But we did enough upgrading on our own once we got some different owners and got the philosophy changed a little bit. And they said, hey, we've got to put some money into this place if we want to have it survive. So they did. So we did a lot of automation in the coating plant, gosh. And that took a couple years to finish up and fine-tune. I enjoyed every minute of it, every long hour of it.

You were getting a new challenge.

Sure, yeah. So I had my own challenge up here, rather than go to Mississippi. Yeah.

What was the worst thing about working for the mill?

I think the worst thing was—for me anyway—was this management versus union. I hated to see that. Not because I'm a manager; I hated to see it because people work side by side; they depend on each other, and they have to do this and can't come to any kind of resolution. That's probably the worst thing for me. I hated to see that, you know. I hated to see the strikes because it just tore people apart. And eventually they got back together, but I don't think they ever got back together as much as they used to be. That was probably my worst experience.

Yeah, I think it's a little better with the new system where you don't have this adversity, yeah. We could even see it in our safety committees. I was a member of the safety committee for gosh, I don't know how many years, and I could see it there. And I said, guys, gals, come on. You know, we're a committee; we're supposed to come up with something; we're supposed to work together; we can't be like this. And I'm not blaming one side or the other; one side was as bad as the other. The management side was just as bad as the union side, and vice versa, on some of the issues. And there could have been a lot more give and take. Unfortunately, that's the way it worked out.

A lot of people say it was hard, hard work, but the check came every Friday.
Oh yeah, yeah. Absolutely. And long hours sometimes, yeah.

I think we've covered everything on my list—unless there's something for posterity?
Well, like I say, I enjoyed my years down there and made a lot of friends, and I would do it again.

So do you have Horseshoe Club reunions?
No, unfortunately. That's too bad. That's the other regret. Oh, gosh.

Delbert Herndon

Interview April 26, 2006

Rosie started at the mill in 1927, so it was just great to talk to him.

> Yeah, I was born in '29. My older brother's passed on, but I've still got one brother alive that worked at Sullivan—

All right. You must have a clock that they gave you…for retirement…

> Actually, up in the top of the garage, I built a little room up there, and I've got my retirement present up there. It's a reloading set. I did a lot of hunting—I loaded 30.06, 30.30, 357 magnum, .38 for handguns. I'm 76 years old. Yesterday was my 51st wedding anniversary.

That's a long time. What year were you hired at the mill?

> 1947.

And how old were you when you were hired?

> Eighteen. I went to work as soon as it was legal to go to work.

Did you have people in your family that were already working there?

> My dad worked there.

How long did your dad work there?

> He only worked 10 years after we came out here from the Midwest and ran into the tail end of the Second World War. Finally made enough money on the farm to get off the farm and get out of debt, and we moved to Oregon.

Why did you want to work at the mill?

> It was good money. It was good pay, and it was a good job. I liked my work there. It was a real good job when I first started. Along towards the end, they forgot we were people and started figuring we were numbers, but back in the old days—what we call the old days—it was a good place to work. There was about 1,400 people working there when I started, and about 500 when I retired. They shut down so much of it. At one time I spent two years as maintenance foreman up in the sawmill and grinder room. And they shut them down.

So do you remember your first day on the job?

Yes. The first day they had me report at midnight, on the graveyard shift. I went down to No. 2 paper machine, and it was what they call haying out. It was making hay; it wasn't making paper. A lot of breaks and trouble with it. And the foreman on the job told me to stand back out of the way until they got it running, then he'd show me my job. Well, before they got it running, the person who was supposed to work there came in—and they said they needed a man down in the supercalender department, bucking rolls. That's where I stayed for four years. At that time they hired most of their maintenance people out of the mill. They didn't hire outside maintenance people; they hired them inside, and they put bids on the board. After you got a few whiskers on the job, you could bid on these jobs and have a pretty good chance of getting them.

And I did not care for shift work. You'd just get used to sleeping in the daytime, and they'd change you, and you'd work in the daytime and try to sleep nights. So they had a bid for a pipefitter, and I bid on. They had a bid for millwright helper I bid on. And then they had a bid for a welder's helper, which I bid on. And I never heard a thing from any of them, and about two or three weeks later, the mill plant engineer and the master mechanic stopped down—I was working swing shift, and they were on their way out of the mill. And they said they wanted to talk to me about that bid I'd put in for, and I said, which one? They kind of hemmed and hawed a little bit, and then they said the welding bid. They asked me if I was a welder at all, and I said no, I went to high school back in Kansas, and we had a pretty good-sized vocational ag shop and a carbide generator and an electric arc machine. I can strike an arc, and I can light a torch, and that's about as far as my welding has gone. And I'll never forget, John Moak, who was the plant engineer at the time, said, "Good, we've got too damn many shipyard welders anyway. We'll teach you our way. You've got the job; report down to the welding shop." So I went into maintenance then—I was a laborer then, of course, in the union. But actually, the welding shop and the pipe shop were run by lead men. And when I first took it over, I took over as a lead man, but—

That's hourly.

Yeah, it was still hourly pay. About a year later they made all those jobs salaried, so I went on salary then—

So after four years of rotating shifts, this job got you onto days. Were you happy?

Very happy, very happy to get off of that shift work.

And your wife?

She was very happy.

Shift work doesn't sound like very much fun.

A lot of those guys worked it their whole life down there, and it didn't seem to bother them, but it did me.

Can you tell me what year you retired and how may years you put in over there?

I put in just shy of 39 years, and I retired in 1986, 20 years ago.

And the next day, how did you feel? Was it weird, after 39 years?

No, not really. I've got eight acres out here, and I keep busy.

Do you remember what your pay was when you first got hired?

A dollar and seven cents an hour. That was pretty good money then.

As a roll bucker?

Yeah. Started out bucking rolls in the supercalender.

That's heavy, heavy work.

Well, on the other paper machines they had these little dollies that you had to push. But on supercalenders they had electric dollies that you could sit there and push a button and just steer them.

In 1947?

Because a roll of newsprint would weigh maybe 1,000 pounds. A roll of that coated paper off the supercalenders—the magazine paper, the slick stuff—it would weigh maybe 2,500 pounds. It was just too much to handle by hand, so they gave us electric carts—but you still had to push them on and off—

That's a lot of weight to move. So did the supercalender have a crew of four or five?

In the supercalender department, they had three calenders. After the paper would come off the machine with this clay on it, then it would run through this supercalender to put the shine to it, like magazine paper got. And there would be an operator and a helper on it. And then right behind it there would be a winder with a winder operator and a helper. And then there was three sets of those machines and winders. And two roll bucks took care of both, of all three machines.

It's a different setup than a regular paper machine.

Yeah. It wasn't like a paper machine, where you had your crews all down the line.

Did you wife work at the mill at all?

No, no. Willis McDaniels—his wife was my cousin. She worked there, and my sister worked there for a while.

Now what did your sister do?

The women, at that time the only place they worked was up in toweling interfold. They made toweling on—No. 4 mainly made the toweling, No. 4 paper machine. They cut it into rolls about so wide, so big around. Probably four-foot diameter.

So that wasn't the blue mechanic toweling?

Some of it. Some of it was blue, and some of it white. My dad worked up there in that department, too. He would load these big rolls on the machines, and then the women would sit there, and as they come off, this machine would cut them and fold them so that, you know, when you pull one towel down, the next one starts. And then every so often, it would kick one up--it counted them—and then they would band them and pack them in boxes.

What's the difference between finishing, converting and interfold?

Interfold was strictly toweling. The finishing department, they would take these rolls and cut them. They put them on a cutter machine, and they'd come out in sheets. And instead of by the roll, they'd sell it by the ream. Some of them were big three-foot square, four-foot square reams. Some of them were just eight and a half by 11. But that's what the finishing department mostly did.

And converting? Is that the same difference?

Well, they called the toweling, the interfold machine, the converting department.

Where did packaging happen?

Well, the toweling packaged their own. And sheet finishing packaged their own.

So were your sister and your dad working in the department at the same time?

Yes, they were.

Any other members of your family ever work there?

No. Like I say, I had two brothers right next door there at Sullivan Electric Plant, but the man who hired me was a cousin of my mother's. He did the hiring for the mill at that time.

There were a lot of kind of clans in the mill.

At one time, half the little town of Willamette worked in that mill, and they were all related some way. Yeah. There were a lot of people related to each other.

How did they train you? People were there who were trained on ships?

Yeah, see, this was right at the end of the war—and a lot of the welders that we got—Glen Jubb, he had welded in a shipyard; he had also welded in the Navy. And he came back, and they hired him. And I guess John Moak made that remark that he didn't want any more shipyard welders, but Glen was a much,

much better welder than I ever could have been. He was a real good welder. But after so long a time, you could get out of the helpers class by passing a test.

Then you went from that helpers class to a C, and then so many years and you went up to a B. And those were simple. To get to the A welder, top class, you had to pass a written test, and you had to pass welding tests. You had to be certified that you could weld a pipe up, and then they'd cut [coupons] out of it, strips out of it, and they'd take those strips and bend them and stretch them, and I don't know what all they did to them. But they sent them into a lab, and—

Checked your welds.

You had to be a welder in order to get an A rating. You had to know how to weld.

So this is union classifications?

Well, no, those classifications were actually just in the mill. But we did work on the union contract--pulp and paper workers' contract. At one time—it wasn't too long after they started Clackamas Community College, I was on the board for welders up there. And I really enjoyed that. I'd go up there and be able to teach those kids how to run a bead.

You were still working at the mill then.

Yeah, I was working at the mill then. In fact, I was still in the welding shop at that time.

You were doing two jobs. Well, and eight acres make three.

Yeah, three.

Apparently your kids didn't work at the mill.

No, no. None of them.

How come?

Well, it wasn't that good a place to work by the time they grew up, really. I think they were out of school when I retired, but the girls both went off to college, and the boy went to work down here in a feed mill in Aurora, and he got caught in an auger down there and lost part of his leg. So after he got over that and got a prosthesis he could walk on, he went over to Clackamas for two years and became an electrician. He liked that much better. It's a good job. And he's got a real good job now.

People talk about their kids working at the mill as a summer job, when they were in school.

Actually, I went to work there—I was going to work long enough to buy me a car. That's the real reason that I started there. But it was a good place to work then.

Roy said he didn't want his kids working there because he was management by then.

Well, where Roy was and I was, foremen—he was on paper machines and I was on maintenance—but we were still just one notch above hourly people. That's probably the worst job in the mill. All the crew has got something that's bothering them, and they come along and gripe to you, and if things don't go quite right, you've got your boss and his boss and his boss, and they come along, and they don't gripe to you; they *tell* you. Like the superintendent on nine told me one time—it was three or four o'clock in the morning, and we'd been down all day and most of the night, and he says, "When that machine's down, this is your job right here." And at eight o'clock in the morning I got to come home. There was several times that I worked 24 hours--as a foreman. As a lead man you couldn't work them more than 16. Yeah, you could not work them more than 16 hours. In fact, unless you could finish it in 16, after 14 you had to call in another crew.

You said you worked in the sawmill and the grinding rooms. What happened?

The mill was real good. They had management classes that you could sign up for and take. And the lead man at that time in the welding shop was Frank Draniger. Before I went to work in the mill, I delivered his "Oregonian" paper every morning when I was going to high school. But he was getting up there in years. Aaron Winagar, the assistant lead man, was one year younger. We're going to need a new lead man here one of these days. Well, I took several courses. They were free; what the heck? One night a week for four or five weeks. And it paid off. When the job opened up, I got the job as lead man, and then a year later, like I say, I made foreman. I worked there for—I don't know— two or three years as foreman. The maintenance foreman in the wood mill, Mill A grinder room, quit and went down to Toledo and went to work in a mill down there. And they just put me up there. I worked up there for two years, and then they had some efficiency experts come in; you've probably heard something about those before. Well, they weren't very efficient, but we had to make out a list of all the jobs we did. Well, a maintenance job in a paper mill—you might do one job and never do it again. But we had to make out a list, and we had to estimate the hours, and if we missed the hours, we had to tell these efficiency people why we missed the hours and what happened, you know. It was a hassle, to say the least.

When was this, about?

Well, let's see. That would have been about—I took over the welding shop in 1963, Christmas Day '63, because the lead man took a week's vacation—at that time there was eight foot of water in my welding shop. That was the year of the big flood.

And that's the week you took over?

That's when I took over the welding shop. So when those efficiency people came in would have been in the late 1960s somewhere. Anyway, Paul Odeon was the foreman in the machine shop, one of the nicest men you'll ever meet in your life. And he told them what they could do with their job. He didn't want to take that white-haired so-and-so that—and that just wasn't like Paul, because he just wasn't that type of person. But he'd had all he could take from him. The guy was a little, short, white-haired efficiency expert. He said "Figure out my vacation, my retirement." He had about a year to go. He retired about a year early. So they went to his lead man, who had two years to go to retirement, and said, "Well, we'll make you the foreman." And "No," he said, "if that's too tough for Paul, it's too tough for me. Figure out my retirement." So there they sat without a foreman, no boss in the machine shop. And I went down, and a mill master mechanic came up, and he talked to me about going down and taking over the machine shop. And I said, I don't know anything about machine work. I read a little bit about millwrighting, and I know how to weld, but I'm no machinist. He says, you don't have to be; you know the paper work. So I went down and ran the machine shop for two years. And that was probably the best job I ever had in the mill; I liked that job.

I really did. It was a real good job. And then the foreman out in Mill C, one of the Mill C paper machines, Ed Karbonski, was going to retire, and that was over No. 9. I had one, two, three, four, No. 9, and No. 7 was still running part-time then, and mixer room. That was all considered Mill C area. When I first went in, I had probably somewhere between 15 and 20 millwrights for that area. And I was supposed to work with Ed for I think three or four months before he retired. He worked with me a week and got sick and never come back, so I kind of had to learn that one on my own too. That was when the foremanship got tough on those machines, because when they're not running, they're not making any money. And when they're not running, you're going to be there and find out why they're not running, or do something about it.

Sometimes you're the bad guy instead of one of the guys...

Yeah, but that's—that was probably the worst job I ever had in the mill.

But that isn't the one you retired from?

Yup. That is the one I retired from.

Tell me about the eight feet of water in the welding shop.

They said we would have maybe three or four feet in the welding shop. They estimated it. So we took all our tool boxes and put them up on the tables. A lot of good that did. And then we sandbagged right out behind the welding shop—we had two big air compressors out there—and we sandbagged around them and put sump pumps in them, trying to keep them dry. Some of the more important stuff that we felt that needed to stay dry if we could, we'd sandbag and put sump pumps in them. We lost them all anyway, eventually. But we did that for a couple of days before we shut down for Christmas shutdown. Well, we were shut down anyway; when the water got up that high, they were shut down. But—of course, when we went back in, we had those logs to cut up and get out of there. And you could see where they had come through. And a bunch of mud all over everything.

They had quite a debate about [whether to start the mill again], I understand. Whether it was worth—a lot of the motors we took out and took them upstairs, so they were dry. And of course, the pumps, you just take them apart, clean them out and put them back together. But the motors that we didn't get out, they had to take them out then and send them into Portland. Some of the big electric companies in there would put them in their ovens and dry them and bring them back. I'd hate to even estimate what it cost to dry all those motors and get everything ready to run again. But there was quite a debate whether it was worth it or not, and they finally, fortunately, decided to run it again.

So you stepped right into cleanup from the flood? Not a typical day at work.

When I was a lead man in the welding shop, I was secretary of the union also. Of course, I had to get out of the union when I went on salary, but I don't know if that made any difference, whether they put me on salary or not, but at least I figured it did a little bit.

You mean they wanted to get the union troublemaker out and promote him up?

Yeah.

Wasn't it hard to make that decision to go to the other side?

It was. It was real hard. I agonized over it for quite a while. The first offer they made me I turned down. I said I'll stay where I'm at. But then they came back with an offer that I couldn't hardly refuse, so—same way with the retirement offer when they made it. I couldn't refuse that.

So you must have seen layoffs and strikes. Do you have any memories about those?

Some, but not as much as you might think. A lot of it was just by attrition, people retiring, and they just didn't replace them. I lost one welder, Charlie Cherry. He went over to Bend and started his own company over there, and I guess he did real well. But at that time he was the only one that I actually lost. The rest of them, as they retired, they just didn't hire new ones.

Can you talk about what the long strike was like?

I was a foreman at that time, and they kept the mill running with salaried people. See, Crown owned two mills in Louisiana and one up in New York, and two, three over in Washington, and one down here on the coast in Oregon. And they would bring a few salaried people from each one of them, and we ran the mill. They [the paper machines] were down for—not very long, maybe a couple weeks, I think. And then they decided to start them up and run them.

What kind of quality paper did they get out of that?

Actually, we did all right. They got some pretty good paper makers. Roy worked through it. He was working on the machines. They had some pretty good paper makers, not only ours, but the ones they brought up from Louisiana. Almost everybody in the mill, our mill, had started at the bottom and worked their way up. They didn't hire much outside help, except in the base pay position.

I think Rosie said that Mr. Zellerbach would come and shake everybody's hand.

Right. When J. D. Zellerbach himself was alive, he'd come in through that mill, and out of the 1,400-some people, he probably knew the first name of half of them. I don't know how he did it. I can't, I couldn't do it, but he did. He knew us by first name. You were a person then. You weren't a number.

Roy said the worst part about the strike being over was when the hourlies had to take over from the supervisors, and it got a little tense. Did you see any of that?

Yeah, I saw some of it. There was two brothers; one was a welder and the other was a millwright. And he wasn't in that picture, but he hired on as a welder later. And they were pretty bitter over it. But as a whole there wasn't a lot of it, no. But there was some. Yeah, you're bound to have some that are bitter over it. Yeah, they would throw rocks at you when you went to work and that kind of stuff.

Oh, crossing the picket line? They must have had a picket line.

Oh yes, they did.

Twenty-four hours a day?

Yeah. First one, I was in the union. They had two strikes while I was there. The first one only lasted about two weeks. One of the salesmen used to come into the welding shop; I got to know him pretty well. And after I'd been off work for two or three days—see, I'd been married before, and I had two other girls; I was supporting them, buying the farm, raising three kids of ours, and things were pretty tough at times. So I went into the shop there in Portland where he worked and told him I was looking for a job. And he said "What do you mean? Did you quit down there?" I said "No, they went out on strike." He said, "I hadn't heard; I was going out there tomorrow. Maybe I better save myself a trip." And my wife and I—for some reason we were in Portland. But before we got back home, he had called and left a message to call him. I called him up, and told me to go down to Murty Brothers' trucking outfit. They had a job for a welder down there. So I went to work for about two weeks for Murty Brothers.

But you were happy to go back?

Yeah, I was. I could have stayed, but at that time I was a lead man in the welding ship, and I was making—well, it's an extra dime over the A rate. And I felt real good when I left. The boss in there came out and told me I was fired. And I said, what did I do wrong? He said, nothing. And then he got to laughing. And I knew something was up. So he said, they settled your strike. Your wife called, and you can go back to work tomorrow morning. But he quoted me a wage one penny higher than what I was getting. I could stay right there. I says, no, I'm not going to drive to Portland every day and back for a penny an hour, not with fighting that traffic.

That would have been nice...

It made me feel good, yeah.

Okay, when did the paper machines shut down? Do you remember?

It started in the '60s, '70s, '80s, along in there. They were pretty much all down by '90.

Down to three machines?

Yeah. There was 10 machines there. Well, there was only nine of them running when I went to work there. They had shut down the old No. 8, which originally was put together in a building someplace up on the Clackamas River. It was the first paper machine west of the Mississippi River. And I've often thought, I hope somebody took that out of there and put it in a museum someplace, because that's where it belonged. It was a little old machine about, oh, maybe 60 inches wide, and it would run 80 feet a minute, maybe. Not like five and six and nine down there now; they're all—let's see, five is 105, and six is 110, nine is 110 wide, and they run about 2,000 feet a minute. So they're making more paper now than we did with nine paper machines then. I've heard it was on the Clackamas River, is all I ever heard. But it was supposedly the first paper machine on the west coast, the west half of the United States, in fact.

It was interesting—the papers that we made—No. 10 mainly made lettuce crate packing, for when they were—they were pads about maybe a quarter of an inch think. They weren't sheets of paper; they were pads, really. But they used them for packing lettuce when they were harvesting lettuce. And No. 7, for years they ran what they call pineapple mulch. Black tarry paper, and they had a wheel on it with little shoes that went around, and it put a white paint mark every so often. They'd roll that out on the ground and put a pineapple plant for each one of those marks. Nine at that time, when I first started, was mostly making yellow telephone directory paper. That's an awful lot of telephone books. They'd run day and night for months on end with nothing but yellow telephone directory.

George had some Crezon at home, a scrap of that. And he had some pineapple mulch.

That was made on seven too, the old Crezon paper. Oh, 10 or seven, either one. They were both pretty slow.

You worked in the sawmill and grinding rooms—why did they quit those operations?

When they built the new mill down here at the coast, down at Wauna—they put in a big pulp mill. They made the pulp and baled it. They could make it cheaper and ship it up here and beat it up again than we could make it here. We had problems with Wauna pulp, all kinds of problems. They'd buy a bunch from Canada, smoothed right out. You'd think they'd wise up, but they kept trying to run Wauna pulp for years.

Henry said that that Wauna pulp kept getting plastic in it…

Yeah, make a mess.

Where would plastic come from in a pulp mill? I don't understand.

They did a lot of recycling, paper recycling. And that's where they get the plastic mixed in with it.

Okay. Sullivan didn't interact with the mill very much?

No, no. It was a separate building. I'd been over there and gone through it, but I didn't know that much about it.

How about the locks? Was that a separate world, too?

That was pretty much. I didn't have much to do with them. The tugboat operators, of course, did—like Milton—and we lost one of those here a couple weeks ago. Whitey Van Domelen passed away here a couple weeks ago. He was on the tugboats.

I guess Paul Miken put in some time on the river.

Right. He put in some time. He kind of bounced around kind of like I did in maintenance, only he was in operation. He was in the sawmill for a while. He was boss up there for a while after—what's his name? I can't think of the name now, but he was boss up there anyway. Benefield.

You've still got your good memory. So they decided to close down the pulp, the grinding room, all those turbines.

They closed down our Mill B. That's the acid plant. They closed it first, and I was happy to see that closed. They melted sulfur, burnt it, and then ran it through these coolers and made sulfurous acid out of it, and filled these big towers with lime rock, and water dripped out through it, and it would come out sulfurous acid, and then as soon as it hit the air, it was sulfuric acid. And every Tuesday they shut it down for repairs. All these pipes in there were lead. Glen Jubb and I were the lead burners when it was running, and we did all the welding of the lead pipes, and that wasn't a good job. I worked in there 10, 12 hours and come out of there with a nosebleed.

So was that before safety committees?

Yeah, that was before hearing aids, you know, earplugs, or earmuffs, and we did have safety shoes. They bought us one pair of safety shoes; it was a steel cap over it.

Who finally thought of safety?

I don't know. It finally caught up with us. As far as I can remember, we had two fatalities while I was working there, in the 39, 40 years I was there. One of them was an electrician who got into some high-powered stuff, and the other one worked on the river. He was at the Pulp Siding, and he'd gotten in the bight of a cable—But as far as I can remember, that's the only two fatalities we had in that mill while I worked there. We had some pretty serious accidents, but—

I asked Paul Miken what he remembered about his first day at work, and he said it was so loud.

It was loud. Here they are. A little hearing aid. I'm about half deaf.

I was just thinking you're the only one I've talked to that doesn't have them.

I've got them. I just don't wear them all the time.

Did you work in Mill A when the barking drum was going?

It was loud. I didn't have to work right there all the time, but I was in and out of there while it was running while I was foreman up there. Once a week they shut it down, and we'd work repair day, and weld it all back together. The thing was falling apart.

So you had regular weekly maintenance on the sulfur lead pipelines?

Not all of them. On the paper machines, they would run them until their what they call the Fourdrinier wire was getting bad. They'd need a new wire, why then they'd shut them down, and we'd do all the maintenance we could. And while they were changing the wire and the felts and whatever else the paper makers had to do, the maintenance crew would go in. We had one guy, Walt Riggs, that had a little box on his belt. I called it his stethoscope, but earmuffs and a prod, and he'd go around and check the bearings and gearboxes, and he did a good job. He'd tell us, now you've got a few weeks on this one yet, but you'd better watch it, keep an eye on it. So he'd say, the next wire you'd better change out that gearbox, or you'd better put a new bearing in there. He did a pretty good job keeping us out of trouble. We didn't always get out of trouble; sometimes they'd go out in the middle of the night. But he did a pretty fair job of it. I've got one roller out of the roller bearing on a supercalender, bottom supercalender. One roller that big around, and there's a double row of those around that.

Salmon would go out in the lunch boxes: Did you witness any of this?

Oh yeah. We were over in the grinder room one time—they had one spot over there where the grating was loose, and of course the guys that worked there knew it, but nobody was supposed to know it. They'd pull that off and snag for salmon out of that hole and put it down. So I happened to look out—on the back side of Sullivan they had some round portholes—and here was a salmon going up the wall. Somebody had snagged one out of the tailrace and was pulling it up the wall into the Sullivan Electric Plant. [laughing]

Just going past the window.

Yeah, salmon going up, crawling up the wall. We were far enough away we couldn't see the cord he had on it, the rope or whatever he was pulling it with. But yeah, they got pretty strict on it towards the end there, but there for years and years, anybody wanted a salmon... When I was working in the welding shop, they'd come down there with these hooks. They couldn't get a triple hook big enough, so they'd buy single hooks. And the hook would be, from the eye to the bottom, would be maybe six inches, and the eye, the curved part, would be that far across. And they'd buy three of them, and then you'd cut the eye of two of them and then braze all three of them together so you had one eye with a hook, no matter which way he came from.

Photo of welding shop crew celebrating at the West Linn Inn, with Del Herndon, center, standing, white circle by his pocket protector.

Perfect for snagging.

I made up a lot of them for some of the boys. So any time I needed a fish—all I had to do was say I wanted a fish, and—

So a braze would hold a 65-pound salmon?

Oh yeah. Oh yeah.

Okay. Now you already told me your favorite job there was that job where you—

Run the machine shop.

—run the machine shop. But generally, what did you like best about working at the mill?

I think it was people, really. We had a bunch of good people. They got pretty strict on it—but when I first started there was quite a bit of horseplay going on. And everybody just seemed to get along real well.

By horseplay, you mean—?

One day I went to the locker at the shop, and I had high-top steel-capped boots that I wore down there. And little Shorty McClanathan then was a painter, and I remember him and I was always kidding one another about something. I come in one morning, and my shoes were bright yellow. Or you might go over to pick up your lunch bucket setting on a loading table, and it would be soldered to the table. But nobody seemed to, you know, get mad about it. And on repair day up in the sawmill, somebody would get a shovel of sawdust down the back of their neck. Invariably somebody would get it.

Not an accident.

No, no.

Now they recommend that you take time out and play games at work because it's good for productivity.

Well, I always thought it was, but they got pretty strict about it. They were afraid somebody was going to go too far and somebody get hurt or something. Whenever they'd turn a paper machine down, they had a cleanup crew that would come in with their high-pressure hoses and wash their pits out and wash everything down. We had one guy that was a master at it. He'd be down in that pit, and he'd look out of the corner of his eye, and he'd hit that corner and it would come right back out at you. And if you've ever done any electric arc welding, then you get your gloves wet and your feet are standing in water, you go to put a new electrode in that stinger and man, it would rattle your teeth.

I was down on six one time, and I'd been shocked several times, and it was getting on my nerves a little bit. And I asked him real nice to kind of watch it a little bit. And he watched it all right; as soon as I turned my back I got it again. All over the mill, you know, out there on that island, you couldn't get a fire truck or anything in there—we had our own fire system— and on the wall we had a hose coiled up in a rack, inch-and-a-half fire hose. I just went over and peeled it out of the rack and turned it on. I didn't get wet any more that day. I put him down in the corner of that thing and held him there for a little while with that inch-and-a-half fire hose, and he was a little careful after that. It wouldn't have bothered me, you know, if it hadn't been that you had to put a new rod in that electrode in your stinger every so often.

We heard about the squirting, and a lot of horseplay on the river. A lot of falling in, being pushed… You can't be like that any more.

No, no, can't have fun. Yeah, you've got to be serious. You can't have any fun at it.

We heard an English guy bought the company, and that's when a lot of people bailed.

Goldschmidt [sic]? Well, he probably did me a favor, because I retired in '84 eight years early, with the same amount I would have got had I been 62. And they paid me a supplement to my Social Security. They figured out what I would get from Social Security, and they paid that until I reached 62, until Social Security took over. And they paid all my insurance and medical and everything up until 65, until Medicare took over. So if I'd have worked till 62, I wouldn't have got any more money. I couldn't afford to—it's one of those jobs you just couldn't afford to turn down.

People think that when they retire, they're going to have 20 years. A lot of people are having lots longer than that.

Well, I've know an awful lot of them that went out of that mill, though, that went home, and if they went home and sat down, they'd be gone in a year. If they had something to do, a hobby to keep them busy, then they had good long lives. Well, Frank Zaniger, he didn't last. Him and his wife were both killed in a car wreck less than a year after he retired. We had an old blacksmith, and like I say, I used to do a lot of hunting. I've got an old '49 Willys Jeep out there in the shop, and the blacksmith had a hip bothering him; he couldn't hardly get around and walk. Didn't know what he was going to do. And I said "Well, Fred, I said, why don't you go hunting with us this year, and I'll take my Jeep and haul you down there, and I know where there's a great big rock, and I'll set you on that rock." And so we did, and he got his deer all right. The next year I asked him if he's going hunting. Nope. Going to retire—I don't remember how soon now, but he was going to retire pretty quick. At that time deer hunting season was three weeks long up there in eastern Oregon. And he says "I'll bypass it this year. Next year I'll go up there and I'll spend a whole week camped up there." He didn't make it. He died about a week before he went deer hunting. Darn heart trouble.

Well, people that we've talked to are busy. They went right into other things.

I did a lot of welding on the side, putting on trailer hitches, for several years, a little extra.

That's pretty good for a welder's assistant who'd never welded.

Stayed in the business for a while, 'til insurance finally run me out. I just couldn't make enough to pay the insurance and the income tax.

So you did have your own small business, your own welding business.

Yeah.

> *Elnora: How many years? You had it from the time the kids were little until they were almost grown.*

Yeah. That's the reason that big 40 by 40 shop is over there. It's full of a boat and couple of cars and a Jeep now, but that was my welding shop.

Do you remember any fires in the mill?

Yeah. I was on the fire department, and we had a volunteer department in the mill just like you would in some of these small towns. The worst one I remember was up in the hog fuel bin, up above the boilers. It got up there and got to burning down underneath that hog fuel. It was tough to put out.

Did you relate to the West Linn Inn in any way?

Well, a lot of us would stop and have coffee in the morning there before we went to work. I did sometimes, sometimes not. But they held all of their banquets there, pin presentations for years' service, and that's where they put on their schools that I was talking about before. Management schools, they'd put them on right there. It was anybody that I guess they felt they wanted to get into management, and they wanted to better their job a little bit, why all you had to do was sign up for it. And it didn't cost you anything; the mill put it on.

People tell me about Paper School.

Paper School was different—I never did go to that. I think it was a Crown class there.

I heard that they built the Inn in a very short period of time, to house strikebreakers.

Right. That's what it was built for originally. The mill itself built it. It was on company property. And there was a strike, and they brought their strikebreakers in, and instead of making them cross the picket line, they just housed them right there. The last two strikes, that big motel down there across from the Oregon City Shopping Center, Crown built that one in the last two strikes. You might just say they did (laughing). That's where they housed all the workers that they brought in from outside. They paid a lot of money to that place.

All the extra supervisors?

Yeah.

When was the really long strike?

The one where the Inn was built? That was way early. I don't know when that was.

But the one where you were management—

That was the last one they had.

Was that in the 70s?

When did you and the girls take that cruise?

> *Elnora: Let me see. It was on Diane's 21st birthday. Let me see; it would be '75.*

I was on strike and working anywhere from 12 to 20 hours a day, and I sent mom and the two girls on a cruise to Mexico and back. I'd come home, and I'd be so dead tired anyway, I'd just fall into bed. I didn't see them or talk to them anyway, so it didn't make any difference.

So, didn't they still need the welding shop, the machine shop, the millwrights, the electricians? I mean, how—

After that strike, they broke up the welding shop, and they put a welder out in each millwright crew, or two welders in each millwright crew or whatever. And that's the way we ran it during the strike. If they needed welding done, I did the welding. But if they need millwright work done, I did millwright work, too. After the strike was over, it had worked so well that they went to that kind of a job; did away with it, practically. I don't know; they had two or three welders in the shop to build projects that come in all the time. But as far as maintenance welding, that was all done out in the— and I taught all the millwrights that were there how to light a torch and burn, you know, a cutting torch. But I had—what the heck was his name? 'Tonto', we called him. He was a union officer, worked for me and the millwrights. And No. 7 had a breakdown one night.

I called him in to fix it. There was a guard there, and it wouldn't come off. So I'm going to have to cut it. So I went to the office, and I called every welder in the crew. Couldn't get a soul. So I went back out and told him. Said I can't get any welders. None? Nope. I went back to the office, I went through the whole list. I couldn't get a welder. And I says, I ain't going to stay here all night; we're going down to the shop and get a set of tanks and a cutting torch. You or me, one's going to cut that thing off. I cut it off. As a foreman, I couldn't use a tool, but I did that night. He didn't complain either. He wanted to go home too.

We've heard stories about people that get in trouble for trying to help out...

Yeah. I never did get in trouble over it, but I was pretty careful.

A lot of people talk about the people being the best part of working there…

We had a bowling league just in the mill. Five guys on a team, you know, and maybe five off of this machine and five off of that machine, and a bunch of millwrights. And then there was a bunch of us, one or two, it wasn't enough to make up. And we called ourselves the Misfits. And we had a little team of our own.

Did you go over to Gladstone Bowl then?

Yup.

Somebody told me that the West Linn Inn had had a bowling alley in the basement.

They'd got four lanes in the basement. I didn't know it either. I've never seen them, but that's what I hear. The Oregon City Elks Club has got, I think, four or six bowling lanes in their basement, too.

The food was supposed to be pretty good at the Inn.

Yeah, yeah. They had good food, there; they really did.

What was the bad thing about working at the mill?

Getting up in the middle of the night and getting a crew in. Oh, I hated to call a crew in at night. A lot of times I wouldn't have to go in. It depends on what was broken down, how serious it was. If it was pretty serious I always went in, but a lot of times I would just call a crew in and I'd go back to bed. And that bothered me, calling a crew.

According to Margaret Witherspoon, sometimes they didn't answer their phone.

I've run into that. Her husband worked for me. Yeah, old Eddie was one of my better millwrights too. You know, I called one guy one night—I knew he wasn't home. It was three o'clock in the morning, and boy! She wasn't lying to me that time; I know. I could tell by her voice.

Now, did you live in West Linn?

Well, I finished one year of high school up at West Linn High School. Before that I lived in Kansas. I was 15 when I come out here; my folks moved out here.

So where did you live in West Linn?

I lived out in Willamette. My folks lived out in Willamette, just the other side of the old box factory out there. I don't know if you know where that's at or not.

Gerber Blades?

Well, there was a road that went off up there, and from that road back towards town my folks owned four acres in there, up on the hill. There's about six or eight houses on it now, or maybe more than that.

There were service stations right down by the mill with ads that said "Mill worker's special: leave your car with a note on it, we'll fix it before you get off shift."

I think they only had one there.

> *Elnora: When you worked there, there was only one. There might have been more earlier, but there was just one there.*

WE'LL TAKE YOUR CAR
When you go to work, service it carefully and when the whistle blows, it will be ready and waiting for you.
Paper Mill Employees, drive in and let us show you how convenient our service is. No walking, no waiting. We are right at your front door to take care of your car.

Mobil Service
MEAD OSWALD
Phone 2-7781
At the West End of the Oregon City-West Linn Bridge

Advertisement for Mobile Station at entrance to the mill.

Did you drive your car to work?

Eventually, when I worked long enough to buy a car. I rode the bus before that. They used to have a bus running from Willamette to Oregon City and back.

This is after the trolley closed down?

Yeah, yeah, right.

> *Elnora: The trolley ran for a long time, because I rode it to high school. Oh, you mean from Willamette. The one that went Oregon City to Portland, it was there for a long, long time.*

Is there anything else that you really want to share?

No.

Thirty-nine years.

> *Elnora: It's a long time. I'm kind of like Mrs. Witherspoon; I didn't like it when they got called in the middle of the night. Or when they would call me in the middle of the night—he was still at the mill—and say, "I want to talk to Delbert." And I would say, "Is this Crown Zellerbach?" And they said yes, and I said, "Well, he's there. He's wearing his radio. Have you tried him?" "Well, no; we just figured he was at home." And I said, "Well, no, he hasn't come home yet. He'd better be at the mill!" Then they'd call him on his radio, and he'd say, "Yeah, I got a call. And I was right there; I don't know why they didn't call me first." I've gone after him in the mornings when he worked 24. I would be there. He would call me and say he was getting off, and I'd say, "You stay right there; I'm coming to get you." I wouldn't let him drive home, because if you relaxed in*

the car, you'd go to sleep. I'd just say, "You leave your truck there, and I'll take you when you're ready to go tomorrow, but you're not driving home."

Can you imagine somebody nowadays working 24 hours straight?

> *Elnora: No, I cannot. Most of them don't want to work eight hours.*

My longest shift was 26.

When you were management?

No, I was on the clock then.

> *Elnora: Why did they work you 26 hours?*

To finish a job.

> *Elnora: That was before me, because I don't remember him having more than 24.*

They had three small sulfur burners in Mill B, and they were changing over to one big one. They had to take those small ones out, cut them to pieces and take them out of there. Then the new one was all built, but we had to slide it over into place and hook the pipes up and everything. And there was, I don't know, two or three of us in there with cutting torches to cut those three little ones out. Oh, a stinking mess.

George Droz work talked about one time they put him down in this ceramic thing when they were cracking out the inside of it to replace the ceramic materials.

That would have been one of the digesters.

He said that was the job from hell—being lowered down through this little hole...

> *Elnora: I couldn't imagine—I'm claustrophobic—they could not have got me in some of these places they put Delbert into. And I'm smaller than he is, and they would get him down in there, and I don't even know how he got small enough to get through some of those holes.*

Oh, it would be scary. And they wouldn't put me in anything like that. Ed's dad worked up there with my dad.

He said that if everyone he was related to went home sick, they'd have to shut the mill.

One of the foremen in Mill C—Gerkman, Ernie Gerkman was his name. And he delighted in calling me in at night. It just made his day if he could call me in at night. And he'd heckle me all the time I was there about being up in the middle of the night. He retired before I did. So after he retired, I got called in one night about three o'clock in the morning. I called him up, and I says, Ernie, I just wanted you to know I'm in the mill. He said some harsh words to me.

> *Elnora: Did you tell her about all the jokes you played in the welding shop? And they used to weld stuff to the ceiling, and tool boxes and stuff. Honest.*

I didn't hear that one.

Mary Lefebvre

Interview May 1, 2006

Mary L-e-f-e-b-v-r-e. Okay, could you tell us your name and your age?

My name is Mary Lefebvre, and I'm 52.

When did you start at the mill?

I started in the mill in 1974, October 8th.

And you're still working there?

Still working there, uh-huh. Different capacity, but still there.

Why was it memorable?

The date sticks in my mind. I don't know why, but it does. You know, looking back, you don't realize when you're young the decisions you make have a long-term effect. I was working in a dentist's office and disliked it. I've always been physical. As a child I worked out in the fields, worked summers, and it was physical work. And I didn't like being in an office, where I am now, but my father was a salesman in the plywood industry, and I worked in mills before. A friend of mine —I was over at his house, he and his wife—we were talking about jobs, and I wasn't happy where I was. And he said, "Hey, my dad works at the mill, and they have to hire women now. So you know, if you're looking for something for a couple of years, the money's good; the hours are so-so, but the money's good, and short-term it's great." So I drove down there – that was a huge step for me—I drove down to the mill and saw personnel. His name was John, they called him Mac. And I went and saw Mac and filled out an application. "Yeah, yeah, they're hiring, but…", you know, he'd call me back.

So I waited about a week or so, and no contact, and I called him. "Oh yeah, I'll get back to you; I'll get back to you." And this started probably in August, and I was talking to my dad about it, and he said, "Then you bug him." he said, "you go down there once a week; you call him once or twice a week. If this is what you want to do, you know, then do it." So I called Mac twice a week, went down there once a week and saw him. And finally he said, "Okay," he said, "you're hired." And years later, years later he had gotten out of HR, out of personnel, and I saw him walking out or something, he said, "You know," he said, "when I said that I was going to hire you," he said, "I never thought you'd last." He said,

"I hired you just to get you off my back. And," he said, "you fooled me." He said, "You surprised me." And he was a nice man; he was a really nice guy, but he hired me based on the fact that he knew I wasn't going to stay, that I would never, ever make it. We laughed about that.

But the first night I worked, it was 8 p.m. to 4 a.m. They had the big barges, and they filled them full of clay, which is a dry additive—makes a coating, it's dry. They just piped it into the barges. They had this great big hose, and you would hoe it into a big suction tube, a miserable job. Miserable, dusty, it was white powder. I did that off and on. We were on "extra board" it was called, where if there was an opening, they'd have you come in. The only two days I worked that following January was Superbowl Sunday and one of the playoff games. You know, people would call in sick, and they'd call in the extra people. I did some training, at that point I did work up in sheet finishing for a little bit, and that's long since gone. They converted the rolls of paper into sheet stock, and you'd package it. I was a helper; I didn't operate any of the machinery because I didn't have any seniority. And it got really slow, and they put me on the paper machines.

They had an area called Mill C, which had the uncoated machines. It was hard work. You can see by the picture, I was smaller at the time, and I hated it; I hated it. It was a real a good-old-boy network. And they didn't really want the women in there. Some of them were okay, but most of them, especially the ones down in that area, they didn't want women down there. I worked down there a little bit after the training. And then they put me up on five and six, which is now one and two, the paper machines, the coated paper machines, the bigger ones, and I trained there for a couple of weeks, intimidated and scared to death. These rolls were eight, nine thousand pounds, and you're moving them from here to there. They had transporters; it's like backing up a trailer. It was a front end where you had this long—I have no idea how long it was—probably 150, 160-inch trailer. And you had to manipulate and maneuver this transporter. Needless to say, I can now back a trailer up. You have to make, you know, 90-degree corners with this thing, and it's huge and it's heavy. It was just humiliating, but everybody went through that. So I trained on the paper machines, and I knew I wasn't going to make it. At that point I was so intimidated. When I first started, and I guess I was naïve, I didn't expect much resistance. I didn't really think about it.

Can you remember the very first day?

Not totally, I can't. I got lost; I know that. The people where I worked—it was the clay plant again on the barge—they were nice men, and they just walked me through it. I mean, all you did was hoe this powdered clay into a big suction tube. There is not much to it. It was pretty physical. And I remember you worked 30 minutes on, 30 minutes off. You had to have two people because of the safety issues. The barge was on the water, and only one person could work at one time.

And I worked with a young kid, and he was okay. But I just couldn't believe that they would pay you to sit half the time, because there was really nothing else you could do, you know. But I just remember being so tired.

Did you have to wear some sort of a ventilator, mask?

Oh, no.

So there wasn't any safety—

I don't think it's toxic. No. Got blisters. That's changed so much, but no, I didn't wear a ventilator or mask.

But you didn't have that job very long?

No. That was really an extra board job. Just the new people coming in would do that. I guess it was great in the summer, because people would fish off the barges.

In the half-hour that they weren't working, they could fish?

Yeah, they kept their lines out.

So where was the barge? In the locks? There aren't any fish in the locks.

It was not in the locks. It's on the other side of the locks. They tied it up—boy, it's hard to remember—on the river side. I mean, it was on the river; that's why you had to have two people, in case somebody fell in—you walked up a gangplank to the barge.

I'm trying to visualize. Was this in by Sullivan, in that bay area, or by the—

It's across from Smurfit. Towards the bridge side. Almost at the mouth of the locks, if I'm remembering correctly. But it's been a long time ago; I don't quite remember.

Did they give you any equipment?

A big hoe, that was it. If you wanted gloves, you got your own.

So you alternated between that and backing up the transporter—how long did it take you to bid on something else?

Well, let's see. I hoed clay, worked in sheet finishing some, and then on the paper machines. Probably most of the first year. Like I said, Mill C wasn't like this; all the machines were smaller, but more physical. Then I went on the coated machines with the transporters, and I will never forget—uh, I hadn't been on the paper machines for a while, and I was really happy because, like I said, it was intimidating. They called me up to work graveyard shift, night shift, and I was so angry. When I found out where I was working, I was so mad I couldn't hardly stand it. I just said okay, if they want me to kill somebody, fine. If they don't care, fine. And at that point I really lost my fear. That was really kind of a milestone. And once I lost my fear, I did fine; I did okay. Very, very physical.

Did anybody know you were afraid? You don't look afraid in your picture.

No, no. Well that point, I'd been there several years. No, I don't think so. I think they just felt I wasn't very good at what I was doing. I don't think they knew the reason. And I wasn't about to tell anybody, you know. But after that, things went much better. I was probably one of the first five women hired after they had to start hiring women. And I think a lot of people were very resentful because they had to hire women, and they had to start hiring more blacks, and they had to start hiring minorities. And a lot of the men weren't happy with that. There were a few women that were a holdover from World War II that were still there, but they were elderly—probably my age now—elderly, and did a lot of the clean-up. So a lot of those women were there before most of the men came.

So was this a Civil Rights Act—

Equal opportunity, yeah. They were mandated they had to start hiring women.

And how did the union absorb that?

I don't know. They had no option, and I wasn't involved in the union at that point, other than having to join. I wasn't really aware of their thoughts on it. But the people who accepted you accepted you whether they were union or not, so I don't think that was really an issue. Some people were nicer than others, so that's okay, too. Looking back you see things and you notice things at the time—I think I was so intent on just trying to do the job that I didn't pay a lot of attention. But looking back—there are several other women, one who still works there, andwe talk about that. And you look back and you go wow, I'm so glad I didn't know that at the time; I'm glad I wasn't aware.

Letting yourself be unconscious about all the hostility or whatever was the way you made it?

I don't think anybody was overtly hostile; they couldn't be. And I don't know from a management standpoint how they addressed that. But I didn't really notice any overt hostility. A lot of prejudices, and a lot of the gentlemen were 40s, 50s, 60s, raised a certain way. Women just didn't do what we were doing. They just didn't. But for the most part, people were really quite kind.

So did you bid your way to Roy's area?

No, I was still on extra board when I worked for Roy. I never bid on the paper machines. I ended up working with Roy quite a while, and you would get on a shift, and with vacations and illnesses, you would be on a shift months at a time, particularly during the summer. And boy, they wouldn't let you go, either, because if you lost somebody, you wouldn't get them back; they'd go on another shift. But probably after a year, year and a half, I was trained in the supercalenders, and then I stayed there. It was a little less physically grueling, not quite as intense, and I took the bid on the supercalenders at that point.

What was your title on the supercalender?

Well, you were promoted up. You started out bucking rolls. What a roll buck did is, you had transporters, and the winderman would run the rolls, cut them to sizes, and then they'd come down off the table, and the transporter was level with the upended table. You'd move the rolls onto the transporter and block them. It was all physical. You had to break the rolls apart and then take them to head or push them on the transporter onto the header.

How much did you weigh then?

Probably my lowest weight at the time was probably between 110 and 115, and that was after working huge volumes of overtime, 60, 70 hours a week. I didn't eat well on graveyard anyway. So yeah, I lost too much weight, that one summer in particular. I didn't let that happen again, because I didn't feel that well. But a lot of hours, and it was hot during the summer, it was so hot.

You had to drink a lot of water.

Drank a lot of water. So anyway, I bid on the supers, and eventually when I left there, I was operator on the No. 9—well, it's off of the No. 9 paper machine, but it was a Wärtsilä winder. You ran the paper off the 9 paper machine, because every paper machine had their own stack. They'd vary a little bit, but—

And how many machines were still running then? All of them?

No, no, because they shut Mill C down—I think by that time they just had three. They might have had four…so my last job in the mill was a permanent winder operator on the Wärtsilä winder. They had converted No. 9 from a non-coated to a coated machine, and then they put in a new supercalender and a new winder. The stack was from Germany, and the calender was from Germany, and the winder was from Finland, I believe. I was helper when they started that up at that time, and that was quite an experience too. I was running winder. I was having some issues with the winder; I remember that. This was probably after it had been running several months, and just electrical issues and huge problems. The electrician on the shift was a nice man. And I was getting behind and was getting frustrated. He looked at me and he says, "I don't know what to do." He said, "I'm sorry," he said, "the instructions are in Finnish," he said, "I don't know what to do." We laughed about that years later. But we got it working.

I was in the mill probably eight and a half years. April 15th—you know, I have to backtrack on the year—one of the people in the planning office was retiring. A lady who worked in there, her husband was one of the foremen I had worked for. They used to take a lot of internal bids on salaried jobs. And they asked the foremen if people out there would be able to do this job. And said he thought I would be able to do it. I look back now and I don't know how we functioned, because it was all manual. Everything was written down by hand. But when you were running winder, you'd have to program…where planning would, "—we need X amount of rolls for this order of this size, and X amount of rolls for this order of this size," and they would try and trim the winder to be as efficient as possible.

It was expected that we would have to go to the header, see how many rolls were needed, adjust our cuts, make sure the orders were met. I guess I was always able to do that, and the one foreman, the husband of this woman, he had real high expectations. If you didn't have it done, he wanted to know why. And I didn't have time. He said, "You shut it down, you make time; this is what you need to do." So he was a pretty good teacher, and you'd better know what you needed, down to the letter. He wanted to know where you were at all times. So that was pretty good training, and anyway, I got the job in the shipping/planning office, and that's where I've been.

So you were on graveyard for the eight and a half years?

Oh no, you rotated. A lot of people liked it; I didn't. My daughter was born in 1980, and it worked out fairly well. The kids' dad was home, and he would help and stay with her when I was working nights, and get her to the sitter in the morning. She was about five, and I just got to thinking, what am I going to do? You couldn't work like that with a child in school. And so this opportunity came up, and it was really a lifesaver, because I knew I couldn't function and take care of her and be there when she needed me to be there. Actually, I kind of missed it, because you had days off in the middle of the week. You know, as far as appointments and going places during the week, you wouldn't run into crowds.

What about the sleeping?

Horrible. You force yourself to sleep. I could never take naps before I worked rotating shift, but I had to learn to sleep when I could. Graveyard was probably the worst, going home. It took me probably a year before I could sleep well; it took a long time. And you know, you get four, five, six hours a night—I was just wasted all the time. Again, some people liked it. Some people tried it, they loved it. I never adjusted that well to it. It was very physically hard on a person.

I don't see how you could work that hard with only five hours of sleep.

They have adjusted, and I think that they've done a lot of work, trying to find the best situation. The rotating shifts we were on was very physically tiring. You worked five days of days, and two and a half days off, started swing shift, and you worked five days of swing, then you had two and two-thirds days off and started graveyard, and then you had 24 hours off and started days again. That's what was so physically hard. The schedules they have now—and I think a lot of mills are adopting this—is, they run 12-hour shifts, and—since I haven't worked it, I'm not real sure how it works, but you end up with two four or five-day weekends a month. The days are kind of long, but you get the days off. You get the time to recoup. And there was just no time to recoup. I don't know how these people did it for 40 years; I really don't. It's just physically very tiring.

A lot of them bid on jobs that weren't rotating, just so they could get out of that.

There weren't that many jobs that didn't rotate; the majority of the jobs did rotate. And talking about the protection and the hearing protection: when I started it was optional. Then they wanted you to wear hearing protection. And if you refused, you would sign a waiver. They had just started that probably several years after I started, that you signed a waiver. That way, you couldn't go back in 30 years and claim loss of hearing. They had had some settlements based on that.

You know, it was—I don't want to say ignorance—just not having the knowledge of what this long-term noise level does to a person. Toward the end, hearing protection was not an option; it was mandatory, and they would check. Safety shoes were always mandatory. Eye protection was not, and that's another thing that had changed. When you were doing certain jobs, you needed to have eye protection. They had goggles and equipment, at least, that you could use if you needed to. That's probably one of the biggest changes, is the safety. Not that I don't think people were not safe; I just think the more you know, the more you can apply. And I think OSHA stepped in, and you realize that sometimes it's just not worth it; you know, that a person's safety is really a priority. I think that was probably one of the biggest changes that I've seen since I've been here.

When you went up to planning, did you miss the old crew at all?

Missed the camaraderie. Oh yes. And I never minded the physical work; that was never really an issue. The work itself was great. Seriously, I'd probably still be doing it if it weren't for the hours. And again, the camaraderie. You became very close with the people you worked with, and especially after I'd been there a while and was on a permanent shift. You would work one helper, one winder, and one stack operator. The winderman and stack operator really formed some long-lasting relationships. The people in the lab rotated on a different schedule. They didn't want—this is what I heard—they didn't want the people in the lab becoming too close to any one shift on the machines or the supers; not show any favoritism in letting things get by.

But we would have these marvelous potlucks. Oh my gosh, marvelous potlucks. You'd all make sure everybody had a chance to eat and, you know, you'd watch a person's job while he was cooking. And I worked with a gentleman by the name of Tony Cristofaro—Italian, obviously—oh my God, that man could cook. He would make these wonderful things. They were probably three or four times a year, huge potlucks. We would feed everybody.

How many?

Oh my gosh -- 20 in the supers; at least that many on the machines and the lab people.

Was it just the people that shared your shift that you became buddies with?

Oh no, no, because if somebody would leave on a shift and the next person to move up was on a different shift, everybody would rotate or move, depending on where you were in the pecking order. But especially working on extra board you became fairly close to a lot of people and obviously had your favorites. So you knew everybody, and I think working under those conditions, you became close to the shift you were working with. You could be on that shift for an extended period of time if nobody retired or got fired or got ill for any length of time.

So you worked there for 14 years before it changed to Simpson?

I was on maternity leave with my youngest daughter—she was born in 1990— when Simpson bought.

Do you remember the Englishman?

He's the one who purchased from Crown Zellerbach. Then it was James River. I remember all my dates by the ages my children were... I'm trying to think. I don't remember the year.

I think a lot of people retired early in '85 or so, just before the big change.

It might have been right about there; I think there were quite a few people that did retire. But it was Crown Zellerbach, James River, Simpson. That's probably '85, '86, '87, because we weren't James River too long before Simpson bought us out.

Do you know Henry Herwig?

Henry was in the coating plant. He might have been up in technical. But he was in charge of the coating plant, for most of the time I remember.

Can you tell me what you mean by the people in the labs?

On shift, they had a lab, right, off the paper machines, that would test the paper run, test profile tests, check various strength—I don't remember all the tests—but there were three people in the lab. Every reel you'd tear a sample, and they had an air tube, you know, like Costco has. You'd stick them in the tube, and they'd go to the lab, and they'd run their tests, and if there was an issue, they'd come out. You'd go in a couple times a shift, or the stack operators would send you in: "I need to see this, this result right away." And they had three people in there, and they'd worked together for a long time, most of these people.

So, I'm thinking there was a strike in the '70s that you would have seen.

It was an eerie, eerie feeling. I was in the shift that closed the plant down. It was just so surreal, because you come off shift and you shut everything down, and I don't think we knew until sometime during the night that it was going to happen. I think they were hoping that it wouldn't. You walk out, and it's just quiet. You walk up the hill and get in your car, and the pickets are out. I stayed home. I enjoyed it a great deal. I shouldn't say that, because they're very destructive, strikes; horribly destructive, and nobody wins. Nobody ever wins. And honestly, I don't really even know what it was about. I can't remember the specifics.

How long did it last, do you remember?

Seven, eight, nine months, yeah. I was married at the time. He worked out of town, so I was actually alone quite a bit, home alone most of the time. We'd just bought a little house. I stayed home and played house; it was actually kind of nice. You know, I'd not done that before, but they tell me you should always keep busy. And in '85?—'84 to '85 there was another strike. I had just been on four months in the office, I was salaried at that point, no longer union. Nobody knew how to run several of the pieces of equipment out there but me. Nobody knew, and so I was in charge of training a lot of people. And I'll never forget a gentleman by the name of Bud Atwood. I got very close to Bud; I hadn't worked with him before. He was a foreman. Nice man. But I had run the Wärtsilä winder, which is off the supercalenders. And they'd put in a new winder off of nine, which was similar; not the same, but similar. They had an old junk reel, practicing and playing with it. I was getting the feel of it, you know; it was coming along okay, and he said "Oh, what's this knob?" And he reached over and turned the unwind off, which is kind of the tension. The reel absolutely exploded; it was like confetti. Absolutely exploded. He looked at me and said "Well, I've got things to do," and took off. I remember that; that was funny. He left me alone after that, hands off.

So this was when you were in the office, but you got out to help train people?

During strikes, they would have the salaried personnel come in and run the machines. And at that point it was Crown Zellerbach. We had people from St. Francisville and Bogalusa, Louisiana, San Francisco; there's a mill in L.A., several in Washington. But all the salaried people would come in and work strike duty. We did that for nine months, and that was grueling. That was ten 12-hour shifts with five days off, and that was grueling.

You had two long strikes then?

Two very long strikes, yeah. They're so destructive. People don't forget, which is unfortunate. They don't forget those.

People have talked about when they got it so that they didn't have to work 16 hours straight.

Oh yeah, there's definitely a lot of validity on both sides; that's what's so unfortunate on the strikes. There's no reason that people can't come to terms.

Roy said he was in the basement of the old mill, where you can hear water trickling, and there were nutria down there, and all the lights went out.

Oh yes. I never went down in the basements. It was really a horror movie dungeon, horrible, all the stories. I still don't go down there without no need for it.

What kind of stories?

Oh, just dumb things people would find down there, like the nutria; they're rodents. I don't like rodents. Funniest thing that ever happened: they used to have what's called a clock alley, where you go over the locks, the bridge. There are swinging doors that we'd keep open in the summer or spring, and then the clock alley was just a big U. I was in the office at the time and I'd just headed out the door, and here comes this thing, walked across the locks, through clock alley and down the dock. It was just hideous; it was an overgrown rat.

It crossed on the bridge?

Yeah. It was awful

You know, there was one story about a muskrat in somebody's locker.

Oh yeah. There was a supervisor on the paper machines. Everybody disliked this gentleman; one person in particular really disliked him. I was in the supers at the time, and I see this gentleman doing something. He had killed a nutria and stuck it in front of the supervisor's door. Big animal; they didn't like it very much

That sounds threatening. That's not just horseplay—

Well, it wasn't meant to be horseplay. I've got to tell you this story. I have never forgotten this. When I first started, there were no facilities for the women downstairs, no bathrooms. If you wanted to go to the bathroom, you had to go all, all the way, halfway across, you know, up the stairs, over into the women's locker room. I was getting off graveyard, and I was going to clean up. I was walking up the stairs, and I believe it was one of the Kunz brothers, and he was washing down the walls. He looked at me and he said, "You know, I am so glad they have women working here." Then he looked right at me, and he said, "They don't spit on the walls." And that just cracked me up: they don't spit on the walls!

You're a civilizing influence...

You know, I think in a way we were a civilizing influence. I think it did tone down a lot of behavior. A lot of the gentlemen had daughters our age, sisters our age, and I'd heard that, again, two or three years into this. I think we did. I will never forget—a lot of "Playboys", a lot of "Hustlers", a lot of magazines, and you know, they'd stick these magazines in your locker; I mean really raunchy magazines. There were probably eight, 10 of us in the mill at the time. Some of it was really offensive; some of it you just let go because you didn't want to be offended.

You have to pick your fights.

You do have to pick your fights, and if it was just bare breasts—whatever, it was fine. One of the gals was drawing sweaters on the bare-chested women with black felt pen. It got to be really out of control, really inappropriate, even in any circumstances. One of the gals whose name was Susie brought in a "Playgirl"; they'd just come up with "Playgirl". Oh, my God! You would have thought the sky had fallen. Oh, these gentlemen were absolutely appalled and offended. From that point on, the pictures were down. And I thought, how clever of her to do that. But the mill manager wouldn't pay any attention. I mean, he tried, but his heart wasn't in it. You know, boys will be boys. But I think, yeah, part of it you let go, and you pick your fights. When you were embarrassed to go into a room, that's going too far. We all had a good chuckle over that. But that's what it took: one of the gals had to bring in pictures of naked men for them to realize things were far enough along as far as equal rights that they couldn't have one without the other. So they were really forced into at least taking the magazine underground.

It sounds like a harassment suit--

Again, you pick your fights. And I do think the women of my generation, you just tolerated it more because that's what you did: you just tolerated, right or wrong. But I do think we set a standard.

Wherever you went in society, it would be the same.

This was across the country. This wasn't, I think, anything specific; maybe a little exaggerated because it was a very male-dominated industry. There were no women on the floor, none. When I got hired, I think there were five that had been hired. Several of them didn't last long. There were two that had stayed, and then I was hired. And then there were several in between. The following February and March they hired more, and several of them stayed. But it's hard work. It's hard work for men; it was hard work for women. And there were things still, physically, things that I couldn't do, and I was concerned about that. There wasn't a lot that I couldn't do, but you made up for it; if somebody helped you, you helped them twice as much. So people weren't offended at helping me, or having to help you; people wanted to help.

You didn't want to be beholden. I didn't want to give them an excuse not to keep me because I couldn't do the job. That's how a lot of us, especially the smaller ones at the time, I can't do this; I need some help. And next time they needed help, or even if they didn't need help, you were right there. So I think that changed some of the tone in the department, and the supercalenders particularly, where I had some issues with just physical lifting. You didn't just stay in your own little area. There's a lot of testosterone. We had several of the older gentlemen, stack operators; they were in charge of the winderman and the helper. And a lot of them said to several of the other gals they'd rather work with us, because we didn't argue, we didn't fight back; we just did our job, did what was asked of us. And I think that was kind of refreshing. I think there was a lot of testosterone going on.

The women had a better attitude?

I think. I think maybe better, more accepting, maybe.

Roy said, "One day they sent me this girl down to work for me..." Do you remember that day?

I do. I do remember that. Roy was a nice man, and he never treated me with anything other than respect and kindness, and his daughter-in-law I graduated from high school with, so there was a small connection at that time. I'd never realized that he really didn't want me on his shift, but I do remember when he came up to me and he said, "You know, I didn't want you; I didn't think you could do the job." And he said, "You proved me wrong." And that was such a neat thing for him to say. I think that really shows what a secure and wonderful man he is. He was very good to work for, such a gentleman. And that's one of those things you don't forget, that somebody would say that. He is a nice, just a genuinely nice man, yeah. I do remember that.

How long did it take him to tell you that?

I don't know how long. Probably three or four months, two or three months. That was very nice. And that was a nice time. But Roy ran a very good shift, he really did. I think his people liked him. It was a good environment to work in.

Do you want to describe what you learned to do in Planning? When you got that job, there were people you didn't know?

...half a dozen, and I was familiar with them. Running the winder, I was familiar with the people in there, because they put together the programs, the machine programs. Things have changed so drastically. We had barges at the time, and I actually didn't do that. What I would do was, when they'd get the machine runs ready, or programs ready, they would have a roll list. It was all manually typed, where it was going. And you would have lot numbers, which was a load, and some you kept in the mill, and some you put on the barge. So I kind of distinguished what was going to stay at the mill and what was going to go, and ran the runs around, passed them out to the machines, every station.

In the morning, you'd go out and collect. The sheets went to the header, and the header would fill them in manually, roll number by roll number. You would collect those, and replace them if, you know, you needed 20 rolls and they'd only made 15, you had to make another lot for five rolls. You'd add them up—and they did have a computer attached, I think to the Crown Zellerbach name company—and then enter them into the Crown Zellerbach system. I think we did that. I think they had just thrown a computer in the office. Everything was very manual. And they had a transportation department that was in Portland, and you would tell the transportation department what you needed as far as trucks and things like that. And they would schedule the trucks in, and—

So was that after they made the change from barging to trucking?

No, that was Crown Zellerbach. That was what they did when Crown was there; they still barged a lot of paper. And they had the terminals at Front Avenue.

What's there now?

Western Transportation; I think it's still there. I don't know if that's what it's called any more. But it's way down Front Avenue.

Do you remember when they decided to quit barging and go just to trucks?

That was when they shut the mill down last time. We barged with Crown, JR and Simpson. And Simpson contracted with Western Transportation. Before, they were all part of the same company. The Traffic Manager was Simpson, who I worked for, in various stages as the jobs evolved. They kept him on. Katie was in first grade; that would be—is that '96?—'95 or '96 I think when they shut the mill down. And the Traffic Manager stayed, and I left in October, and he stayed to finish things up. When Ron Stern bought the mill then, it wasn't cost-effective to barge, the cost of paying somebody to do that. And at that point they quit barging. It was just not cost-effective.

I heard that it was because of high longshoreman labor costs.

That's part of it, but at that time too the real estate property on Front Avenue—I mean, you can see all the condos on Front Avenue now; they're inching their way down. The property values and the rent became much, much higher, and that had just started when Simpson sold it.

So Western had to charge more also just because of overhead, not just labor?

Oh no, overhead too. You bet. Western Transportation may have owned their buildings, but the warehouse we were in was not owned by Western; it was leased.

Do you remember what they paid you when you first started there?

You know, I want to say four dollars an hour, $4.35. It was a good wage. For a woman particularly it was a good wage. Oh yeah, great wage.

Western Transportation, a subsidiary of Crown Zellerbach, at the lower mouth of Willamette Falls Navigation Canal with a barge load for the mill.

And you had to go through the new rigorous rehiring process in '97?

I actually did not. I was one of the few that did not. Nobody else knew how to do my job; I think they were kind of stuck. Particularly in the planning area and the traffic area, which had evolved into more of a traffic area, billing and those kinds of things, there weren't a lot of people who did what I did. There were actually two of us, three of us, and then they didn't hire the one gal back.

Did you ever not consider going back?

No, no, I didn't. I didn't.

The die was cast.

The die was cast, yeah. No, it was an exciting time to start up. It was wonderful. You put in a lot of hours, but you were doing something valuable. I think it's really important to be able to feel like they're being valued. Oh, we put in some long hours, but it was great. It's exciting times.

That's the way Henry talks about it, too.

Oh, it was exciting! I think Henry stayed…

He retired in '98.

Yeah, but as far as the startup goes, they kept probably eight or 10; they needed them for cleaning up and—

Oh, just sort of bodies on site when the mill was closed?

Yeah. I think they shut the mill down around the first of September, and I left toward the end of October. As your jobs and things got finished, you know, whatever projects, there was a lot of cleanup to do. And they still needed people there to run some things.

I remember the lay-off was staggered. They came out in batches.

Yeah, it was hard. By the time I left, I was glad it was over. It was such a stress at the time. It was just the roller coaster: somebody's going to buy us, you know, somebody's going to buy us. I was relieved when the decision was made, as horrible as that sounds. I was relieved that somebody had made a decision. Emotionally, it was horrible. I think I vegged for about a month; it was just physically exhausting. And I've kept in touch with a lot of the people, and the IS guy, he and I were good friends, and he'd keep me posted. So I was going to wait until January, take a couple of classes, bone up on my math and maybe take an English class, and just general classes. And he called and said, "Don't look." He said, "I think we've got it. I think we've got it." That was around the first of January, end of December. And then Ron started to purchase the mill in February, I think. Or it was the letter of intent or whatever it was. So I went back to work in February, first of February.

So you had the holidays.

I did have the holidays. It was wonderful. I really miss that. You know, working full-time, you miss that leisurely pace, the ability to have the time. I had great fun. Both the kids were in grade school. And one day a week I was in the morning in one class and the afternoon in the other; it was great. It was so much fun. I enjoyed that.

And you could afford it because—did the union help, or unemployment…?

You know, I never took unemployment, and I guess I could have. But my ex-husband at the time was working, and we did okay. Come January, I was going to have to—you know, I'd probably have until February or March before it might have become an issue. But it was nice.

One time the guys talk about that made them want to retire early was when the efficiency experts came. Do you remember that?

Oh, yes!

What caused that?

I think it was just corporate trying to find ways of cutting money, trying to save money. I mean, bless their hearts, but until you work at a job, you don't…we went through several of those.

Multiple rounds.

Multiple rounds. I think with every new regime, they always have this bright idea. I mean, history repeats itself. I think that's regardless of where you are. But yeah, we had several of those. Call it what you like, it's basically the same thing. I think that's one thing that has improved over the years: I do not think the employees are as suspicious as they once were. I think they were justified in being very suspicious. I think the way it was handled was pretty dehumanizing. You have somebody with a college education and a white hat and a clipboard, come in and try to figure out what you're doing. You spend two or three hours with that person, and you can't make a judgment. It was unfortunate. I remember the last time they did that I was in the office, don't get me wrong, they were very nice people. I mean definitely nerds; nice people. Well, you could cut here, you could cut there, and you could do this, and you need to do that. I'm looking at these people going, "Do you have any idea how long the three of us spend on the phone every day?" Oh—no. They hadn't checked out phone calls. I think the Traffic Manager was on the phone three hours a day. I was on the phone anywhere from two to three hours. And the other gal's on two to three hours a day. Well, they were going to cut a person—we were on the phone a total of like eight or nine hours a day between the three of us. I do remember that. But think of the management at that time. It was a real hard, heavy-handed management style. I do remember that.

There seems to have been quite a gap between hourly and the managers as far as the clubs and parties.

Being a mother of a small child, I really don't know all of the perks. There were some benefits, definitely. I think animosity was promoted from both sides, though. There were several of the union leaders that were so hardcore, and I think one begat the other. I don't think there was any chance of any kind of reconciliation anywhere, on both sides. It was not just a one-way street. But yeah, at the time—and I think you find this nonexistent in corporate situations, or minimized—but, the managers and the supervisors had their own shower and had their own restrooms, and—there were parking spots; it became a real status issue.

And, presumably, they didn't spit on the walls?

The managers? Well, you don't know that. They were men, most of them. They might have. But you look back, and it's really kind of petty.

I wonder if any of them ever worked 26 hours straight, like Del Herndon.

Some of the managers did, you bet. If they had issues—if they were in charge of that machine and they had issues, you bet they did. You didn't go home. That was a little bit later, probably when I got in the office, that the expectation of the managers…that you be there. You need to be there.

I know the managers for years worked up through the ranks.

A lot of them did. I think where I really noticed a lot of animosity is when they would bring in these young college-educated kids. And it makes me chuckle. My dad was a salesman for the phone company. And he didn't even have a high school education, but he worked his way up. I think he lied about the high school education. He had a lot of disdain for these young new salesmen who had a college education, who thought they knew maybe a little bit more because they went to college. The good ones in the mill took the abuse and learned from it, but there were those few who never really got the fact that these guys would make or break you.

Just because you worked in a mill didn't mean you weren't intelligent. Highly intelligent men worked down there; highly intelligent men. Very good at what they did, very good people. The longer I was there, I think the attitude changed. But you know, you don't treat people in a certain manner and expect to be treated with respect. There were some of the foremen and some of the managers who did have a lot of respect. Those were the times, and you didn't have to like the person. And I don't think a manager—a good manager—expects that. But you needed to respect them. And if you lost respect, it's miserable.

If you're working in an unrespected situation.

Oh yeah. And one thing too—we had a foreman for the supercalenders and a foreman for the paper machines, and a lot of times those two were in an adversarial position, because it was the supercalender's fault that you were ruining the paper machine's paper, and the paper machine's paper was bad, and it wasn't your fault. Oh, highly adversarial. Oh, I can remember two foremen nose to nose, screaming at each other. Those were both foremen manager positions. But that's changed. I mean, I think the paper machines are now more of a fluid line.

So you had short hair when you were in that area.

I didn't right to begin with; I had longer hair.

Roy was talking about hair in the nips.

When I got there it was mandatory. Your hair was up. At that point I don't know if somebody had gotten their hair caught or not. That was mandatory; no loose clothing. Everything had to be closer to your body, and no long hair. Your hair had to be contained. Several years jewelry was banned if you were working. No wedding rings; you had to take your wedding rings off.

Roy was talking about a biker guy with an earring that you worked with?

Yes, yes, I remember Terry well. It wasn't long and dangly. He could keep his earring. It was just a post. Terry was wonderful, but scary looking, scary guy.

But nice, calm, a lot of tattoos...we worked together quite a while. I don't think he came back after the next shutdown. But he was probably pretty close to retirement age; he might have retired before then, but I don't think so.

So did you get to know the people on the barges?

A little bit—when I was in the office, they would load the barges until 11 o'clock, 10:30, and then you'd cut everything off, you know, no more paper. You'd reconcile what was on the barge with your paper work. I got to know the tug crews because they would be waiting for me, I needed to work with them in that respect. You know, we're going to be a half-hour later; we're going to—so I knew them a little bit, not much. I didn't work with them closely, but I knew several of them.

Why did you have to be done by 11?

Because they had to get down the river.

They had to be somewhere.

They had to be somewhere. They had to get down the river, and then they had to come back and spot the pulp—the barge with the pulp on it.

Do you remember what hours the locks were open then?

They might have had somebody there 'til midnight. I know in the summer they did.

...

Oh yeah, you get a lot of pleasure craft going up and down.

'Til midnight?

I think so. I'd be working on the paper machines, and they had big doors that opened right close to the locks, and you'd stick your head out now and again to get some fresh air. A huge yacht—oh, my gosh, you'd just be going "What am I doing here?" But that would be eight, nine o'clock at night.

It's fascinating. The locks are fascinating. They're just amazing.

So did you have any high water stories?

Yeah, the last one I was with Simpson... I don't remember any big floods. There were some scares, but nothing like what we had. We were living here at the time. I had to go in. The day before it peaked, I was going to go the freeway, because this [road] had flooded, and there was water across the road down there, and I stayed home. But I came in the next day because the water receded. They had a dike break up in Molalla, and if you know where the bridge is going to the mill, the water was up to the bottom of the bridge. It was incredible. I don't remember if water got on the dock or not. I think in '64 it had. But it was amazing. The lower parking lot was flooded. Most of it was.

Now, this is—are you talking about '96?

Ninety-six, yeah.

Okay. you said it was Simpson then.

It was Simpson. We had two right in a row within two years. Then maybe they shut us down the following fall…The water was just incredible. It makes you feel mortal.

You didn't have to go down and shut the waterproof doors.

No, I didn't. They had other people do that—at that point, though, they'd gotten pretty good. There were enough people there who'd done it often enough that it was fairly efficient. Pulled a lot of the motors, and by that time, too, it wasn't as intense since we only had three machines, as opposed to nine or ten at the time, in earlier years. But I know there was some concern that they may not start up. I do remember—I don't know how long we were down, how long the machines were down. I remember for at least a week they brought a lot of the more senior gentlemen back, and they were cutting up wood and the logjams and things that had accumulated, debris—I know they wouldn't allow them to keep any of the wood of any of the trees because it was contaminated. So they were throwing it back in the river. Some of the guys wanted to take it home for firewood, but it was contaminated; they wouldn't let them.

What was it contaminated by?

Sewage, maybe? I remember I had a little battery-operated Coleman lantern, and I brought that in and sat at my desk, because in the office there was only two, maybe three small windows up high. I was sitting at the opposite end at the wall at the time. And I brought my little lantern and would work.

So your office -- has it been called planning forever?

Yeah, since Bill Gray took over, we also have customer service. So it's traffic, planning and customer service. And before then it was just planning and traffic, basically.

And it's on the island?

Yes. The mill's from here to the end of the house away, or less, yeah.

Okay, and it's upstairs, or …

No, no, we're downstairs. We're right—actually, right across the locks and over the bridge, that's the door to our office. Right by security.

Okay. What has made this career worthwhile?

The thing that I think I remember most about working in the mill—not so much in the office, but on the floor—was the camaraderie. It's hard to explain. You were a part of a group, and most of the time quite close. I think shift work tends to make people close. You're either going to become very, very good friends, or you're definitely not. And those relationships are what really pull you through, you know, four o'clock in the morning, and you were up all day. That kind of support—you really worked together as a team. And the little jokes you'd play on each other, and that's what I remember. That's what I enjoy, and that's what got me through.

So is using the word "family" too strong?

No, no, it's really not. Really you are—it's an extension of your family. Shift work is different than any other kind of work, and I think particularly the rotation that we were doing at the time—at least for me—it isolated me from my friends and most of my family. You had two weekends off every three or four months. You couldn't go out Friday night; and I was young—I was 20, 21, 22. You couldn't go out and be out late on a Friday or Saturday night because you had to get up at six in the morning. You couldn't stay out too late. You couldn't have a drink because you had to go to work at midnight.

And so I was pretty isolated. My ex-husband at the time worked out of town. The first four or five years we were married, he worked over on Black Butte Ranch. So it was 'two cats and alone' most of the time. So yeah, they were really part of my family. I think a lot of people felt that way. With the shift work you're fairly isolated.

Isn't it interesting that it can feel like that when you're with them, and you never see them outside of work?

I didn't socialize much outside of work; a little bit, but not much. Part of the reason I was uncomfortable—some things were said but not many. I think a lot of the wives didn't appreciate women working down there, and especially, you know being young. That was made clear at times. Some of the wives were fine; some of them weren't. Nothing overt was said, but it was known that they really didn't want us around, which is interesting. So I didn't socialize much—the last day of swing shift, every three weeks, we would go out and have a drink after work. That was really probably as much as I socialized with a lot of them.

Did you go to the West Linn Inn?

No, I never did. That wasn't really open that much at night. Proms and things like that. I didn't go to the Inn. I regret never being inside it; I hear it was a wonderful building. I regretted when the city let them tear that down; that was horrible. It had the big veranda porch looking out over the water. And you'd be going down to work, and some of the gentlemen who'd get there early would be sitting out there, just enjoying.

What was the worst thing about working for the mill over the years?

I don't know if I can answer that. I don't look back on that period as anything being horrible.

It sounds like the sleeping was the worst.

Sleeping was tough. The shift work was hard, but—I think you have a tendency to forget some of the more unpleasant situations. There were isolated incidents with people, but in general, everybody was really decent.

It's fascinating, though—another gal I think still works at the mill, her and I were at about the same level, and we were in a position of moderate authority, very moderate. We would have these young helpers that were men, and really had more issues with the young men than the older men. And that was very frustrating to me. When I was running winder and dealing with those issues, that was really frustrating. You were constantly being questioned, constantly being challenged, by the younger, 20, 21-year-old men. Interesting. I don't know, it was just—"Do what I tell you to do. Shut up." You didn't say it, but you wanted to. And I did, with a kid, I just said, "Because I said so." I'd had enough. "I said so; do it." I don't have any huge negative memories.

Presentation pinning, five-year pins? Did you get in on that?

I never went to one of the pin dinners. I never went. But I think after 10 years they eliminated it.

Why didn't you go?

We lived in Woodburn, and it was too far to drive, and I just, never had the energy to go. I have my pin. And there's a bowl over there; that's my 10-year bowl. Yeah, you had several things you could pick.

So did you have any close calls, you know, see anybody get hurt, fires?

Um, one fire.

What happened?

Well, I was on the paper machines, so we were isolated a little bit. You know, I don't even remember what started it; it wasn't serious. I saw several close calls. It was a long time ago, but frightening. Supercalenders are tall, and the paper threads through them, and then on the bottom it's wound up. And if it breaks, the helper and the winderman, everybody, would go in there and put their rump up against it to stop the reel from spinning.

And—off the story—one of the foremen had a pair of polyester pants on one time, and it burned him. But the coating would come off, and you'd have these little white [indicates the back of her pants]—and you'd walk around. But you'd put your rump up against them and kind of go back and forth, because if you stayed in one spot it would get really hot, because it was going, you know, 1,800 feet a minute. If you didn't, the paper would unwind as it was spinning, and it would flop. One guy tried to stop it by himself; nobody was close enough. And he got sucked under. He was really lucky; he didn't try and raise his head. Because—I mean the force is—as it would loosen, it would kind of egg-shape flop. Does that make any sense? And he was stunned. He was off for a little while. He was stunned. That was frightening, because you couldn't see him. After a while you knew he was under there. You couldn't get it to stop; once it started to loosen up, you just allowed it to stop on its own. That was frightening. I saw another kid—I hadn't been on the supers long—the winder was running, and I don't know how fast it would go; 12-1,500 feet a minute. And he clipped the edge of the winder with the transporter. I don't exactly remember what happened, but it exploded -- I mean the sheet -- and it flipped him. He got caught somehow, and it flipped him. He was lucky. But these are really the only two serious accidents.

They had a lot of horseplay. But apparently they knew when to do it and when to work?
That was always kind of fun, picking on the crew you worked with, yeah.

When the fire happened, did they come and put it out, or is there just an extinguisher?
Honestly, I don't remember. I know it was more smoke than fire. It smelled really bad. It was obviously a dangerous situation, but it was never a critical situation.

So how many different kinds of papers do you remember?
I didn't work much down at the other end. There were a lot. Down at Mill C they'd make—I remember canary yellow, and there was pink and blue and—vibrant pink—blues, greens, and they made some newsprint. They made—called it Avon paper—I think it was that Avon tissue paper. Made it on little tiny rolls. It was called Avon paper. They also made, uh, Crezon. And that was on No. 7. And they'd make like two reels a night or something. But you know, they used that for backing on siding.

Somebody who restores historic homes says they used to nail it on the walls because you could paper over it …
It seems to me, and my dad was aware of the product, it was some backing. They used it. But it was heavy. It was cardboard almost heavy.

Are you the only one in your family that worked at that mill? Your son? Daughter?
No, no desire either. It's good summer money, but no, I'm fine if they don't.

Roy said that he actively discouraged his boys from working there.

It wasn't just the managers' kids; it was all employees' children could apply. And I don't think there's any preferential treatment, but—

So there are whole clans of related people.

Fascinating. Oh yeah, families, brothers, three or four brothers. That's changed some. But I think the whole nature of industry has changed. You will more than likely never work at the place where your grandfather worked and your father worked. That doesn't happen any more. There were sometimes three generations working. At the same time. That's not going to happen any more.

Ed said if his relatives stayed home, they'd have to shut the mill down.

The Kunz. A lot of Kunz worked there. Everybody was a cousin, yeah. A lot of families. It was, I think, a good living. It was a good wage. And the working conditions were not the best; part of that is just the nature of industry...

But if you were in the shipyards or any manufacturing—

It's the same. A lot of those people get close. It was a real community situation.

Is there anything else that you have in your heart about Crown Zellerbach?

Oh my gosh, I don't know. I'll remember them when you leave. It was so long ago.

No regrets, though?

No. If I had to do it all over again, I probably wouldn't stay; I probably wouldn't. A lot of uncertainty. You know, bless the mill's heart, I think 20 out of the 30 years I've been there, they were going to shut it down every year. So I probably wouldn't have stayed had I known; worked a couple of years and done something else. But then you look at industry: what else would you have done? I don't know.

Do you know somebody named Mary Kenney?

That's me.

That was your maiden name?

Kenney, K-e-n-n-e-y.

Somebody remembered your maiden name.

Kenney is my maiden name; isn't that funny? And there's a Margaret Horner. And she was just a doll. She, I think, had been there since World War II. I have no idea if she's even still alive. Does Horner sound familiar?

Jim Bonn

Interview June 13, 2006

I have a picture here of the supercalender department. [under James River]

This is a great picture. Is this the room that you worked in?

Mm-hmm, that's where I worked. Right.- and this is a supercalender, and this is a winder that follows it. This is a winder, and then these two are winders also. And there's a big machine called a supercalender in line with each one of these machines. So this makes three supercalenders. This is what they called the salvage winder. When you had a roll of paper that had a defect in it, they would put it to this winder and cut it to a different size and cut the defect out. The reel ran through the stack, and—I don't know if you can notice the gloss or the shine—anyway, that's what the supercalender did; it put the high gloss on the sheet of paper. And then the winder, it was a machine that took these reels of paper and made them into rolls of paper for the customer. I was basically rank and file. I come in when I was about 19 years old, and I started working. And you worked the extra board, and you did this and that. And I gradually worked up to a shift foreman. Then from a shift foreman I went on to be assistant superintendent in the supercalenders.

My boss gave me that when I retired. I have a clock. Here's another picture you might want. This is my father, myself and two sons. This in the background is No. 9 paper machine, which—I don't know what number they call it at this time, but I think it's different since the Canadian people come in.

They renumbered?

Correct.

So these are your two sons?

Correct.

Do they still work there?

Oh, no. One of them runs the computer programs at Portland State University. Another one works for the City of Oregon City in the parks department. I probably am not going to be a very good interview. I hope I can answer your questions, but I'm not the most eloquent person in the world.

Three generations of Bonn males, posing in the West Linn mill.

One thing that I thought might be challenging—almost everybody suffered hearing loss.

Oh, I have that.

If you could tell us your name and your age.

I'm Jim Bonn, and I will be 74 years old on the 12th of September of this year. I was hired at Crown Zellerbach in September of 1951. That's a long time.

How long did you work there?

A little over 40 years.

Could you tell us when you retired?

In July of 1990 was my last day to work. I did have accumulated time that took me up for another eight months before I actually drew my first Social Security and retirement check. Yeah, 58 years old.

Fifty-eight. That's early. Why did you apply?

I applied because I needed a job, and I believe my first job I worked in what they called the beater/mixer room, and I was running—or trying to run—a beater, I guess you'd call it. You're probably not familiar with the word, but it was a large machine that took rejected paper that wasn't usable, and you run it through the beater, and it ground it up into a pulp, and then it was reused in the system.

It had some chemicals that went in there with it to help break it up?

Well, not really. There already were chemicals in the sheet, and at times they would place dye in it to change the color or whiten it up, or whatever it needed.

How did they train you for that?

Well, a fellow would come in and say, "There it is. Watch this guy." And that's what you do. There wasn't a whole lot of training.

Did they have a hierarchy, where you started as a helper?

Down at the mill they had what they called an extra board. People that were new, they weren't permanent jobs that they were placed in. They were placed in jobs where somebody may be on vacation or off sick, whatever. You were just a fill-in. So you may work two or three days at that position and then move on to another extra board position where somebody was absent or whatever. You would work in this area until you accumulated a little bit of seniority or there was an opening in a department and you could apply. They would interview you, and if you already had previous experience there, why, that always helped you establish yourself in that department.

So your dad worked there?

Oh, yeah.

Did your grandfather or—

No, no. Just my father. My father, he was basically a logger. He worked all through the Depression, and he worked in logging camps. And at the start of World War II, why, there were more lucrative paying jobs at that time. And he went into the shipyards in Portland. He worked there, and then about 1945, why, he was looking for a permanent place to work, and he applied at the mill. He went to work in the ground wood department. From there he moved on into the sulfite department, and he was the cook in the sulfite department up until a point in time when that particular department was closed down because of contamination they were putting into the river.

That's really hard, working in the grinding room—not very healthy?

Well, I don't know. He lived to be 88, but he was a very strong person, and he worked in the woods, you know. He was used to hard work. Of course, he was born and raised on a farm, so I guess back in that era, why, hard work was the name of the game, and a lot of people did that.

So he was probably a 'cook' when you applied for the extra board job?

Yes, he was. From there, why, I worked also ground wood, or what they called chippers, or paper machines and different areas, and finally in the supercalenders, where I wound up getting a finger hold and was able to work in there for the next 35 years, or 38 years, or whatever it would be.

What was the first job you bid on after you had enough time on the extra board?

Well, basically the supercalenders was the first job where I had an opportunity to get in a full-time position. I was a roll buck. That was the very bottom of the ladder. Initially, when I started in to the supercalenders, they had two paper machines capable of running coated paper. But they did not have enough orders to keep both paper machines on coated paper full-time, so one paper machine was on coated paper all the time, and No. 5 paper machine was on the coated paper when orders would allow it to go to coated paper.

When I started there they just had one paper machine running coated paper full-time in the supercalenders. At a later date—it was probably three years or so—they established enough orders that they could put both machines on coated paper full-time, and then that instantly gave me a nice jack-up in the seniority ladder. I went from a roll buck to the senior supercalender helper, which in that day and age was a little bit of a raise in pay.

Do you remember what you got paid when you started to work there on the extra board?

Oh, gosh. I remember the first year that I worked there that I made about $4,500 for 12 months—so as far as my hourly rate, I don't recall it.

A lot of overtime in there?

Anytime I could get it, yeah.

I know a lot of the guys talk about really long shifts. Were you on shift work then?

Yes, I was on shift work for the majority of the time I worked there.

How was that for you?

Well, I guess it's just something you get used to, and I didn't mind it. At that time of my life I liked to fish, and I could go fishing where I didn't have to be surrounded by people on the weekend. I had a lot more freedom to have the river to myself.

Get to the hole first.

Yeah, that's right. Yeah, salmon was always a big thing. There was a lot of salmon fishermen in the mill, and in the spring of the year, why, that's when a lot of the people took their vacations. They would spend it on the river. There were fishermen that, they would anchor their boat, and then there would be somebody in that boat day and night. They would get into a spot that they felt was a very productive area, and they would not leave that spot. There would be boats shuttled back and forth, taking people to work or whatever. This was pretty common.

Does that relate to 'hog lines' like we know today?

Mm-hmm. There would be specific areas, especially late in the season, up near the mill itself, up near the falls, that the fish would get congregated up in those areas. And you had a very good chance of picking up a fish. They would go in there, and they would have a top on the boat, and they would have a heater in the boat, and sleeping bags and whatever it took to get by. And then they would work it in shifts. That's how it was done. I never did do that, but then I wasn't, I guess, that obsessed with catching a salmon.

Rosie Schultze showing off his Meier and Frank Derby prize-winning salmon. Circa 1955.

Did you ever work a river job?

I never worked on the river at all. Paper machines, ground wood and beater/mixers, supercalenders were the areas that I worked in. No, I never did work on the river.

So you didn't work with clay barges or logs?

No, I never did. Henry Herwig, he was the superintendent in the coating plant. And I knew Henry very well. We fished together also.

Did you ever work with Rosie?

When I first met Rosie, I had been taken off of my job as a winder operator, and I was asked if I would go into a relief foreman's position, which was strange to me at the time. I accepted the offer, and that was the first I met and worked with Rosie. He was working straight days, but he would come in and relieve paper machine foremen when they were on vacation or such. And I was lucky enough to work with Rosie through several of those periods. He was wonderful to work with, you know. He was always so willing to help, and maybe more so with a very green or very inexperienced person. He was wonderful to work with. Yeah, I always used to cherish working with him.

Do you know how much of your career was hourly?

About 18 years, and about 22 years on salary.

Whew. So during strikes, were you hourly or supervisory?

On the first strike that the mill had—I guess it was back in the late '60s—I was hourly. And from there on, why—the two succeeding strikes I was on salary.

How did you feel about going from hourly to salary?

Well, monetarily, salary paid better, and that was the basic reason I accepted it. It gives you a few options; a little better retirement and things of that nature.

Did you ever feel it was going to cost you any friendships on the crews, or—

No, I don't think so. I got along with my people very well. I made a lot of friends and hopefully still have them. John Hanthorn was always a very good friend. His brother Ted Hanthorn—Ted, I believe, just retired recently, maybe in the last year or year and a half. He was not old enough to retire when all of the changes started happening down there, and anyway, he just recently retired. Well, you met John—John is quite tall, very slender. He would be about 80 years old, I believe. And Ted, he's even bigger than John; he's about six foot six, six foot seven. Those two fellows were the absolutely the hardest-working people you have ever seen. They were wonderful people.

It sounds like for a long, long time, morale was pretty darn high, and people were proud of the work they did.

I believe people at the mill, they were quite proud of—especially once they were established in their department. A lot of the people, because of the way they could advance, they could bid on a position that maybe they'd had previous experience with, or something that they wanted to do. And they could maybe be a welder, or they could be a plumber, or they could work in maintenance. There would be different things that they could do, and there were some excellent, excellent people down there.

I don't know if you've ever heard of Richard Buse or not. He died about a year ago, and he was a fireman down at the mill. Oh, he was a jewel. He was just really a great guy. As one of his sons told me once, if a rock could talk, he would make friends with it.

You started in '51 and worked your way off the extra board. About how long did that take?

Oh, I suppose—I would think about three years; three years that I worked on the extra board prior to going into the supercalenders. I believe that would be pretty close.

When you got to supercalenders, did you just feel it was going to be your place?

When I started in the supercalenders, the supercalenders were practically brand new. I believe they started in about 1946, but this department, it was basically a new department. Everything was bright and new and shiny and clean. And compared to a lot of places in the mill, which were kind of dark, dark dungeons, you might say, it really stood out. Anyway, I enjoyed it my whole time, really.

You were still hourly in '64, then.

I was working by the hour in the 1964 flood—I did not have a lot of seniority, 12 years or so, which was minimal at that time—and for cleanup, basically, this is what they brought people in to do. In the sub-basements and the main basements, the mill was full of sand and mud. And they brought in Bobcats and different machinery, shovels, and they worked and worked to just basically clean up the place. There were areas of the mill that were a lot more disturbed or destructed by that than others. My department, in the supercalenders, there was about nine feet of water in the basement just below, underneath the supercalenders. Of course, with all this water, all the damage that was done to the electrical systems, the motors—you know, we'd heard rumors that the Crown Zellerbach Corporation was wondering if they would even restart the mill because of all the damage.

But they did.

Yeah, they did. They started it, which was great for a whole lot of people, about 1,500 of them at that time. I believe it was shut down for—oh, six weeks, two months. It was a long period of time that it was down. There was just a tremendous amount of damage. And then, I don't recall what machines started first after this, but they weren't all started up immediately. One would get to the point where they could start it up and it would be going, and eventually another one would get going. It would depend on the amount of damage and what all they were pushing at. But anyway, it took a while to get things going—quite a while.

So the low-seniority guys, which included you, were down there cleaning out the sand?

Anybody that wanted a job at that time—I always wanted a job—if you wanted to work, you listed your name, and you would come in if there was a position for you. And you'd go ahead and you'd grab a shovel or whatever they wished you to do, and you went ahead and did that. But it was very, very, very cold in the mill then. It was extremely cold. Basically, on the upper end of the mill, like in the supercalenders and underneath the paper machines, I never seen any logs; just lots of debris, lots and lots and lots of mud; lots of mud.

On the night that it flooded, were you working?

No, I was not. We knew the river was extremely high, and they had a lot of maintenance people in pulling motors. Prior to the shutdown, maintenance was just basically working 24 hours a day. Those people—electricians, maintenance, anybody in that area—they worked to remove motors to try to salvage them so they wouldn't be ruined. And they got a lot of them. I don't recall what position I was in at that time as far as whether I was on days off, or what.

During the strike of the '60s, do you remember what the issues were?

During the strike of the '60s—it was very short; as I recall, about two weeks—it was basically the International Brotherhood of Paper Workers versus a new AWPPW that wanted to take over. It was just a clash between those two unions that created the strike. There were people that wanted to keep the old International, and then there was the people that wanted the new one. And the new one did win out. There was a lot of hard feeling over that; there were some people that crossed the picket line, and they were not well received for a while. But in most cases I think that it wore off, and they let bygones be bygones. I can remember walking the picket line there, and one of the—what was it back in those days?—anyway, there was television cameras down there, and I can recall them being there and watching the people on the picket line.

Down by the West Linn Inn somewhere?

Right. It was right by the West Linn Inn where we were walking the picket line, and I was thinking it was KPTV—well, it's Channel 12 now, but I can't recall what the numbers were at that time. But anyway, right, it was while the old West Linn mill [Inn] was still standing.

Did you ever go in the West Linn Inn?

Probably 80 percent of the days that I worked, I walked through there going to work. I walked through the old Inn. They had a pretty magnificent area; it had a huge stone fireplace or rock fireplace and beautiful stone floors. It was a real magnificent old building. Yeah, the West Linn Inn was the place to stop. And some people had breakfast, had coffee, or sit down in the main part of the Inn and just sit down and relax and read a paper or whatever. It was pretty magnificent. Thinking back on it, I believe at the time—you seen it so many times you didn't really feel that it was that much different. But looking back on it now, it was something else. A lot of granite—oh, it was a beautiful old building.

It was torn down but I don't really know why.

I imagine that the structure itself probably had some rotting conditions and whatever. But I believe the base of it, or the area that was being used all the time—the restaurant and the areas and such—they were pretty substantial. But the rooms, I believe, were getting in pretty sad shape, and the stairways.

Jerry Herrmann, when he was going to school, he used to work for me. He worked for me in the summer months when he was still a student. My wife and I walked over to the community college quite a bit, and I would run into him over there. People change so much after they get some age on you, and a couple times we were walking, and Jerry Herrmann, he let out a holler at me, and I'd finally figure out who it was, and we'd have a chat. So it was kind of nice.

Was your father the first one in your family to work there?

To my knowledge, my father was the first person in my family that did work in the mill. I never heard him say any bad things about it, so I—but he'd worked so hard all of his life, why, I imagine the jobs that he'd taken in his later years might have been pretty easy for him.

Then did you have any brothers or anybody else that worked there?

My brother Wayne, when he was going to school, he worked there a couple of summers, yes. He worked in the extra board as a fill-in.

I've heard a lot of college kids and high school kids would work summers there.

That's right; a lot of the college kids did that. And it was a good deal for them. They could pay their way through school and have a job and all of these things. It worked out very well for them.

And your sons? Did they work summers for college?

No, my one son, Dennis, he graduated from Clackamas Community College with a degree in horticulture. At that point in time, to get a job that paid anything in horticulture wasn't working very well. So he went down there. He worked there for quite a while until he finally did get a horticulture job. He started working for the City of Oregon City; he's in the Parks Department there. And my youngest son, Eric, why, he worked a couple of years in the supercalenders. He worked there full-time for a couple of years, and then he wanted to complete an education, and went to Oregon State. He's got a bachelor's degree in computer engineering. Anyway, he has a very good job where he's at now. He worked for the State for, oh, six, eight years, and now he heads up the Portland State computer department, and he's been there for a few years. He's married and has two little boys.

So do you think he would ever recommend that his boys take a job down there?

I don't think so; not in this day and age. I don't think Eric would even consider his boys taking a job there. When the owners that are there now—they call it the West Linn Paper Company, I believe—they called me and they asked me if I would come down and help out on the training and such as that. I told them I would for a short period of time—I would do it for three months or 90 days. And then I would not work beyond that. And I did that. Things are tremendously different. The wages, amount of money people earned then versus now; so many job eliminations. I would imagine it's real hard to make a real, real competitive product now. I guess I shouldn't say that, but that's the way it seems to me.

I think they're down to about 250 people, and only running three machines.

I've heard that 250 people—and I know a number of people still working, and of course it's getting smaller all the time too. But well, they've always had good days, and they've always had bad days, no matter what era it is.

I talked with Mary Lefebvre.

Yes, I know Mary. Mary Lefebvre used to work for me. You know, Mary Lefebvre and another young lady by the name of Peggy Word—I don't know if that's her name at this point in time; she may have been divorced. But they both worked for me, and they could work alongside of the very best men I had. They were just outstanding workers. They were going to show anybody, everybody, they could do it. And they could do it; they were excellent. I see Mary Lefebvre once in a while, and she's always extremely friendly, and she's an outstanding young lady. Oh boy, could she work, though. Very frail, very thin—and Peggy Word was probably six feet tall and weighed 120 pounds, but she was a working machine. It's just amazing.

I haven't heard about her yet. Was this the '70s?

Well, I believe Peggy is still there. Yeah. No, those two ladies, they were outstanding; just amazing. No matter what job you put them on, why, you knew it was going to go well. They would do it, and there would be no problem. They were excellent.

And no problem with the men?

Well, 99 percent of the men, they did an adequate job. Then I had a few John Hanthorns and a few Ted Hanthorns that just did outstanding jobs. And most of the guys were pretty good. Sometimes absenteeism was pretty bad—individuals, that is; not everybody. But basically people were pretty darn good.

How did they deal with the women when they first came in?

Well, how did they deal with the women? I suppose there was a few love affairs that came along the line. That's a hard question to answer for me, but I think that—in general I think they dealt with them fine. But I think maybe they were a little bit astounded by what they could do, too—what the gals could do, because they did well.

What about any strikes when you were a supervisor?

The strikes when I was a supervisor, they were extremely long hours and extremely hard work on both of them that I went through—seven months on one and—I forget—about six months on the other. You worked 10 days in a row, 12 hours a day, and then you had five days off. I worked in the supercalender department, and so I run that department, and they would bring in people from the offices in San Francisco, or people with salaried positions in logging camps, or whatever, with no experience at all—never seen the machines in their lives— and it was a real chore. You had to train everybody, and in some cases you'd work your 10 days in a row, and then when you came back after your five days off you might have half of your crew that was already trained decide that they didn't want any part of that, and you never seen them again. So here was another bunch of green people you had to train. It was not a lot of fun; not a lot of fun.

How did you feel when the strike ended?

Yahoo! When the strike ended, it was yahoo! I was just so glad it was over—it hit so hard on everybody, you know. The people, the hourly workers, they depend on that for their living, and it was very difficult for them. And although us that did work, we were making some awfully big paychecks, it really wasn't worth it; not to me, anyway.

Roy said when the strike ended—it was a very awkward moment.

Yes, I remember when the hourly persons came on to relieve the salaried and/or people taking positions, and there were some confrontations. No blood spilled, but lots of words exchanged. So it wasn't a very smooth transaction.

That's too bad, really, because then it takes a while to smooth the feathers down again.

Oh, yeah. There were some people—very few—that even held it against the salaried people because they were bound to cross; either that, or they didn't have a job. But I do remember one thing: John Hanthorn—I don't know what position he held in the union anyway—and as the hourly people were crossing picket lines, they'd take your picture. They had a camera, and they'd snap your picture. Anyway, apparently they snapped my picture, and they posted my picture on the bulletin board at the union hall. And John Hanthorn, he seen my picture on there, and he ripped it off and he said, "That man has the right to cross. He's a salaried man. Do not put any salaried people on that bulletin board!" But anyway, that was the kind of a guy he was.

Maybe I'll have to go talk to him. He sounds like he has a lot of integrity.

Oh, you bet! Yeah.

Do you have any memories about the locks or the Corps of Engineers, or PGE people?

I never was associated with the locks or PGE—just walking across the bridge going to work or going home was as close as I got to them. The locks were running 24/7, and there was a lot of activity. In those days the locks were used to ship the paper into Portland. They had barges, and they would load the barges, and the tugs would take them into Portland. And then the empty barges would come back. The locks were used extensively by the mill.

And the clay barges would come down?

Well, the clay barges did not use the locks. They didn't have to use the locks; just about where the head of the locks are now would be where the clay barges were unloaded. They unloaded the barges with a vacuum hose, a large vacuum tube. They had a large hoe, and they would pull the clay to the tube, and it would be sucked into the silo. And that's how it was done.

Sounds dusty.

Dusty and pretty monotonous.

Were there any fires that you remember in the mill when you were there?

I'm sure there were fires. I know there were fires and problems here and there. We had power outages—you would be in the middle of swing shift, working from four in the afternoon until midnight—why, you'd have an electrical storm that would come through and knock all the power out. Of course, when the power goes out, all the machines go down, period. It's totally black in the mill, you know. There were some emergency lighting that operated adequately, but in most cases, it failed also.

So in that case, why, you'd go to your locker and try to come up with a flashlight so you could find your way around. But in most cases it would be intervals of half an hour, or maybe an hour and a half, before things were up and trying to get running again. After something like that, it was quite a chore to get all the machines going again. And of course you could do a lot of damage to the supercalender. If power went out—why, the supercalender has what they call filled rolls; they're rolls that are comparatively soft if you compare them to a piece of steel. And if a wad of paper goes through, it'll leave a large indentation in that roll. Then that roll has to be changed, which is a pretty good, pretty hard, job, and also maintenance has to come in; they have to do the changing. So when you—as we called it—knocked out a stack, a supervisor was always on the phone to get a crew in to change the rolls and to get it up and running again. But that was just all part of the game.

So it would take quite a while for the machine to stop. Wouldn't it snarl?

Very much so, you'd make lots of snarls, when you lost your power you'd lose your friction or the huge brake that's holding back the roll, and it would wrinkle everything up and if you were lucky it would happen during a reel change. If you happened to be shut down at the right moment and you were changing reels, why then you could get by without damage, but it was pretty common to knock out the stack during a power outage.

That wrecked paper would go to the beaters?

The amount of paper that really got wrecked wasn't very much because it would tear the stack up so bad that it would snap off on top, so the paper that was damaged might be several hundred pounds but it was minimal to what you might think.

Did you wear rubber boots when you worked on the supercalenders?

The supercalenders was a very dry, very neat, clean place, no rubber boots required. Nothing like groundwood. It would be hot in there in the summertime and comfortable in the wintertime. The paper machines would be hotter because all of their dryer sections were all steam heated, that dry the paper. The paper machines would be very hot.

What about the humidity?

The humidity was very high. In later years in the supercalenders we installed sound booths. Each supercalender, each winder position would have a sound booth that really cut down the noise. Each sound booth was refrigerated or air-conditioned, so it was a huge change from my first thirty years.

Did you see very many injuries? Were some of the smaller injuries covered up?

Individual departments counted their injuries separately. Our supercalender department worked 00 days without an injury, for instance, but as you say a lot of things were covered up--there would be minor injuries and of course you wouldn't really count it unless it was a time-lost injury. Otherwise, if you had to lose working time then it would be counted, but there were times when they would bring a person back in and let them just sit around to avoid a lost-time injury. Not often, but it did happen. The supervisor would not be able to do that on his own—maybe an injury would happen and maybe the next day the supervisor would see how he was and say, "Would you feel like coming back in and just maybe doing something at a desk", and if they said yes, well that's the way it went. If they said no, well it was a lost-time injury. That did happen

The internet shows photos of hands missing fingers from the old days, but not many men said they witnessed that type of injury.

I have a small injury here in my hand from when I was a winder helper, I got my hand caught in a nip point and tore it up a little bit—that wasn't a lost-time injury, either. Just a bunch of stitches—I can't straighten it out any more—but at that time I was just asked if I would keep coming back to work, and I didn't have to do a job, but I had four or five days or whatever and then I was back doing my regular job. Those things happened to avoid a lost-time injury.

When you worked in groundwood, what was that like?

I only worked in groundwood a very few days and it was just plain heavy, hard work. The chunks of wood were about 125 lbs. apiece and you filled up a cart, it was kind of on a railroad track they had little tracks that the cart run in and you filled up the cart up, I suppose the carts were six feet long and five feet high, and you'd fill it up with wood and push it to the grinder and there was a grinder man there and he would unload the cart and he would stuff it into the pocket that goes down to the grinder—it was heavy, hard work.

1951 photo courtesy of West Linn Paper Company archive

Did they still have flumes then?

I'm not familiar with that area, but the flumes were there—they come in from the sawmill. Some wood was taken by a conveyor belt to what they called the chippers, which was a very large wheel that had knives in it that turned, and you would stand as this large belt come up by the chippers...you know what a picaroon is? A picaroon probably has a hardwood handle about three feet long, and the metal head has a sharp point on it and as the wood came by you you'd reach out and you'd kick that into that block of wood, and draw it back into the pocket that would drop it down into the chipper. And the chippers were probably the nosiest machine you could ever hear in your life. When you come out of there you'd have hearing protection on but it would hardly do any good. Boy it was terrible. It was really vicious—it was somethin' else.

Louder than the debarking machines?

Oh yeah. It was terrible. The old wood chippers, I suppose the wheel was 12 feet in diameter and there were three of them, and they were grinding this wood constantly and having never worked in the sawmill, I don't know whether it was louder than the debarker, but it was terrible.

They had ear protection then?

When I started—the hearing protection wasn't really required, I think it was more or less an individual thing. In later years it became a required thing. If you were the shift foreman and you seen an employee without it, why he had to go get it, he had to get it on, and he had to wear it. That was it. They had it very available, you could wear an ear-muff type protector, or you could wear the type that plugs right into your ear...there was four or five six different types, it was all free.

You had your finger in a nip. Would you call that a close call?

It was a close call for my finger. There was a roll-lowering cable, and I was a helper, and there was a pan, or like a bridge, from the reel of paper coming out of the winder—a roll-lowering cable came up to the same height as the reel and the winder would be, and this table would lay over like this, and there was a hand-held bridge that you had to drop in place so the roll wouldn't drop down into that little area, and anyway I'll try to shorten the story, the reel had already came out of the winder and up against the roll stops on the roll-lowering table.

I'm trying to relate this in a message that you might be able to interpret—but we'd reach over and grab the pan, and I always was a little bit ahead of myself, I'd already pushed the down button to make it go down and I had my hand under this table, and as the table sat down there was a roller there, and it sat right down on my hand and it tore it up. But you know, that was a bad habit I had—I was a little bit ahead of myself on everything, and that was a bad habit. I never did it again.

Did they stop the machine?

No. In the mill, the laboratories all had a first aid station. I went into the first aid station and pretty soon my foreman was there, and "What the heck did you do?" and then they sent me over to the hospital and cleaned it up, and that was it.

Did you socialize outside of work?

We socialized with a few, not a lot. We had special friends-- they're still special friends.

Rosie told me about the supervisors' Horseshoe Club and Christmas parties…

I participated in the Horseshoe Club I believe one time, and as far as going to the Christmas parties I was never a drinker, so it really didn't fit my style—I guess I'm more like Roy Paradis than a lot.

Have you been going to the reunions?

I have attended three or four of the picnics at Boyd's farm out past Highland. Did you attend?

Yes. Olaf invited me.

When 12 o'clock comes, they eat, that's for sure.

What was the worst thing about working at the paper mill?

Would a person say, graveyard shift? Or the two long strikes that I went through? That would probably be the worst things.

Did you ever think you were going to stay there that long?

Initially I never thought I would stay there that long. I'd been there a few years and I decided I was going to go to college part-time and try to get a degree in teaching. I've always loved athletics—baseball, basketball, and I did a lot of wrestling in high school, and I wanted to be a coach—that was my dream or my wish, but I attended Portland State for a time, but trying to work full time with a family was difficult, so I finally gave it up.

But I have a daughter that is a teacher, I have a daughter-in-law that is a teacher. My son Eric has an outstanding job at Portland State, and I have a daughter, Kathleen, who is completing her doctorate degree, so my kids are making up… I've got a little bit of pride there.

Was the mill your first job?

Well, as a kid, I worked in the woods. I worked behind a Cat, setting chokers, and things of that nature, more of less a summertime job. I was born and raised on a farm and of course you're always working at something there, and I'd worked in the Woodburn canneries for a couple of seasons, and such as that, but I guess basically it was the first permanent type job that I ever held.

Do you remember interviewing for it?

In those days, Crown Zellerbach, most of their employees came out of the small little town of Willamette. The personnel man, he lived in Willamette, and if you lived in Willamette you had a job in the mill automatically. I went down and just applied for a job—there was no interview, you just applied for the job—they needed people and they put you to work. I didn't live in Willamette, but there was a lot of people from Willamette that did work in the mill.

How many years did you work there?

Forty years and some months.

What was the best thing about working for the mill?

The best thing? I believe probably the people that you worked with. There was a lot of good people. I suppose the next best thing was your monthly paycheck.

Rosie said Mr. Zellerbach said if you worked at the mill five years you'd have a job for life.

I guess I have heard that. I guess he was quite an individual. One story about one of our mill managers, Ray Dupuis, by name. The mill used to provide the mill managers with a residence, right there close to the mill. Anyway, one of the machines, maybe 5 or 6, was haying out, and when they'd hay out, the steam would come out and it would make an awful blast--a lotta noise, and it kept waking Ray Dupuis up and he came down and said "shut that blankety-blank thing down, I can't sleep!" So they did!

Somebody said it was a bad day when Zellerbach sold to the Englishman.

Yeah, that completely ruined the Crown Zellerbach Corporation. That sent the mill into a kind of spiral. 'Course I suppose the mill would have a hard time competing in this day and age with the new mills that run twice as fast or twice as wide and they don't take any more people to run, so it's pretty hard to get in there and go dollar for dollar with them and try to come out on top. So the only thing they can do any more is specialty papers and such.

Did you know about the other papers that were made in the old days?

No. 7 ran Crezon. And that pineapple paper was a very very dirty job. I never did witness it but I've heard a lot of stories.

Anything else for the good of the order?

You've asked me a lot of questions and hopefully I've answered some of them correctly, as I've seen it through my own eyes.

John Hanthorn

Interview June 28, 2006

I'm going to ask you if you would give your name and your age, please.

My name? John. I'm 81 years old as of June 24th.

I wouldn't have thought it. Do you remember the year you were hired at Crown Zellerbach?

I was hired in 1947 at Crown Zellerbach. It's a funny thing happened at my hiring. I went up there and put an application in, "Well, we're not taking any applications right now," he said, "but we'll give you a call." About two weeks went by. So I thought well, I'll go down there and find out what they're doing, what's coming on. So I went down and asked him about it, and he said, "Well, have you put in an application?" And I says yes, I have. So he says, "Well, I'll look through the application pile here." And he opened the drawer, and he picks out a pile about six inches high of applications.

And so he's going through them, and here I'm clear at the bottom of the pile. And he says, "Yeah," he says, "I've got a job for you," he says. "It'll be down in the sawmill." And he says, "I'll take you down and show you where you're going to be going and everything." And I says fine. So then we were just going out the door, and he was going to take me down there and show me where to go, and the phone rang. He says, "I think I've got something better for you." He says, "They need four people down there in the supercalender." And I says it's all right with me; I didn't know one place from another. So we went down there, and I've been working in there ever since.

It would have been horrible to go down to the wood plant.

Oh, yeah. I'm glad I didn't, because I went down there and seen the things they had to do down there.

It's hard work.

Oh, it is hard work down there in the sawmill.

Of course, it's all hard work, but some is a little more miserable than others.

That's for darn sure.

So what was your first job title?

My first job was a roll buck. I started out as a roll buck, and then I moved up along from job to job as I got a little experience, I guess.

You weren't on the extra board; you went right to a real job as a roll buck.

Oh, no, no, I never did. I started in that department, and I ended in that department.

Wow. What year did you retire?

What year did I retire? Nineteen eighty-seven—yeah. It was 40 years, anyway, from the time I started. Is that 40 years?

That is 40 years. So you retired after it changed from Crown Z to something else?

Yeah, it became James River. When I retired that was James River, yeah. But I think it was practically the same as Crown Zellerbach yet; the same benefits and everything. -

A lot of times in a takeover you lose retirement or something. But you didn't?

No, I didn't get as much as I thought I was going to get, but it was better than—I guess you could say a kick in the eye, or a stick in the eye, or something like that.

Okay. So first job was roll buck. And how old were you then?

Twenty-two or 23.

Had you had other jobs before you went to the mill?

Well, I got out of the service—see, I didn't get out when most everybody else did because I signed up—I was in the regular Navy. And I didn't get out right away. So then when I did get out, I had money to buy a car, but I couldn't buy a car; there wasn't none around to buy. So then I ended up buying a motorcycle, and for a year I didn't do anything. I just run my motorcycle, and we went down to California, another guy and I, and rode them all around. Well, we come back, and my sister lived in Klamath Falls. Her husband at that time was working at building some kind of a building. Well, we stopped at my sister's place. And her husband was building, I think it was a starch plant—[starch] made out of potatoes. But they needed some ironworkers, and the kid I was with, he was an ironworker, so he stayed there and worked, and I came on home, and that's when I finally started working in the mill.

So you were a footloose bachelor.

Yeah, I was. But I was still living at home, though.

Were you born in West Linn or Oregon City, or—

I was born in West Linn, raised in Clackamas area—the old Clackamas area, not Clackamas today. The Clackamas I know today covers a lot of ground, but there was just that little bitty burg of Clackamas. And then we moved over to the Milwaukie area, and—well, there and here [Gladstone] is where I've been all my life.

When you went to work at the mill, you were living in Milwaukie and you commuted?

I had to drive to work. Well, I think 99 percent of the people did drive to work, didn't they?

There were a whole lot of mill workers' houses pretty walkable distance to the mill.

Well, everybody I knew that was working in the mill—they came all the way from Canby and everywhere else. Woodburn.

Ole lives here in Gladstone—Olaf Anderson.

I never did know where he lived; I still don't. Well, I talked to him when I was in the mill all the time, you know, at different times, but never did ask him where he lived. I figured, well, it was none of my business.

What about mill work made you want to apply there?

Well, there was a guy, he was hauling lumber—well, it wasn't lumber; it was cants. They called them cants, and they hauled it to a little mill over here off of 82nd somewhere. For about two weeks, maybe, I helped him unload his trucks. And then when we got to where we were going, they were unloaded for us. So, probably only making a dollar an hour; less than that, maybe. Oh, I'm sure it was less than that. But other than that, after I come back from California, from riding the motorcycles, I don't think I worked anywhere else. Well, I was drawing my 52/20; that's what I was doing.

What's that?

Fifty-two twenty? When you first got out of the service, they gave you $20 for 52 weeks.

Twenty dollars total?

Twenty dollars a week, total, for 52 weeks. So that's what I was getting by on.

A single guy could get by on that, probably.

Well, yeah, gave mom half of that just because I was staying at home.

That's really nice.

Well, I didn't have a lot of spending money, but I worked once in a while and picked up some.

Then when the 52 weeks ran out, you had to get a job?

Yes, I had to. Well, that's when I was getting right down near the bottom there, scraping the bottom of the barrel then. In fact, Richard Buse -- did you interview him? Oh, no; that's right; he's gone. His wife worked in there where I signed up for my—they called it 52/20, but it was kind of like, oh, unemployment. And you'd come in and sign up for it.

So you decided to go to the mill. Do you remember how much the pay was?

Do I remember how much the pay was? I think it was $1.27. No, it was less than that. I don't remember what the pay was, but when I can start to remember, it was $1.27 and a half. They always get that half in there; I don't know how they did that.

Half a cent?

Well, when you get a percentage raise, you know.

That's right; that's why we need pennies.

I guess that's where they get the saying every penny counts, or something to that effect.

So when you started as a roll bucker, do you remember how long you did that?

Well, actually, when I was a roll bucker, I bucked rolls for about two weeks. But that first day when they brought me down there—and I went to work immediately when they brought me down there—I had my street clothes on then—so I worked that first eight hours. And while I was there learning the job, trying to figure out what to—you know, someone showed me, but I had to learn fast, because they brought down three more guys to take on the following shifts. And I was supposed to show them what they had to do. Well, one of them came in at midnight, and then one came in the next—or swing shift—and one came in on graveyard.

So when it was time to move up the ladder, then I was moved up ahead of them because my seniority started before theirs did. My next job, after about two weeks I went to helping on one of the winders. I can't remember how long I worked on one of them. But then they had a little rewinder they had to run, and it was taking rolls that were damaged, or something was wrong—they were cut down to a different size or something like that—and I ran that for I don't know how long, until I moved up again. I moved up to helping on one of the supercalenders. I have no idea how long I helped on one of them.

Do you remember who your boss was on that first supercalender job?

My first job on the supercalenders, I had a boss named Britt. I can't remember his first name, but he didn't stay very long. He came from a mill in Wisconsin, I believe. In fact, all four of the supervisors at that time came from different mills in the east. Well, they weren't supervisors; they were kind of like a straw boss. But he was my first one. And then from then on, I can't remember who was second or third or fourth, but—I know who they are, but I don't know what order they came in, you know.

Were you working there when Rosie was there?

Yeah, I worked there when Rosie Schultze was there, but he was always on the paper machines, and I didn't see too much of him—only when he'd come out and talk and looked around and seeing how things were going. But I never did speak to him, not very much anyway. Everybody liked him.

I think of supercalenders and the paper machines as being together, but they weren't?

No. They made the paper for us, for the supercalenders. The machines made the paper for us, and then we finished it. We had to work together, but one was in, you might say, this building, and we were over in this building. But they were all hooked together.

How did a roll get from paper machine to supercalender? Is that what roll buckers did?

Well, at one time, yes. But at the very beginning, they had the cranes that followed a track and went around corners and everything like that and hauled them right out to you. They were roll buckers on the machine, the paper machine itself, that brought them out.

So then the crane actually positioned—

Followed a track up here. Thing's up here, and the control is down here. It was on the end of a control, and you could work it, and it followed this track out. It came out one door and then another one, and around a corner.

And it would just lower it right into position over the supercalender?

Yeah. They brought it right into where we could get it with our cranes. Then they'd set it down there. Well, when they left, then we could move our cranes back and forth and pick up these rolls as they came out, or as we needed them.

Cranes inside the building. I don't think anybody's talked about that before.

Well, they were overhead cranes that run on tracks.

They had the cranes as long as you could remember, or people did it at first?

Well, it got to the point when them cranes weren't operating good enough, and they were kind of dangerous. One time I come in at, oh, it was midnight, and there was a big roll of paper, and a crane laying on the floor. And there was a lot of millwrights in there at that time trying to get it squared away. But after that time they got carts, big carts, pulled with a little electric-type tractor. And they loaded them on them carts then and brought them out that way. It was a much better operation that way than them cranes ever were.

Did the carts have some sort of lift system, or did they still have to have something overhead to pick it up?

Oh, at the paper machines themselves, they picked it up and set it right on the cart. Then when they came out to us out in the supers, they came around and parked it right where we could pick it up. They'd pick it up and set it on the floor, and picked up an empty spool to take back with them.

When you say 'they'd pick it up', you're not talking about men actually picking it up?

Oh, no. Our overhead cranes on ours—they were separate from this here one that went around the corners and all that.

Okay, you had to have that overhead crane on the supers.

Oh, yeah. The cranes on the supers couldn't move the length of the building—back and forth the length of the building. If them overhead cranes were coming out, they had a couple of little ropes that you pulled. If you wanted to go straight by, you'd pull it this way and it would go straight by. And if you wanted to make a turn and come into where our cranes are, you'd pull it the other way, and the track up there would move. And there was a corner track and a straight track. Well, after that crane got back out and was back out in the paper machines, we had to make sure that the track was in the right position for us to move back and forth, because our crane was moving back and forth this way.

That track was put in—when they built the building itself, actually—they built that building just for the supercalenders only. And it was built in 1946—'45 or '46. The first paper that was made was made just before July the Fourth. The contractor had got some special incentive to make paper by July the Fourth. Well, they made paper by July Fourth, but they only made a very little bit on the paper machine itself.

A technicality.

Yes. And then they were down for the Fourth of July weekend, so they had I don't know how many days—about three or four days—to finish up the work they had to do get it to running. Well, as soon as they got it running, they got some paper on the supers. And as soon as they got paper on the supers, then they got it on the winders. And as soon as they got it on the winders and cut it up to the rolls they wanted, it was dumped off, and I had to be there to haul it away.

It sounds like when you started, they had just opened the supercalender operation. And so they had to people it with people from afar.

Yeah, it was the first calender operation on the west coast. Yeah, then they got another one down in California someplace. But that was a different outfit, a different mill. They had to have clay to make the paper. The paper on the machine starts out just like newsprint, but as it comes through the machine it gets dried, and then the clay coating is put on the supercalender paper. When they built the supercalender department, actually the machine was there for the paper machine. It was there, but it had to be completely overhauled to make the coated paper. And when they made the coated paper, they had to have the clay barges coming up and dumping in the silos. So—

Who knew how to convert a paper machine into a supercalender?

Well, they didn't convert the paper machine into a supercalender. It was still a paper machine. But as it came through the paper machine, it had to go through a coater—it was called a coater. That's where they put the clay on. And then it went through some more dryers to dry that clay onto the basic paper itself. And then after it was all rolled up, it came out to the supercalenders. I could show you the diagram.

The coating was put on, dried, and then went to the place where it's ironed and polished.

The supercalender—all it is a big iron with a lot of heat and some pressure. But it just makes it nice and shiny.

People say, "This machine had 14 nips," you know. I'm trying to imagine all of the rolling and turning that would happen.

Well, the super had, I think, 10 nips. Nine or 10. Went over the top—one, two, three—I can't remember. It's too late for me to remember now.

Oh, well. Anybody ever get caught in a nip, or injuries on the job?

Well, we had one guy that got his hand in a nip, and he couldn't get it out. I can't remember how it actually happened now, but it burnt all the skin and everything off clear right down almost to the bone. He ended up being a boss in there later on. Well, he still had movement of his hand, but not as much as it was originally. And he was quite an artist. He could sit down and make a caricature picture of you, and you'd look at it and say, oh, I know who that is. He could do that to almost anybody—

The one that did that sketch of the cavemen working on the paper machine? You know which one I'm talking about?

Yeah. He didn't do that. He didn't make anything for anybody; he was a caricature picture drawer, or artist—in fact, I thought he was even better than the guy that was doing it. That's just my opinion. Melvin. Melvin Vergara. I think his folks came from the Philippines.

Well, there probably weren't very many jobs for artists, compared to jobs for mill workers.

Well, I don't think he really cared about it. I don't think he did.

Employee cartoon art, courtesy of Rosie Schultze.

He just did it for fun?

Yeah. I think if he would have went down there to Disneyland, they'd have put [him] to work immediately. Of course he probably wouldn't ever get an interview, maybe.

Okay, about your first day. Did you actually stay to train the other new guys?

Oh, no. They came down, and they watched what I was doing while I was doing it, and that was their introduction to it.

You started right on days?

Yeah, I was on the day shift. I came down and found out about my application, and that's when—it was around 10 o'clock in the morning, 10, 11 in the morning, and he took me down to the mill before noon and showed me where it was at, and I worked until four o'clock. That was the end of my first day. But during that time, these other three people came down and tried to find out what they had to do.

You were a trainer?

Yeah, I was training myself. I suppose I showed them what they had to do, but I don't know; maybe I was telling them wrong.

Did other people in your family also go to work at the mill?

Well, my older brother worked there—I don't know how long he worked there, but he didn't work there very long. That's when it was still a newsprint paper machine, and he was working on the machine that made the coated paper. But he was still making newsprint at that time. Then I got a job there, and after I worked there for a while, I had another brother that came down and went to work for a short time, and he says this isn't for me, and he left. Then my youngest brother, he came down, and he got a job, and he was like me; he stayed there for 47 years.

Oh, really?—what was his name?

Ted. Yeah, my brother Cliff, he worked there maybe six months. Then I worked there for, like I said, 40 years. And well, I had a brother that didn't come near the mill.

You had a lot of brothers.

Well, there was five of us boys. So then my brother Tom, he came down and he didn't like it at all, so he quit within a couple of weeks, I think. Then Ted came along—he was the youngest of us—so he stayed, hung in there.

So he worked there after you were gone?

Yeah. He just retired last year. He's 17 years younger than I am. He was born in '37, and I was born in '25.

Was it kind of nice to have your brother at the mill, or did you ever even see him?

Well, he was always on another shift. As far as I knew, he didn't work there.

Did you have to do rotating shifts? How did that go?

Not good. Ask my wife. She didn't like rotating shifts. I could have went to millwrighting, but I didn't see the handwriting on the wall, so. Yeah, after I was there for—I don't know; I was running a winder at that time, and I had a helper. And he says, "Come on, John, they want nine millwrights, trainees." And then he says, "Let's go to millwright." He says, "This isn't for me." Well, I figured I was halfway up the ladder then. Why should I stop and start at the bottom again? So I didn't go, and he did. Well, as time went by, he went up the ladder faster than I ever could dream of.

What was his name?

He lived in Vancouver. I don't know what his name is any more.

It seems there's a lot of opportunity in a big organization like that.

I would have moved along and probably—but it was something that was different. And I'm sure that 'different' is what mostly held me back, because all nine of them people that did finally get them jobs, they moved along way faster than me, monetarilywise and everything.

Millwrighting was a good job.

It was.

It seems like people were engineers in their heart to really do the best at that. Like Ole.

Well, you know, as for being engineers, there was some of the people we had in that mill that had no education at all hardly, that were really good at that type of work. They had a supervisor that was way up the ladder, way up, and the first time they took one of the king rolls out of the supercalenders, they were putting it back in and the supervisor says, "Well, that's got to be bolted down." And this guy told him, he says, "No," he says, "you can't." He says, "You'll tear the whole place apart." Well, it's a good thing they didn't. He wanted to have it bolted down some way, because the way that's set in there, the weight of it was the foundation. I mean, it was kind of like a saddle and it set in there. And, well, he's gone now too. In fact, he's been gone for quite a few years. His name was Tito, Tito DiCenzo.

That was the boss, or the other guy?

No, that was the millwright. The boss was the guy he had an argument with—but I remember the first time they ever tried to change rolls in one of them supercalenders. They had one heckuva time because it was all new to them, too; they'd never had anything like this before. And they'd pull it out, and they took it and set it aside, and had a new one ready to put in. They put the bearing housings on and everything, and they went to put it in. It wouldn't fit. It wouldn't fit. They got four bolts on one end, and the other end didn't line up.

Well, they took it out again and looked at it, and put it back in again. These four bolts lined up right, but these—hey, we're way off here. Nobody could figure out what it was. Then one of the millwrights that was helping, he was looking at it, and he gave that bearing housing a shove because he was mad, I guess, about something—he was really teed off—and it moved. Just the bearing housing moved. They found out, hey, that's a floating bearing housing on there; we can move it anywhere we want to.

So they got the one end in, worked fine. This one here was off, so they shifted it in a little bit, because they had about that much room that that bearing could float in there. When you got it in the bearing housing, it was snug the way it was supposed to be. But it was this idea that it floated in there. One end was fixed, and the other end wasn't.

Sounds like it didn't come with an instruction manual.

No, no, no. None of that stuff did.

They were just trying to get it running, when you were hired.

Oh, yeah. There was times I went to work there, couldn't go home. Couldn't go anywhere, you know. You had to stay there, because they didn't know when they was even going to get the paper machine to running. Days on end I would go in there and do nothing all night long, all day long.

For pay?

Yeah, they paid.

It would be pretty boring, though.

Well, I got to wander around the mill and find out what was going on in the rest of the mill, though. Yeah. I went and seen parts of the mill that I probably wouldn't have seen, you know, if I had to be working all that time.

Mm-hmm.

Everything was all brand new.

Photo of grinding room, courtesy of Ken Cameron.

There's this big room down by where the balcony is, above the grinding room.

Oh, yeah. I saw most of the mill, but I had nobody to go along with me and tell me what they were doing or what it was all about. So I looked down there where the grinders were, and I thought to myself, boy, that's one place I don't ever want to work. I think that's where I was headed when they wanted to take me down to the sawmill. I didn't want no part of that after I found out what it was.

There's a supervisor at the mill now that tells a story about the day he got hired. He said, God, I thought these people are so nice; they gave me a pair of rubber boots just, you know, for free.

That reminds me of something. They had one of these guys came and got a job. They did the same thing: they gave him boots, and they gave him hard hats and everything. Took him down there, and he worked there that day. When he left, he took everything—all that stuff they gave him—and left and never did come back. [laughing]

Okay. Did any of your kids work at the mill?

No. I wouldn't have let them.

How come?

I don't know; it's no life. In fact, the kids always wanted me to take them down to the mill. I'm glad I couldn't even find time.

So you stayed on that supercalender for 40 years?

Well, they had three supercalenders. They changed No. 9 over to coated paper, too. I think it's No. 3 machine now. They've changed all them numbers on them machines. Well, anyway, they bought a new one from Germany, brand spanking new. And the four guys, the four of us—when I brought them into the mill and showed them how to buck rolls—we were the first four to run that. We didn't—nobody showed us anything—it was altogether different then. It was so much different than the other three that it was just like night and day. And so what we would do is, at the end of our shift, when one of the other guys come in, we'd tell them everything that we knew or found out about it. And then when they went off shift, during your eight hours maybe you learned something that you didn't know before and didn't know whether they knew it or not, well, you'd tell them about it. So we just all learned by "by gosh and by golly."

Gosh, how can you even make paper under that condition?

Well, basically, they showed us how to get it started, but that's about all. There was other little things that you could do, and it made life easier. But if you didn't learn them little things, little tricks of the trade I guess you'd call it, it could be a miserable job.

Did you have quotas?

No.

No production quotas or anything like that?

They would like to have had, but you can't keep producing when you're having problems. Because we'd get all kinds of things coming from the paper machines. Did anyone ever tell you about hair cuts?

I don't think so.

Well, a little piece of hair from your head could get into the stock of the paper machine before it gets on the wire, and it goes down through the paper machine, and it runs through there fine. And they rolled it up and sent it out to us. Well, under the pressure of them rolls when it goes through them nips, BANG, depending on where the hair is. If it's near the edge, you're darn sure it's going to break. And then it damages the rolls, so then you've got to shut down and change rolls.

Well, the millwrights used to change all the rolls. Later on they got to thinking, well, that's crazy; we've got four people here standing around and doing nothing now. So instead of having the millwrights come in in the middle of the night, they had the four operators—the operator and his helper and the winderman and his helper—and they had to change the rolls. So you're down during that time, even if the millwrights come in, and they could probably have done it faster than you could have. But this new supercalender, the one I was telling you about where we had to learn and tell each other everything—they had a beautiful way of setting that thing. They had the rolls in a rack right there behind the stack. And if you had to change two rolls, you took this one out, put it on a cart and got rid of it; picked up the other one out of the rack and set it right over and set it right in there. You didn't have to—on the old ones you had to move this and take this out and take that out, and—

Pretty slick, huh?

Yeah, boy, this new one was a slick job for changing rolls. It was a good machine.

So a hair cut essentially made a tear in the paper?

Well, yeah, I've seen them that long. They may be curled or half-curled or whatever.

And that would wreck the roll?

Oh, sure. When it breaks like that, it all wads up, and these are filled rolls; they're not steel rolls.

You'll have to explain that part because I don't know what a filled roll is.

Well, a filled roll is a roll that's made out of cotton. Some of them were made out of denim. They have a big shaft—this is how they make them. They put on a ring about so big around with a hole in the center, and they put this piece of denim, all these denims [motion of round layers going on], and they apply so much pressure to it, and then they put a whole mess more on and put pressure to it until they get that thing all the way filled up. When they get it all filled up, they cap it, and then they put a cap on the end to hold it so it'll stay there. Then they take them and put them in a lathe and turn it down until it's all smooth, all the way across the roll. It's as hard as this. I don't see anything around here right now that's that hard. But you can't dent it.

When that paper wads up and goes through them nips, it puts a big dent in there. Well, on the coated paper, that leaves a mark on the paper itself, and the printers don't want that. So you have to change them rolls. They take them out, take them over to the lathe, smooth them all down again, and then put them back in the stack.

How many times can they do that before you have to throw the roll away?

Well, they start out at 21 inches, I think, size in diameter. And they run them down to around 15 and half or 16 inches, somewhere right around there, in diameter. Then they have to take them and send them back to the factory. Sometimes they take off, oh an eighth of an inch, sometimes maybe a quarter of an inch, something like that.

Depending on the dent?

Yeah, how big, yeah.

Well—but it's steel we're talking about, right, on the outside of the denim?

No.

It's just denim?

Well, you had a shaft about that big around. Then you've got pieces of denim that they put over that shaft.

And then they press them from—like this?

No, not like that. They're standing on end. And they press them down.

Okay.

You keep piling it up until you get from the bottom end of the roll to the top end.

So the dent is actually in the denim? Okay. I understand now.

Well, the paper, when it breaks it gets wadded up because it's running, say, like a thousand feet a minute or more. And when it breaks, one end goes down through the stack, and the other just wads up and goes through in a big chunk. And when it does that, well, that's when you get all the dents in the roll.

Do they still make them in denim and cotton like that?

Yeah, but they've got a new type now, and I don't know what it's made out of. They got them in after I left.

But it's still impressionable, soft?

Yeah, but it'll work itself out. They've got them so that if you keep running it, it'll smooth out. It's not rubber. It's some kind of a plastic of some type or something.

So it's subjected to heat and water? And it holds up under it?

Has been, as far as I know. You'd have to talk to somebody that knows more about them than I do.

What was the length of the roll on your—

A hundred and fifty-six inches, 157. I think 157 inches long. Well, it's got to be more then—12 feet would be 144. Well, a little over. Then you have to have some on the end for—maybe three or four inches on each end, because you can't run out clear.

What was the longest you ever had to stay at work?

How many hours I had to put in without—? I think about 18, because they always got somebody in to relieve me of my job, except when they had to call somebody on the spur of the moment and have them come in. But most of the time you knew that 16 hours would be the most you'd ever have to work.

Somebody said that they remember real clearly when the union got the 12-hour limit.

Yeah, oh yeah. That was before the 12-hour limit, yeah. Okay. They could work you as long as they wanted to at one time, as long as you were able to stand up, I guess. But 16 hours was the most I ever had to work. Actually, 18 hours, because there was a couple times when my eight-hour shift was over with, something would happen, and nobody could come in. They couldn't get hold of nobody for the next shift, so I'd have to work that next shift, which would be making 16 hours. Well, by that time they were trying to get somebody else to come in a couple hours early, or something. But they're doing this, they're compensating for one person there that can't make it. I'd have to work 18, and then he'd come in and work maybe 16, because maybe by that time they got somebody. There was something else I was going to tell you.

About hours?

Yeah, the hours. They got to the point where you got down to 12 hours; that was the most they could work you at one shot. But they got by that by—I don't remember. Well, I never did work too many overtime hours anyway.

Were you always hourly?

Yes, yes. I could have went to salary, but I didn't want to. In fact, I turned it down three times. I just didn't want to. I'd probably have been better off today if I would have.

Some of them did it for day shift.

Well, I did it as a relief time, you know, when the regular foreman was off on vacation or something. I was the relief foreman. But I didn't want it permanent.

This leads me to the question about strikes. You must have been there for the—

First one.

Do you know what year that was, about?

Hmm. I don't remember.

Did you get another job when you were on strike?

The first strike was nine months. I had a pickup truck, and I had a chainsaw, and I was going out in the woods and cutting up wood and bringing it in and selling it. I made out pretty good; fair, anyway. We got by. Then the next strike came along, and I forget how long it was. It was near the same amount of time. But I made out like a bandit on that one. I went to work for the longshoremen. I made more money working for the longshoremen than I would have if the strike went on forever. In fact, I was kind of disappointed I had to go on back to work. Yeah, I had to go back because the longshoremen only hired me for during the strike. The longshoremen was good that way.

You're the first one I've heard of that got another union position to tide you over.

It couldn't have been better.

So the union had a strike benefit or something. You didn't have to take that because you had a job?

Well, they gave it to me anyway.

The strike benefit?

Everybody had a strike benefit.

And you had a job at the longshore?

Sure.

That was a good deal...

Well, what did they give you? Two hundred dollars a month, I think it was. It wasn't very much. Two hundred dollars didn't go very far, just like today $500 don't go very far. I don't think anybody turned theirs down.

So they were allowed to work and take the strike benefit, because they'd paid their dues?

Well, I'm sure I wasn't the only one. But the one that bothered me the most was the last strike that we had. The union paid our health benefits for us, with the understanding that if and when we got back to work, we would pay this back to the Clackamas County Physicians Association. Well, I was kind of one of the collectors, and when they were getting their $200 benefit, I always had to be there to collect. I forget how much it was now—their health benefit package was like— I'll say it was $50; I don't remember what it was. Some of them refused to pay it, and here it was for their own benefit, because some of them even used their benefits during that period of time. And some of them paid two or three months on it, and some didn't pay any.

You were in a hard position?

Yeah. I couldn't demand it from them. they volunteered for it: "Sure, we'll take it." They took it. You can't hardly blame them. Some of them were living on a shoestring to start with. The thing is, they could have after they got back and kind of got on their feet again. Even two or three months later, you know, cough up. But they didn't.

So it left the treasury down?

Oh, you ain't a-kidding it did. Well, I guess the Clackamas County Physicians Association, they're the ones that footed the bill.

Interesting. Do you remember the day you went back off strike and back to work?

I only remember the one going back in. Yeah, we started—everything ran smooth with me as far as I was concerned.

Roy said guys went to their job and took over from a supervisor.

I don't know anything about that. No. All I can remember coming back after a strike is coming in and, hey, everything was down, so machines started running, they started bringing paper out to us, and we started running it, and that was it.

And you were happy.

Yeah, yeah, sure. Everyone was—

Except for losing that longshore job.

Yeah.

Okay. Were you there in 1964, at work when the flood came?

Yeah, and I looked out the window all the time. I said, the water's getting higher, it's getting higher, it's getting higher. And pretty soon it was in our basement, and that's where all our electrical equipment was, so we had to shut it down.

So you worked until the last minute?

Oh, yeah, up until the last minute, yeah. Everybody did. I think they wanted to keep it running as long as possible.

Then what did you have to do?

Well, during the flood I didn't do anything. I just hung it out, waited it out. I can't remember, but it doesn't seem like it was a very long time [before] we were back to work again. They had a lot of work to do, I know that. Somebody probably told you how long we were down.

But you didn't get called in to do cleanup?

No, I can't remember going back in. Well, I know a lot of the guys went back and did hosing down, hosing mud out of the building and everything. But I can't remember going back at all. But for the windstorm—oh, power was off. But I remember going up on the roof of the building and watching things going flying by. Sheet roofing, you know, corrugated roofing. They'd go flying by. Up above the mill there were oak trees; you could hear them crashing and banging and falling.

That's pretty exciting.

Yes, but we weren't off very long then. In fact, I don't think we were off hardly at all.

Well, you had your own generators and—

We were getting power from PGE. Oh, yeah, they had grinders up there that could either produce electricity or turn the grinder stones.

So they could generate some of their own if the power went down.

Yeah, but not a heckuva lot, not enough—yeah, they used a lot of electricity down there in that place. Some of the machines, all of them big motors and everything, all running on electricity. Yeah, that windstorm, I don't think they could possibly make electricity on their own.

That might have been a little dangerous up there on the roof, by the way.

Oh, it could have been. But it was blowing across this way, and I'm here. Of course, something could have come flying up there and hit me. I wasn't the only one that would have been hit, though; there were several of us up there.

I hear a lot of stories about salmon.

Oh, yeah. I never did get in on any of that, but they used to do that all the time.

You didn't even get any fish to take home?

No, never did. It was up at the other end of the mill where they were doing all that. We didn't have any access to anything like that.

Yeah, Ole says sometimes the fish would come in through those vent pipes, and they'd go out in the lunchboxes.

The lunchboxes were heavier going out than they were coming in.

What did you like best about working at the mill?

What I liked about the mill? Now, don't laugh. The showers. That's the only thing I missed when I left there. They had showers there with an endless amount of water, and they were good showers, a lot of water. I'll put it like that: the thing I missed the most is the showers, because you could go up there and take a shower every night.

Nobody has ever mentioned the showers before.

> Jean: And he didn't have to clean up the shower.

Where were the showers?

Let me see: you gone up the stairs, and you went through the finishing room there, went down the hallway—it had to be above No. 9 paper machine. I mean, you could walk out of there in clean clothes; you didn't have to walk out—what surprised me is how many people didn't take advantage of it. They had the opportunity, but they didn't take a shower.

They went home hot and sweaty.

Yeah. I couldn't believe it—just to get out of there 15 minutes earlier.

Well, was it like a gym shower or something?

Oh, no. Well, they were shower stalls, two to a stall. And there was probably about six stalls there. And then down on the other end there was probably about six more.

I didn't know they had those for the men. That's really good.

Why, sure. They've still got them.

Really? It kind of perks you up when you get clean, you know.

Well, right after the war, when they made all these new things—the shower rooms used to be, I understood, down below No. 5 and 6 paper machines, someplace down there. Hardly anybody ever used them. But after the war, when they built on, they updated everything, and it was all brand spanking new showers, and all tile and tile floors.

Did you guys get to take breaks?

What—what—what's a break? [teasing]

I'm thinking how hot it must have been in the supercalender building in the summer. A five-minute shower on your break—would be pretty nice.

Well, a lot of guys used to go out and jump in the locks, clothes and all, and come back in and go to work.

Because of the heat?

Oh, yeah. Or you'd get a big rag and wrap it around your neck, with water running down your back.

What was the worst thing about working at the mill?

Shift work. Shift work was by far the worst thing to have to work there. But like I said before, I missed the opportunity by not going to millwright. Then I would have been working more days.

Some of these machinists and millwrights put in the longest, most god-awful hours...

Well, it would happen that something broke down that couldn't be repaired by the operators. Well, they had to get somebody in there to repair it, so it had to be millwrights. Millwrights were, hey, you're stuck, guy. But, you know, if they came in and worked, say, eight hours, they didn't have to come in the next day if they didn't want to, because you can't—you know, you can't work a dead horse.

That's what they got paid the big bucks for, is to be on call too, I think.

Oh, yeah. We had Eddie Paullin. There's a man that never turned down a call. I don't care what time of the day or night or afternoon or when. He never turned it down. One year he made more than the supervisor, the head of the mill. He made more money than the mill manager.

How did he know? Nobody's supposed to know how much people earn, you know, so—

Well, the mill manager talked to him, and he says, you—well, no, that was for supervisors you're talking about; nobody was supposed to know. But everybody knew what I made or what anybody else made, you know, a union person. You couldn't get around it. You got paid for what you did, or sometimes what you didn't do.

How did you and your family work with these rotating shifts?

I don't know. I just did, I guess. I managed it some way. Shift work just—it didn't come easy, but—well, I had complaints about it, but I didn't—

I'm trying to get some information about how the families coped with it, you know.

I don't think the family did. They were here. Well, my wife, she was raising three kids and practically by herself. I was working. When I wasn't working I was home sleeping. If I worked graveyard shift, I was home sleeping when I got off. When I was working swing shift, I got home after 12 o'clock, and I didn't get up until after the kids were gone to school. I didn't see them. So she was, you might say, raising the kids. I had very little to do with it. That's why I say I sure missed it when I didn't go to millwright. But that's water under the bridge now.

You must have really appreciated it when you retired.

I was glad to, finally. But monetarilywise I should have stayed three more years. I was 62 when I retired from the mill.

A lot of the guys were 58.

Yeah, I know. Well, some of them—they either had something else going, or—

Jean: In three years I'll be 80.

Unbelievable. So why did you retire then?

Well, there was that incentive pay. I got in on that. But that didn't go very far because I retired in the middle of the year. That incentive pay went onto what I earned in that year, and so I had to pay income tax on it. I'd have been better off waiting until January the first and then retired.

So they don't give you counseling about this?

No.

But you were glad to be retired?

Oh, I was glad to be retired, yeah. I wouldn't want to go back. Well, I went back one time for nine months. After a company from Canada that started it up, and they were nonunion, absolutely no union at all. And they were picking people to run these jobs that had never been off the street before. So myself and two other guys, they called us back in to help indoctrinate these new people into jobs that they've only seen and watched once in a while. So I worked at that for nine months, but I didn't do anything. I just had to be there and help them out.

Sounds good.

Well, they paid for it,too.

Do you have any stories about the West Linn Inn?

Well, no, I don't have any stories to tell about the West Linn Inn. All I know about the West Linn Inn is that it was a congregating place for everybody that worked in the mill. At shift changes, if you're coming into the mill, you always got there about 10, 15 minutes early or so—it was a place to get the day started, I guess. Oh, it was a beautiful place. They had a bowling alley in there in the basement.

That's what I hear. Two lanes, hand-set your pins.

Yeah, you had to set your own pins there. I've never even been down there to look at it. I wasn't a bowler and I didn't care about bowling, and so I didn't ever go down there and even look at it. But I know there was two lanes. Somebody bought one of the lanes. They were in sections. And he cut them up in sections and made a table out of it, like a table in the kitchen, you know, or something.

Do you go to the Crown Z retirees and Associates' picnic?

Well,—Boyd Ringo?

Uh-huh. Do you go out there?

Oh yeah,—oh, how long have we been going out there anyway? Not long enough. Not all the time I've been retired, though. We missed some. When it first started out, it was just millwrights, electricians, everybody of that type. It was mostly for them.

George told me it was one union, and then they expanded it to take in the others.

Yeah, yeah, they did. Well, it was like you said. It was one union, and then the other one, they expanded it to include them all, because, well, you know, people move on. Hey, we're here. We've got only a small crew here today; we'd better get some more blood in here. So they expanded it and took in the rest of the paper makers and—what did they call them other guys?

Associates.

Anyways. George Droz, he's the guy that puts it on all the time, now that Boyd's gone.

Do you remember when safety became an issue?

Well, I don't think there was anything as safety before. Every so often something would change in the department. You'd come in one day, and you had a broke hole here, and the guardrail around it was only like 36 inches high, and all of a sudden it jumped up to 42 inches or something like that. And so you couldn't accidentally fall in. Different things like that. I mean, it was such a slow process that I didn't hardly notice it. I know when they raised them broke holes, you had to lift them laps up about that much higher to get them down the broke hole. That broke -- you know what a broke is?

I don't. I was going to wait until you got done and then ask you.

Broke is a paper that's—well, it's no good. It's going to be re-beatered, taken down and beatered into slush again and then made paper out of. Well, when you have a crash on the stack, so much paper is lost, and that one bit of paper got thrown down that hole.

That's how it got to the basement.

Yeah.

Nobody told me that part. They said all the paper that had to be recycled went to the basement, to the beaters and digesters or whatever…so there's a hole in the floor?

Yeah. And what made it so bad is, in the supers, that building, when it was made it was a new way of making it. They strung cables from one end of the building to the other, which is a long as a football field, or close to it. And cement was in there, and so it's suspended on these cables so the floor can go like this, can give, you know. And for years: "you can't cut a hole in the floor." "You can't cut a hole in the floor." We tried to get them—especially on the winders or on the supercalenders—there'd be that much paper on a big roll of paper 157 inches wide that you had to stuff down the hole. So you'd kind of make a lap out of it; when it fell down on the floor, you would roll it up, and then two guys, one on one end and one on the other end, and you'd pick it up and throw it down the hole. Well, we tried to get them to cut a hole in the floor. "Oh, we can't do that; can't do that." Well, now today, everything is the same, and they've got all kinds of holes in the floor.

Well, that should make it easier to get the paper downstairs.

Why, sure. Now you just tear it off and let it drop, and it drops right down the hole for you.

So there's a hole by every machine?

Oh, sure. And even paper machines have them. The paper machine—the paper comes through so far, and then it'll go down into the hole and into the slusher. And it just keeps running down there until they get down there and put it the rest of the way through the rest of the machine. On the machines there's probably about two—yeah, there's two there and then one on the winder itself. So when the paper comes down and comes to the first set of dryers, there's a broke hole there. Well, then they have to send it through the rest of the machine, so they send it through the coater. At the end of the coater, it goes down the broke hole. And then after it—see, you should be talking to one of the paper makers and not a supercalender operator.

You're doing really well.

After it gets through the coater and goes down the chute, then they send it through a pull roll that keeps it tight, and then it goes right onto a drum to make a whole big reel of paper. And they've got a broke hole, too. So then there's the winder. They got the one behind the winder, too. Anyway, them paper machines, they had it lucky compared to the supers, because they wouldn't let us put a hole in the floor.

Photos courtesy of West Linn Paper Company archive.

How big are the holes?

Well, they're only, probably, some of them, only maybe that wide. But they went from one end of the machine clear to the other, see?

The same width as the paper.

Yeah. A little wider because there—

No wonder they put guardrails around them. People could fall down in there and get slushed or something, right?

Oh, yeah. Well, and then down below it's on a conveyor belt. There's a conveyor belt going around and around and around. And it carries it down over, and it falls down into a slusher, and it makes mincemeat out of it.

I wouldn't want to go down that conveyor belt.

No, no, you wouldn't want to. You wouldn't last long.

I understand these big machines, if you ever got in trouble, they don't stop fast.

Oh, no.

If it happens, it's going to keep happening for a while no matter how bad it is.

Well, they had that one kid on the paper machine that got his arm in between a guard roll and a rubber backing roll—they called it a backing roll—on the coater, and there must have been about that much room. And he got his whole right arm in there, all the way up in there, and he could have had it taken off or kept, whichever. But it just hangs there like that now. The most he can do is hold a pencil in it.

Hmm. I would call that a disabling injury.

Oh, it was.

We've learned there were lots of little injuries that they tried to prop people up or fix them up so they wouldn't have lost time and have to count it as an injury.

Oh, they'd come in and punch holes in dandy rolls. They had rolls on the machines—or I guess they were suction rolls. I don't know—anyway, they get plugged up with stock, paper, slush stuff all the time. And so they have to keep changing them, say once every three months. I don't know how often they changed them. But they'd take them out, and they'd give you a hand drill and a little drill, and you'd punch this stuff out of these holes. Instead of having a disabling injury, that's what they'd do. They'd give you a job doing nothing.

So there were jobs you could give people that couldn't do much else.

Yeah, you couldn't do anything, but probably you sit there and you—

That's very interesting. I wondered what kind of little things actually happened. You know, Roy—

Paradis? Yeah, well, Roy was on the machines all the time. He worked his way clear to the top, as far as you could get, on paper machines. So he spent a lot of time on them.

Is there anything you want to get on the record?

Well, there's nothing I want to get on the record. The thing is, there's things that I wouldn't—that are memorable—that I'd never mention.

Well, that's not playing fair at all.

I can tell you one incident—about Delbert Herndon. Okay. Shortly after I started there, Delbert was helping on the supercalender, and he was helping a guy by the name of John Wilson. Well, there was nothing to do, so Delbert, he went over and he got in the pile of rolls over there or someplace and laid down to take a nap. He told John just, "I'm going over there and I'll be in there." And he says, "At the end of the shift, wake me up, or tell me when you need any help." So he went over and laid down, and John came by and woke him up, and he said "Okay" and then he went back to sleep again.

And when we come in at midnight, he was still there sleeping. Somebody told us about it. And he finally woke up about two o'clock in the morning, a couple hours later. And oh, he was mad! He went running out of that place. And his wife had been calling up at the mill, wondering where he was at. So I guess he got into a little trouble. I mean at home.

Right. So did you meet Jean [wife] at the mill?

No.

All right, so how did you meet your wife?

The first time didn't mean anything. A friend of mine, it was his car, and we were going out—I don't know; maybe he was taking me home. School had just gotten out for the day. She was in the last year of her high school. And her and this kid were walking home to where they lived. And they had about three miles to go. Well, anyway, we stopped and asked them if they wanted a ride. And "Sure," so they crawled in the back. We took them home, and that's when I found out where she lived.

Then—how long later was it? She was out of high school and went to work in the mill, and I didn't know her yet. I didn't even recognize her as the gal we picked up a year or so before. Then her and her sister and her girlfriend got out of a movie in Oregon City, and they were walking up the street. I don't know whether they were going to catch a bus or what. But anyway, we stopped and we picked them up, and we took them home. And—let's see—what the heck was her name? -- Emma—I don't know whether we took Emma home and then took her and her sister home. But that's when I found out for sure that, hey, you only live about half a mile from where I live, maybe a mile. It was out in the country in Milwaukie. And that's how I met her.

I told her I worked at the mill, and she says, well, she works down at the mill. "Where do you work in the mill?" Well, that's how we really got acquainted, you know. And she didn't believe me that I worked at the mill at first. But she must have changed her mind when she saw me down there working. She had no way of getting back and forth to work from where she lived. And so when I worked day shift, I'd take her in to work when I went. And at nights I could take her home. And on swing shift, maybe sometimes I did. Well, anyway, I had days off in the middle of the week, and she'd work and I could take her in. But anyway, that's how we got together. Sometimes when I was on graveyard shift, if I got up early I could go and pick her up. And after about six, seven, eight months we said "I do."

Well, that's a nice story to end on…

> *Jean:…my sister, we resembled each other so much, and while I went into the house to get ready, my sister went out and was talking to him on the garden swing, and he thought that was me. And I guess he was going to take my sister out.*

Then pretty soon here she's coming out, and oh, my God—I made a mistake.

So you quit working at the mill after you got married, after you got pregnant?

> *Jean: Yeah, when I got pregnant.*

Did they have a rule about that?

> *Jean: I don't know. I just walked.*

They didn't make you quit.

> *Jean: Oh, no. I didn't show—my neighbor lady didn't know I was pregnant when I was six months along. And I finally told her. But I wasn't feeling well, so I just decided I was going to go home. You couldn't fire me after that.*

I was going to ask you if your kids were around here.

> *Jean: Yeah, my daughter's in Fairview. She lives on Blue Lake, if you know where that is. And our other son lives in Oregon City, up by the hospital. And our oldest son is a supervisor to the City of Corvallis.*

Mary Gifford

Interview July 24, 2006

[pre-interview conversation]

Well, when I hired Mary [Lefebvre], we were discussing what to call us, because I'm Mary and she's Mary. So one of the things that was suggested was, well, how about Big Mary and Little Mary? And I says, "I'm not for that at all!" [laughing] So she was Mary Jo. So when people just call her Mary, it takes me a while to figure out who you're talking about, because she's always been Mary Jo to me.

My husband, he's got Alzheimer's, I told you, and well, I know everything about the mill, but he doesn't remember any more, and that's really sad. Because he was in charge of No. 6 paper machine—well, it would be Number 2 paper machine now—and he knew from ground wood to the entire thing, every nut and bolt everywhere.

He was a supervisor. He started when he came home from the military, and he worked his way from being roll buck on the lowest machine up to supervisor of his section of the mill. But my son works there now, so he comes over and he talks. And he's on the same machine Bob was on.

Well, Bob, today when we were talking about it, he says, well, we never had any trouble with the women. Well, there weren't really women until the last few years on the machines…

Could you tell me your name and your age.

My name is Mary. I'm going to be 65 on Sunday. I was 17, right out of high school; I graduated one Friday night, and I started the next Monday at the mill. They had to get a special permit to allow me to work there because you couldn't work at the mill before you were 18, so I had to go through a lot of hoops to be able to work there. I was always so glad that the guy who went to bat for me did. It was a great place to work.

Did you want to work there because your dad was working there, or—?

In the olden days, when I was hired, there were eight high schools in the area, and they took the three top secretarial students from each school and tested them, and then chose from those. And that's how I got my job. I could not believe that I was that lucky. I was really lucky.

You didn't have anybody in your family that worked there at that time?

Not at that time. I have heard since that I had an uncle that worked there—oh, back in the '30s or something, when you didn't even have to wear shoes to work. And then my dad worked there for a short period of time. But no, nobody had worked there for years and years that I even knew. I didn't even know of anybody at the mill.

They didn't have to wear shoes? The women?

Well, no, there was no women working at the mill at that time. Women didn't really come into the mill—oh, gosh, I don't remember when. It was before my time, but they worked in the lab, and they worked—about the time I started they worked up in sheet finishing. But no women around the paper machines at all.

So the men didn't have to wear shoes?

Back in the '30s, Isn't that amazing?

I thought everybody got steel-toed boots when they were hired.

I don't think steel-toed boots were all that old. I suppose in the '50s, '40s maybe, '50s, they probably had steel-toed shoes.

What year were you hired?

Nineteen fifty-nine. June 6, 1959 I started to work.

Do you remember what happened on your first day?

Well, here I was, this shy little 17-year-old going into the business world, and I was just scared to death. Oh, and I was told how to dress: you wear dresses, you wear nylons, you wear either nice flats or heels. And I always wore heels because I was a grownup now. I was hired as mail girl. I ran forms that they used throughout the mill, and I was delivering the mail. And I walked through this one door, and thank heavens this guy was behind me. I don't know; my heel slipped on something and I went straight up into the air and came down like that. And he caught me just before I hit. I thought, oh my gosh—in front of the entire office—and I thought, oh, well, what a way to make an entrance.

That was your first day?

That was one of my first days there. So from there, nothing could have been worse than that, so I did fine.

I started at the mill, like I said, in 1959, in June, and I worked there for almost two years. And then I quit, I got married, and that was an interesting thing. When I came back from my honeymoon, John F. Kennedy was campaigning for the presidency, and he was there on my first day back after my honeymoon, and he wished me well. So that was kind of cool. At the mill. Oh, yeah. At that time there were 1,200 employees; that's a good voting group, you know. So then—

You only took two years off?

Yeah, I got married and had a baby. And then I started just coming back in the summer, doing whatever the office women's jobs were if they needed someone. And I worked in the summers. I had another baby and then I decided after a while, well, hmm, these people are getting more vacation time than I am, because then they started calling me through the year and all this type of thing. And they were really good to me. I had good pay. I could accrue vacation time. And I got some sick—you know, medical, so it was really great. I mean, it just was the best of the best. And then in '72, I think it was—I went back full-time.

At that time there were four openings—there was one in purchasing, one in the lab, one in production and planning, and engineering, I guess it was. Oh, and the time office. And I could take my pick of what I wanted. I loved production and planning. That was where I would be out in the mill. And that was the first of the women getting out in the mill with the men. I'd go out, tell them what I needed, if there was any changes on the orders, these kinds of things. In those days we did roll lists on paper. And I picked those up.

What's a roll list?

As paper goes in roll form across the headers, where they're wrapped and labeled for shipment, they would put the roll number and the weight on a little list.

Like a stock inventory?

Mm-hmm. And each one is like, okay, we need 24 rolls for a full truckload, and we would put a roll list out there. If they wanted a million pounds of paper, say, we'd put X amount of roll lists out at 24 rolls per roll list, so that that would be a shipment. And they'd go to the trucks, and they'd go to however they were being shipped, by truck or by barge.

So that was '72 that you took that job? That lines up pretty well with affirmative action. Was there any problem with you going into the mill?

I kind of felt like I was the token girl. You know, you kind of did in those days. I knew I was going into a man's world, where they'd never had to deal with women before—and some of the guys would just try to shock my socks off. I mean, some of the language I heard, some of the things that they would say to me. And I just acted like I hear it every day of the week, so don't bug me about it. And we all became great friends.

There was one guy that I really had a time with for years and years and years. He got throat cancer and had to quit work. And I was the one person that he had to come and see whenever he came back into the mill. He'd find me. He was just a gruff old codger, but he was a sweetheart. And we were really good friends. And I didn't have any trouble with any of them. Most of them, anything I wanted they'd give me.

And I'm not saying that they were any better to me than they were to the guys that had had my job before me. You know, I'm not saying that at all. It was just that I just always treated them with respect, and that's what I expected of them.

Could you say what jobs women were allowed to do when you first got there, and how that changed? I've heard women worked in converting and finishing, payroll.

Women worked in converting, sheet finishing, payroll, accounting. There were several secretarial jobs at that time, and typists and that kind of thing, and that's what they did. Some statistical typing. And that's pretty basically—but they didn't go out in the mill. Now, of course, the lab, you had to walk through into the mill, but that was it. Sheet finishing, the same thing. You walked through and up the stairs where sheet finishing was, and that was it. You didn't wander around. You didn't go in other places.

Well, that's what I had to do. When I started at 17, I rode with another fellow, and I could walk downstairs into the office, and I couldn't leave until it was time to go home. I had to stay in the office.

And something changed on your 18th birthday?

They think so. I was grown upon my 18th birthday, I guess, and I wasn't before. And it was a labor kind of an issue—you know, around the big machinery and out in the mill proper, you just weren't allowed out there till 18. Now, I don't know how they got around it for the guys, because there were 16, 17-year-olds. I don't know that there was when I was working there, but I know before that there were.

You were the first production person who had to cross that gender line?

Right, mm-hmm.

What was happening in the rest of the mill about women?

You know, I don't know that there was an issue that they were trying to get women in. That didn't start 'til—and I really can't tell you when—but it didn't start till later, much later, because I was the only one for a while, for a number of years.

I've got some old newsletters…and there's a picture of Mary [Lefebvre]…

She worked on the winders. That's when I knew her. She was on the winders.

Did you have to join the union?

We didn't join the union to work in the office. Office help was different. Let me see if I can remember the difference. The men, union, were waged; we were salaried. I don't think the office people ever did join. In fact, there was a big deal when you were waged and you went salaried, because that's what my husband did, and I know of several. They were on the machines, and they were union, and then they were promoted to supervision, and they were non-union at that time; they became salaried.

That must have been a hard decision -- it could change your relationships.

It did for a few of the fellows, but I could see where it didn't in a lot of them. Most of the guys—there were some with chips on their shoulders all the time. I mean, you'd find that anywhere. But pretty much these guys were pretty nice. They knew guys that had worked their way up and went into supervision; they knew that they deserved that. I think the hardest thing was as they would get older and see them bring in these young kids from college that didn't know diddly-squat about the mill, and then to have to start taking orders. That I can see maybe, sometimes. But most of the time, no.

Oh, I remember one time—I was assistant supervision at that time—the assistant mill manager was in charge of production. And every morning at 7 o'clock we'd have our meeting. All the department heads that were in production would meet with him. And I can remember being in there, and one of these young kids out of college was there, and he started to tell this fellow what all was wrong with everything in the mill. And it's like, oh my gosh, I can't believe he's saying this, like his first week here he's—going to straighten everything out. He lasted a while.

Really?

Yeah, he really did. I guess—you know, they go through the tests, and they're smart enough. They just need to let a little experience catch up with their brains. They think they know it all.

Why did you want to work there right out of high school?

Going to work there, I didn't even dream about it, because that was steady, good insurance, good money. It wasn't something that ever entered my mind that I would work there. Once I started working there and I just kept going back—I always felt I was very blessed to have fallen into that, because that's exactly what I did. Actually I stayed with them for 28 years. It was probably the best thing that ever happened to me that way.

Women were entering the workforce in the '70s. But not so much in the early '60s.

Right. I kind of grew away from a lot of the friends I had, because for them, going partying, and let's go out for a drink—and that wasn't my ballgame. And

I had my children young. You know, my first one was born when I was still 19. And 19, 21 and 25 is when I had my children. And I became very serious. I was more serious anyway, and I became very serious early and just stayed with that. I loved my job. I loved every one I had there. When I worked part-time on all the different—in billing, in engineering, in purchasing—a lot of times I'd work like three months at a time, maybe down in the time office or payroll. I just loved them all.

Strike photo courtesy of Mary Gifford

I used to have what I called my bible, and everybody knew "Mary's Bible"; it was a notebook, and I had every job description, down to the nth degree, for each job that I did. So whenever they'd call and ask me, "Well, will you cover here this week or there this week?" I had everything down. Eventually I had it all in my brain, but you know, I had it all written down.

Most places now have a policy and procedures manual or notebook …

In those days they didn't have policy procedures, except on the switchboard you were to be nice. That was, you know, the only thing.

Did you meet your husband at the mill?

Mm-hmm. I was divorced, and he was divorced. Oh, '79 or 80, something like— well, we got married in '80, so it would have been '79 or something like that. On strike duty, actually.

So there was a strike in—

Seventy-eight, '79.

You were picketing?

No, I was working 12-hour shifts—they put me in charge of my shift where they wrapped the rolls and got them ready for shipments, because that carried on with what I was doing in production planning at that time.

When did you go from regular salary to supervisor?

I just gradually went up. I started as the clerk/typist in production planning, and after a year I told my boss I was just so bored. You know, there was no challenge to do that, and so I got a promotion to the next step up. I never did have the job of typing the bills and that type of thing. I just did it as part-time when they needed somebody. From there I think I went to planning the shipments, and from there—well, they put me on a special assignment for a couple of years, because we had like two or three warehouses full of broke. Broke—or rewind rolls—are rolls that they needed to get back in the system in some way. And what they did was go to a broker to buy lots of this paper to ship overseas. And my job for, gosh, the better part of a couple of years maybe, I was out in the different warehouses putting these rolls together, measuring them—most of them had weights on them—but measuring them sizewise and diameterwise to get ready for shipments to put in these—I can't think of what they're called—the shipping vans to send them overseas and put them on—I can't even remember the name of the company that I worked with, but I had a fellow that I worked with trying to ship those.

Wymore wasn't doing your shipping when you were there?

Yeah, Wymore was doing local shipping. In fact, oftentimes he would take these containers into Portland. But generally we used the barge. He did most of our deliveries, yeah.

So barges were still taking the paper, bringing the pulp, for the time you were there?

Oh, all the time I was there. They didn't actually quit using the barges until—let's see, I worked there through Crown Zellerbach and James River and about a year and a half into Simpson. And then that's when I quit. My husband had retired under James River, and then the boss that I had at that time retired at the same time, and I didn't feel like I could quit then, since a new company was coming on board. I felt as a department head I needed to stay and see that there was a transition. It didn't go very well, but I felt responsible to do the best I could do.

That loyalty, I see it a lot. People don't say "they"; it's still "we."

Well, it was really hard, because—and I'm not knocking Simpson—but there were all these 30-somethings that—well, no, they were under 30, because I remember, oh, we had a big major 30 birthday party, and there were six of us turned 30, or something like that. Through James River we had a wonderful computer system. And Simpson came in, they had their own computer system, and they were going to have their computer system, and we were to make it work. And it just never did. But it was an interesting year. There were many times I was at work at 3 o'clock in the morning, and I would go home at 10 o'clock at night. I'd call my husband and say, well, I'm on my way home. And he'd have a hot meal for me. I'd eat and try to converse and couldn't, and went to bed. That went on for about a year. I worked 16-hour days, mainly, for a year.

One thing nice about the "office jobs" was that they probably didn't do shift work.

No, we didn't. Except during the strike, [when] we worked 12-hour day days and we worked 12-hour night nights…

But then, the transition from James River to Simpson…

Well, see, by then I was a supervisor, and that made me where I don't get overtime. So I would work 12- to 16-hour days. I was on call 24 hours a day.

So what was your impression of how the supervisors ran the mill during the long strike?

The strike was very interesting. The first strike lasted six months, and I just had a lot of fun. I always loved doing physical work, and gosh, I went down to 104 pounds. It was great! They brought meals in to us. It was really hard work. But I enjoyed it a lot. I enjoyed the people; we had a great group of people working the first strike. The second strike was no fun at all. It was nine months. By then I was married to my husband now. He worked such long, hard shifts, because he didn't work just the 12. He was there longer than that for the transition because they were having a lot of trouble on the machines.

They had people who didn't know what they were doing on the machines, and the machine work was a lot harder to learn than what I was doing. I was just teaching people how to wrap rolls, and it was repetitive. So we would stay up with whatever the winders were putting out. The machines, it was a whole different ballgame. Oh gosh, there were days that in 12 hours these guys would never even get to sit down, maybe wouldn't even have time to eat their meals.

So it was really hard. You had lab guys. They brought people in from other mills. I can remember I had an accountant from St. Louis work for me, and he could never get the tape machine to work right. And all you did was pull this four-inch tape out and flip it, and slap it on the roll. He could never get that. One day he picked that thing up and threw it clear across the room. And I thought, oh my gosh.

But you know, they were away from home for 10 days in a row, and that was part of what was getting to them. And they were just not used to working like we did. They were in an air-conditioned office. In the summer, working in the mill is very warm. This was days before they had those cool boxes or cool whatever they call them now. I know my son works there now, and of course he's out on the floor a lot. It was hot in those days.

So the men flew home every 10 days?

Mm-hmm. It had to have been very spendy. They used a lot of people from the Portland office, and those were office people again that never got their fingernails dirty, and that kind of thing. And they brought people from—well, let's see: Wauna and Camas were on strike, too, so they had that—but they would bring them from the mills in Texas, from the mills in California. The main business office was in Portland; they'd bring some of those out.

Talk about them flying home: They've mentioned all these supervisors coming, but I thought they just stayed.

Oh, no- I would say probably 50 percent of the people that came for strike duty were from out of town. They would have to fly in. Like if they were coming in on day shift, they'd fly in the night before. And then they'd fly out the morning that they got off. Sometimes the wives would fly in, and they would spend their five days off here and go to the beach, or you know, that kind of thing. And it got for Bob and I, on the second strike, it took him three days of almost solid sleeping to catch up. So we'd stay home on our five days off, when we would go from day shift to night shift.

But when we'd go from night to days, you kind of had like an extra 12 hours in there, so we'd go someplace. We went to Alaska, Mexico many, many times. We'd fly down to Mexico for five days, just to get him away from—because he'd think, oh, I've got to do all this stuff around the house, and I need to—you know—and I'd say, no, you've got to rest. You've got to sleep. And so that's what we did.

So do you remember the dates of the strikes?

The first strike started in July of 1978. I think we'd just spent six weeks in Europe; my two sisters were over there. And we'd just took my mom and spent six weeks over there. I don't think I was home a week before the strike started. So I'm not exactly sure what day we came home, but it was like then and lasted for six weeks. I felt really bad for the fellows who were out on strike.

So it was six weeks or six months?

Six months was the strike.

There was a two-week strike somewhere back there.

I know my husband worked on one—oh gosh, I don't know—in the '60s sometime. That was one of the times that I was off, and so I wasn't involved in that.

I think there was one strike that was caused by kind of the battle of the unions or something.

Well, I think there was one union at one time. And then they split and had two unions. I think one was 166 and one was 48, or something like that. And of course, I wasn't involved in that at all. I didn't have a relative in it, or I wasn't working. I was still in junior high school, probably.

So when they brought you in the meals, where did they come from ?

The first strike, the meals were prepared at the West Linn Inn, and most of them were very good meals. The second strike that was brought in, they had those catered, because by that time they had shut the—I'm thinking; I don't know if I'm wrong—I'm thinking they had the Inn shut down by then. I remember one breakfast shift—we had worked all night, and then they'd bring breakfast in, and I found out later that the cook had a hangover. Gianotti was the mill manager at that time—and they'd bring them in in Styrofoam boxes, and it would be scrambled eggs, bacon, hash browns, toast, you know, that kind of thing. Breakfasts were all pretty much the same. And I can remember opening that box, and six pieces of bacon in there were just limp. They hadn't even been passed over the stove. Well, I'll tell you, when I was off shift, I took that box, and I took it up, and I put it on Gianotti's desk, and I said, this is what they expected us to eat this morning. And I guess you know what hit the fan. That didn't happen again. But most of the time the meals were wonderful. They really were very good meals. And they would bring in snacks, sodas, these kinds of things, that we could have during the day. Like I said earlier, I went down to 104 pounds on the first strike. And the second strike I gained weight, I think. Everything was so good.

Somebody told me that the West Linn Inn came down in '81, or '82. I don't know if that sounds right to you.

That sounds about right.

And so the second strike would have been—

Right after that. They had closed down some of the paper machines. In fact, they had taken out one, two and three. They had left four. They'd taken out seven, eight and 10 by that time. Then when I quit, they only had three machines left, in those days five, six and nine. They're now one, two and three. But that was all that was left. Now, the West Linn Inn many years ago was a major thing. A lot of the guys lived there, in fact, that worked in the mill. My husband said when he started there they would stop for drinks on the way home, or coffee before they'd go in. So the Inn was built, actually, for the mill. But it had a wonderful restaurant that was open to the public.

It did seem like it was a dark place. When I first started there we had Christmas parties, and we'd do Christmas skits and beautiful dances held in the lobby and the big ballroom. And I guess it didn't include the union. We used to have huge crowds. I can remember being in skits the first few years I started. And then the Christmas party got to be a little bit less and a little bit less. And you know, it had to have been very expensive, and they started to become aware of expenses more and more. And I think at that time a mill was standing alone financially instead of as a corporation, maybe, and so West Linn maybe wasn't producing as much as Camas or some of the other bigger mills.

I have pictures of Rosie Schultze when he was Santa.

Oh, I remember the Santa days, yes. The Christmas parties were great fun, but they did become—there was a group of people that drank a lot and became pretty rowdy. I didn't drink, so I probably left earlier than most, and don't know how things went from there. But I do know—I don't think the Christmas parties lasted more than maybe five or six years after I started. And then in the summer they had big picnics too. I don't remember where they'd have them, but huge picnics. And, oh gosh, tons of food and lots of entertainment for the children, and games and that kind of thing. That was great fun.

That was again expensive.

That was expensive, and the union was not included. Well, I always heard, though, that the union put on a nice Christmas party for the union fellows and their kids. And they had a picnic in the summer.

A lot of these guys had incredible other lives at home when they were doing rotating shifts. They had cattle, or they were running a welding shop on the side, or whatever.

My husband had seven rentals at one time, and he worked 16, 12-hour days a lot because of the position where he was, while he was still union, in the mill. And he did building for people and that kind of thing. I don't know how the union guys did it, actually, because in those days it was rotating shifts that were, you know, the eight hours days and then the eight hours swing and the eight hours graveyard. When you would go from graveyard to days, you would get off on graveyard and went in the next morning on days. Oh, I remember those were killers for the guys. That was horrible. Now they've got wonderful shifts. My son works there now, and they work 12-hour days, so you've got two shifts that you're working on. And he works like three days, two nights, five off. Two and two, five off. Two and three and four off. Then they start over again. He loves it, because he hunts and fishes, so it just—it's like having three vacations a month.

So did that just start with West Linn Paper?

Yes. Yeah, when West Linn Paper Company started, they started at different kinds of shifts, and then came into this pretty soon, finding that it really worked. And when West Linn Paper Company came in, too, you sign off for your job. My son is machine tender, and that's the guy that is over whatever machine they're on. He's responsible for that paper from the time it's pulp coming to him until it's loaded on the truck. And see, when we worked there, you know, these guys, some of them were there for eight hours, and they didn't care because they didn't have to buy into anything they did. And that used to just bother the heck out of me. They didn't even have to sign—well, at one time I remember they each had a stamp, and you put your name on the end of that roll. And you were responsible for that roll. If one of the sales people or lab people went in when they were having trouble running paper in a print shop, they could see who put this paper out. But then it got to the point where, well, we can't do that; we won't make these people responsible. And maybe they'd work hard and maybe they wouldn't.

Well, and one of the bad things is, if a paper machine is running poorly, the ones who have to put it through the supers and the winders, they're the ones that get credit for that poor paper. I mean, it comes into their shift. And then you get out of the problem, and then the next shift gets your good paper. And that always was a real bone of contention between supervisors.

When did they start also putting things into Spanish? That's while you were still there?

At the very end. Because most of the Mexicans worked in the ground wood.

Oh, before they closed that. Whew! That's bad work.

Oh, now, my boys loved it. Both my boys worked there, and they started in ground wood, and they loved it because they worked really hard and filled their carts. They filled the cart, and then they could go out and sit on the verandah and watch them fish. And they would tell me that there were even guys that would go down the steps and fish, and they'd maybe be gone a half an hour, and then they'd go back up and fill five or six carts again, and then they could—as long as there was a flow and the continuity, apparently it was okay. I don't know; I never talked to the supervisor about it, but apparently it was okay.

I remember efficiency experts. Efficiency experts weren't the most popular people in the mill. They usually screwed things up more than they helped, actually. Probably mainly because it was like, wham! We're going to put this into effect now, instead of slowly going in, and I think that would probably have been maybe a better way to do it.

Let me ask about your son: why did he want to work there? When did he go?

Well, I have three children—there were all five of us working there at one time. Bob was a supervisor on the paper machines; I was a supervisor in production planning and shipping; Tim, Jeff and Susan did shift work; they were on the extra board. Susan was putting herself through college, working there summers. Tim started the first year, quit, and went to become a welder, and liked the mill better—pay, hours, time off, benefits, that kind of thing. Jeff went to work—he went to college for one year, decided that he wanted to make the big bucks and have his toys and buy a house.

That was their goal—to have all their toys, their boats, their cars, their trucks, and their house before they got married. And their goal was to get married at around 25, so this is what they wanted to do. I mean, I thought this was really something. But Susan worked there—I think she was there three summers. Of course, living at home, she could—you know, I did all the grocery shopping and the cooking and the wash and this, so all she had to do was work. And she worked a lot of splits or doubles. And she had almost enough money when she quit, or when she graduated from college to buy herself a car. We didn't have to put that much in to buy her a car.

Tim, he rented for a while, bought his truck, and then—he loved working at the mill. He liked the guys, he liked the pay; it was great...his time off... He worked a lot of overtime. And he liked it enough that when Simpson shut it down and then West Linn Paper Company started up, they had gone to the supervision and said, we need 35 core people to start. And both him and Jeff were chosen. And people always told me, boy, if we had a crew like your two boys, we wouldn't have any trouble at all. And that's the way they were brought up, to work hard and own in to what you're doing and be responsible.

Jeff at that point went to work at the other paper mill for a little while, the one on this side of the river, Blue Heron. Then when they wanted him to go back over there, he says, you know, I've always wanted to be an electrician; I think I'm going to do that. So he became an apprentice, and now he's a journeyman.

They have electricians at the mill.

Well, he joined the union, and that's how you're training; it's a five-year, very intense training. And you have to give them back so many years, or you have to pay them back the $20,000 or whatever your education would have been to be a—so he is working for Cherry Hill, and he's just kind of stayed with them. They've kept him pretty busy, and he's foreman. And when he finished—well, when he worked at the mill, they'd do testing. And one of the guys that tested him, one of the foremen that tested him, called me, and he says, well, I don't want to give Jeff a job; he really should be in college because he's so bright. And I said, well, you know, that's not our decision to make. That's his decision to make, and if he decides eventually to go to college, then he will. But I said, this is his decision; he's 20 years old; if he wants to work in the mill, and you think he's good enough, then you need to do that.

So when he went to electrical school, he never missed a class in five years. He had classes three nights a week, for two and half hours, three hours a night. He never missed work, and he got the highest grade that had ever been, at that time, listed in the state of Oregon. So that's pretty cool. He just works on big buildings. He's building a hospital right now, up 84. Up I-84 they're building a big hospital, and he's supervisor up there. My dad was an electrician, and he always thought that was kind of neat. And so my dad died 29 years ago, and when Jeff graduated, mother gave him all these tools. Of course, they're nothing that he could use now, but it brought tears to everybody.

How many people were in your department when you started your first full-time position?

When I first started in production planning, let's see—there was me, the typist, supervisor and assistant supervisor. There were eight of us that were in there when we first started. And it was production planning and shipping. So you had the production planning people, and then you had the people that were in shipping. We were under one roof, but we cross-traded until there was me, because then I got where I did go…When I quit there was six. We did our own transportation, and that's how they still have it. We got our own trucks or made arrangements for the railcars, this type of thing. When I started in there, there was an office in Portland that did that. They had, gosh, three or four people in that department in Portland.

You worked with the barge companies too, then? Can you talk about the barging a little bit?

Western Transportation Company. I remember several things about it. On a Saturday once a year, we'd have to go in and inventory this enormous warehouse. It was just a huge job. And another thing, one of my jobs, was to—I'd hop on the barge as it came through the locks, and I'd inventory what was on that barge. I'd have my list pretty much that I knew what I'd ordered down, and I'd have to hop on the barge with my life vest on and count everything and have it all inventoried. If there wasn't that much, I'd jump off sometime as it was going through the locks, or else I'd wait until it docked. During strike duty one time I was a Hyster driver, and I was driving the Hyster at night sometimes. And my job this one night was to unload the barge, and it was snowing. And I started down the ramp, and I slid, and my Hyster just slipped all the way over and just teetered there; scared the socks off me. I had a full load of pulp on the end of it that—gosh, what did they weigh—maybe six tons, something like that, on the front of me. And if I'd slipped clear into the river, there wasn't a soul around, so nobody would have known.

Ooh. Sounds like a safety committee issue.

Yeah. But we didn't load barges in the snowstorm any more.

After you almost fell in?

Yeah, that was really scary. Actually, you know, I don't even know how many people I told that to. I know I didn't tell supervision up in the office, because I think they'd have said, you're not doing that any more. That was kind of fun. One time I was hauling pulp again, down through—it's No.4 paper machine; on this side. And I took out a Fourdrinier. It was kind of spendy, except at that point they had decided that they weren't going to start that machine up again. So it was kind of a moot point at that time, but it was like an oops!—took out, I don't know, a $10,000 piece of equipment.

That was during the strike, because that's the only time you would be doing a job like that?

Absolutely, yeah. That was during the strike, and that was one of the things that made it really fun. I am very good at backing up now. I can go down a long ramp backing up with my car. I do a good job now.

I always ask if you have any stories about the locks.

Oh, yeah. Many times I'd walk across the locks, yeah, on that little bridge or the walkway. Oh, yeah. Yeah, they'd go across those, and—the bad thing was, when we were having a truck loaded, and I needed one more load of paper across there to get that truck out of there, if that bridge is up, you just sit there and wait.

Willamette Falls Navigation Locks and Canal, Crown Zellerbach mill buildings, circa 1950s.

There were several years in a row when a mommy duck would come to the top of the locks—do you remember this?

Oh, absolutely. I remember them opening and closing the locks for the ducks going up and down. And oh, that was so cute. And we had beaver in the locks. I remember that several times. I remember something else swimming through the locks. It wasn't a deer; something bigger. That wouldn't end quite so well. But I remember the ducks and the beaver. I remember it being in the newsletter. I've kind of passed my newsletters on now. I don't have them any more.

How aware were you of safety issues, except when you were doing jobs with the forklift?

Working out in the mill was a real scary thing. Every place you went was a pinch point, an injury waiting to happen. You could cut yourself on the paper knives, you know. Some of this was during strike time, and some—when I'd walk down the dock, you'd have to be aware of the forklifts coming. Some of them weren't as slow as others, and they'd just whip in and get a roll and come out. There were times that you could be pinned against the wall, and I mean that's happened to me several times that it was that close. And I would, you know, whoo! I would think they would see me there, but maybe they didn't. It would be pretty hard to miss; I'd be walking down the dock. But it was just constant. You were just aware of it all the time.

By the time I became supervision, it was drilled into us constantly to watch for things. And then the earplugs and the safety shoes. I hated to wear the safety shoes. The first strike duty, I feel like I almost was ready to lose my job because I refused—I wore tennis shoes instead of safety shoes. Finally I—okay, okay; I'll wear your safety shoes. But one time I caught my foot down in the rail, and I could have had my foot crushed if I hadn't gotten my foot out of my shoe. But that's not something that happens very often. It's more the other way: something falling on you. When I first started there, safety was an issue, but not near like it was as the years went on and they really became aware of things happening. So many scary things could happen.

Okay. I've heard about the nips, like the ingoing nips and the outgoing nips.

Well, it was like rolls going this way, and you get your hand stuck in there. Or if you were throwing broke down the hole or something, you know, that—just so many place things could—I was scared to death having my boys work there, to tell you the truth, because you know a roll—how do you stop a 2,000-pound roll? You don't get in front of it, that's for sure.

People talk about broke holes and you talk about broke and rolls. Are we talking about the actual roll that the paper was on, or are we talking about the paper?

Well, there's several things. Broke rolls: if they knew that they were into a problem, and this reel, say, this reel of paper couldn't be used, sometimes they could take it over in front of the broke chute and use a paper knife and take it off, and it would fall down in the broke hole, and it would be re-pulped. If that was full, they couldn't do that. They would bypass the supercalenders and just run it—make 35-inch rolls into broke rolls, and then those would go in the warehouse to be used when they could use them. Every so often they would say I need, you know, so many tons of broke rolls. And that wasn't something we kept an inventory of, particularly. So the supervisors could tell a forklift driver, go in and bring me 10 rolls of this kind of broke or that kind of broke. Usually you'd try to use like Capistrano or Sonoma or those—you'd try to use the same kind of broke in the same kind of paper that you were making.

...

I used to just relish this kind of sun, and I'd pick berries in it, and didn't think that much about it until I went to work at the mill. That summer my two sisters were still picking berries. I don't know; we were going to go on a picnic or something—so my dad and I drove out to get them. Well, for some reason they wouldn't let them leave until they finished their row. I had on shorts and a little top, you know. So I went out—being an air-conditioned office worker now—I went out to help them pick, and I had heat prostration or whatever you call it. I'll tell you, I don't do sun as well anymore.

Well, I'll tell you, working out in that mill was hot. I don't know how these fellows did it. When we did strike duty and—you know, you worked through the seasons—I can remember we'd put a tarp down in front—you'd wrap the rolls and they roll outside onto the boards, you know, that they put them on. And then the forklift could take them away. Well, that blast of cold that you'd get when that roll would roll out would just freeze your toes, and yet it was really warm from the machinery. But if they shut the mill down in the winter—and they did once in a while; they'd have the mill all down for some reason or other—oh my, it was cold in there. It was just like working in a refrigerator, just freezing. So I guess you had to have something to complain about.

In the summer did you ever get down to the pulp area?

Well actually, for two or three years, we used to do tours of the mill all the time, 10 o'clock in the morning, two in the afternoon. I remember that very distinctly. If—any visitors—we had tours. And usually they had people from the lab take these people on tours. I don't know how it came around that—well, let's just have Mary do the tours. They cut out the two o'clock tour in the afternoon, unless somebody special was there. But I did tours for a number of years, which was the entire mill. So yes, I did get down in ground wood. I don't remember so much that we went up into stock prep. I suppose we hit part of it, because we did go up to the paper machines.

And I got to where I knew enough to get by, or else the people that I was taking on tours didn't know anything, so what I knew was enough. And we usually walked down nine paper machine—well, three, now, it's called—and I could point out all the different points, because you were at the side of it, and that was a good way. And if you went through one and two or five and six then, there was too much activity going on. So that was a good place to stay away from. No. 9 is

No. 3 now. It's the one that's by itself; that was No.9. When I worked there, they rebuilt that. In fact, if you look at the dry end, my name is in the concrete there. Mel Tietz was one of the assistants then, and he said, "Sure, put your name down there." It's under where the reels at the very end are. And the year is on that, and I can't remember what the year was.

That's great. What happened to these machines? Did they scrap them?

I don't remember—when they started taking the machines out, I don't remember any of them that they just tore down and sold for scrap. I don't remember that at all. As far as my recollection goes, they sold them all. They went to other countries. I used to love to walk through the mill. I'm very technically challenged, but it just fascinated me seeing these machines run. No. 8, 10, those were small machines. They were narrow machines. And one of them, we'd made a grade of paper called Crezon. If you'd run 185-pound Crezon, that machine would just [repetitive sound]-- so slow. It would take, gosh, I've forgotten now, how many hours to make a roll of this paper. It just was so slow. And you'd make different colors, different basis weights and that kind of thing. Really heavy. It was interesting.

We used to make soda straw. We used to make that bilious orange paper for McDonalds. I remember one time my boss and I—his name was Al McDonald; he was a real sweetheart—we must not have had much to do to go to this level. We were making pink packet paper that they put the sugars in. So in a reel of paper, which was 140 inches wide by 40-inch diameter, we were trying to figure out how many pink paper packets you could make out of a roll of paper. I don't remember what we came up with. I think we were so dazzled by the number that we finally quit.

Millions.

Millions. It was just huge. But yeah, we used to make 34-pound paper, and that was for some of the special "U.S. News & World Report." "Newsweek" used 34-pound. I don't think they go down that far now, but I know that we used to not like to go under 40-pound on 6, which is now 2. Sixty-pound ran best on 5 and that kind of thing.

So you can't run certain kinds of paper on certain machines?

But they tried. Sometimes they'd hay out for hours. Sometimes it took hours to get a transition, let's say from 80-pound to 70-pound, it would take hours. Now my son says that it's first reel, you've got good paper. So that's great. I mean, they've just done wonderful things.

If you had the chance to go down there now, would you like to go down and tour?

Oh, I would love to. And my son keeps saying, well, I'll take [Dad]—because Bob would like to go through and see—they've done total rebuild of the machine he worked on, and he would like to go down and see it.

You told me your husband was one of the gentlemen that did the group interview up at the [Clackamas County] historical society?

Mm-hmm, right.

How did that come about?

They were doing some oral histories, and with the total transition of the mill—you know, they took out sheet finishing, they took out the wood mill—I mean, it is still there, but they're not running like they were when we were there. And they didn't want to lose that. One of the fellows who became my boss before I was supervisor in production planning and shipping, he came from sheet finishing. He was the supervisor up there. He came down—I don't know; he was there maybe six years, something like that, but I was assistant and he was the supervisor. When Al McDonald retired, Del came in. Del Kennedy.

Where would sheet finishing go? I mean, how did it get phased out…

It just quit. That was another one of the jobs that I did, is I was to order paper from the machine to sheet finishing. And usually orders came in in so many reams of paper, or so much weight. And if it was weight, I had to figure out how many reams did it take to make the weight, and how many reams could I get out of a roll of paper, what size, what basis weight they wanted, and all that kind of thing. So I worked between production planning and sheet finishing that way. And if they made dunnage, which is the stuff that's kind of a nondescript-color paper that you get things wrapped in now—if you order something on the Internet or something, it'll come—and that's what we call dunnage. We would make so much of that and sell it to whoever. I don't even know any more. I don't know that I ever knew.

Does that mean that they just ship big rolls? They don't ship reams anymore?

Right. Oh gosh—in the '80s they quit making—I think sheet finishing was shut down in the '80s, and they quit making reams—we would make whatever sizes they wanted. A lot of print shops just run in sheets instead of rolls. And I think, well, our machinery was old, and it was becoming a very expensive thing to do, and most of your print places, the big ones, were going to rolls.

Was that the same as converting?

Converting is where they made towels, paper towels. Or not paper towels, but the shop towels and stuff. Yeah, totally different. Actually, converting shut down years before that, before sheet finishing did. I'm trying to think. I think I was only up in converting once, so I think they shut down fairly soon after I started.

One of the gentlemen said that one of those machines had come around the Horn. It was so old. It had been in the first mill.

Yes. Mm-hmm.

He thought it was No. 8. He said, "I sure hope somebody saved that, because that's a big piece of history."

They did—I heard where that machine—it's intact, in a museum, and I'm trying to think where it is, if Oregon Historical Society has it, or if it went to the Smithsonian. But I did hear that. Now one thing that we did save—I had Ian, I think it was, call me, and he says the steel Crown Zellerbach sign that was over someplace, do you think they'd want it at the museum? And I said, oh, yes! He says, well, you've got about a half an hour to get here and get it if you want it, before it leaves on a truck to be scrapped. So I got my husband and a couple of other friends that used to work at the mill, and we took the truck and went down and got it and took it up there. And oh, my, that's a heavy son of a gun. It was about so high and so thick, and really long, probably—I don't know—

There used to be the sign on the roof, facing Oregon City, that was like a football field long.

That one they broke down and scrapped. This one—where in the world was this one? Was it over the entrance to the clock alley? Of course, the clock alley used to be in a different place.

What's a clock alley?

The clock alley is where you punched in and out, and they don't even do that any more. They have time sheets that they fill out now. Actually, production planning and shipping took over part of the clock alley, because there was a clock alley and there was a time office. When the production planning and shipping was halfway down the dock, that's where I started. That's one of the reasons that they had to get this special thing so that I could walk down there was because that's where I worked, and I had to walk down the dock. But when they took the time office up to the main office, when they built the down-below part, the nicer part, they redid that office for us. Now there's production planning, and there's a door that leads into the supervisor's office, up into the mill area, through that way. You don't have to go all the way out and around and in like we used to.

So that was payroll alley. Was there a special Christmas bonus?

Well, they used to hand out turkeys like that for Christmas, at Christmas time, some of us would be up in the parking lot.

So you didn't have to bid jobs like the men in the mill did?

No, I didn't. I didn't bid on a job—I understand when the new paper company, West Linn Paper Company, came in, everybody had to bid on their job—office, mill, everything. No, I was just really lucky. I mean, I was just promoted right on up the hill, and I never asked to be promoted except that one time when I said I'm really bored just doing typing and filing by number, you know, that kind of thing. Not enough to fill eight hours. I needed more.

Do you remember your first paycheck, how much an hour you were making?

My first paycheck—I have it somewhere—I don't know where; I haven't been able to find it since we moved, but I know I still have it. My take-home pay for two weeks was $119. Forty dollars I had go directly into the credit union, so I actually would have gotten $159 for two weeks' work.

That's good pay in 1959.

It was! It was wonderful pay in 1959. I liked working out in the mill better than I liked working in the office, just because— I don't know—it seemed more challenging. But it got to the point where I thought, oh, it's payday again. I can't believe they pay me this good salary to do what I absolutely love doing and have so much fun at.

You are a very fortunate person.

I was, very.

What was the worst thing about working at the mill?

The worst thing about working at the mill for me was that last year that I worked under Simpson. And I don't know that that's something that you'd want to put in, but that was really tough. Working for Crown Zellerbach, working for James River, was a pleasure; I loved it. I can't think of anything I didn't like. You know, I'd do pretty much anything they'd ask, and I'd do the best I could, and—

…

Simpson just wanted it their way. They had made it work in a couple of other mills, but those mills didn't have a system anyway, so what they did was come in and just rip ours apart. And I heard that after I did quit that they hired somebody, brought somebody in, a girl in from the supers, and she worked at least a year because they had so much paper they didn't know what they had. With our, the James River, computer system, every roll that went across the header had an I.D. on it, and it was written down. It was somewhere.

With Simpson, they hadn't thought about putting something in their system to take second-class paper, which is paper that you couldn't sell, you know, the broke and the rewinds. They didn't have that in their computer system, how that is handled. So it just went on out on the dock, and they filled one warehouse, and then they filled another warehouse and another warehouse, and they didn't know what they had. So she had to do basically what I did, I told you earlier that I did, for a year or two, was to go through that. And that's when the mill was having real difficult times. But we were running all these paper machines, and that was before the computer system.

You keep saying you quit. Did you not retire?

No, I didn't retire. My husband retired. I wasn't 50 yet, and even though I had 28 years in, I just decided that I wasn't going to be one of these women that continued working after my husband retired. We had planned to do some traveling, which we have done extensively. And I just felt like I wanted to spend the time with him. He's older than I am, and we wanted to do what we could do when we could do it.

So you didn't get any kind of pension?

I have a pension. It's quite small. But I think the hardest thing is that there's no medical. I mean, I get medical through him, so I'm okay that way. And since I did work so long and I did get a good salary, my Social Security is as much as men who worked, or anybody who worked. So I have good Social Security.

That was a big decision. But how great that you got to travel with him.

Mm-hmm. We've had some great trips.

Were you working the extra board, during the '64 flood?

No, '64 flood I wasn't working at all. Probably during that time I was pregnant with my last one, because he was born in '65. And there were periods of time where they didn't need somebody to work—floating is what I called it—and they didn't need somebody, and so maybe I wasn't there for a year, two years, something like that. Now, somewhere we've got pictures of my husband working on the '64 flood, filling sandbags and trying to keep the water from coming up the locks and into the mill. Yeah, he remembers that, doing that. I was going to wear one of my Crown Zellerbach T-shirts. I haven't been able to let any of that stuff go. I still have my jackets and stuff. But they don't fit. My granddaughters were here this summer; they wore them.

Mary was very small, Mary Jo [Lefebvre].

Yeah, she is very small. She worked out in the supers. Actually, she took over one of the jobs that I used to have, planning the shipments. I think she's still doing that. Yeah, I hired her. She was kind of scared to come in. It was a big change for her. But it was days, and at that point she only had Aaron. And then she had two more while she was working for me. She needed something where she could make a little more money, have better hours, and so she came in. She's a very bright girl, and I've always been proud that I brought her in, that she was my choice.

Okay. I always ask people, what's the best thing about working at the mill?

Well, for me, the best thing about working at the mill would be really hard to choose. I loved the people, I loved the work, the paycheck wasn't too shabby, good friends.

…

You know, one of the amazing things about the mill is that there were 10 paper machines, 1,200 people working there when I started in 1959. And I can remember the first time we got a 500-ton day. Oh, my gosh, there were celebrations like you wouldn't believe! I mean, we got these little things that have, you know, the kits that you carry along with scissors and clippers and those kinds of thing in them. We got one of those that said "500 tons," and we got T-shirts that said "500 tons" on them, and all this kind of thing. A five-hundred ton day now is way too low. They're making 700 tons of good, saleable paper.

…

We get invited back every once in a while if—somebody that Bob worked with, if they retire, they always—they used to call Bob "Uncle Bob." He was Uncle Bob to everybody in the mill, the workers, you know. He was pretty well liked. And so we get invited back for those. So we've been back—we were back about a year ago. We didn't get a tour, but you know, as you walk around, you could look and see. Now, when I was doing tours, it was just for people, tourists, that were coming through, you know.

A paper mill is something that most people don't ever get to go through. And you just don't even think about it. So if you're traveling, and here's this paper mill that you can go through, this is great. You know, Bob and I one year just took a drive up into Canada. Well, we used to get pulp from way up there. And we just dropped in, and boy, we were treated like royalty, you know, and got a tour of how they make the pulp that they shipped to us. I was excited about that. I went into a paper mill one time on the east coast where they made seven-pound paper. Well, the lowest we made was 34. And to see this tissue paper was like, oh wow! You guys can keep that on the reel?

Did tourists pay?

No, tourists did not pay to come through the mill. It was just something that we did. And we had beautiful brochures that we handed out to them—I still have some of them—really nice brochures that we'd give to them. So I'll have to look before you leave. I don't know how many of those I have left.

Did you ever get involved with Pulp Siding, Clay Siding—?

I knew all the drivers, the tug drivers. I still know several of them. But I wasn't involved with those guys. I jumped on and off the barge. Once in a while one of them would be riding the barge up as they would go, most generally. But no, I didn't have anything to do with those people. I rarely went clear up the river, where Pulp Siding was

And when you were there, the locks were still working seven days a week, 24 hours a day.

The locks worked 24 hours a day, seven days a week. And in the summertime you'd see the boats going through, and again like the ducks going through and that kind of thing. But I think you have to make an appointment now for the locks to be opened.

I used to be fascinated to sit there and watch the locks work, too. I thought that was just really neat. When we were back in New York, we went on the Erie Canal, and I just thought, wow! Whoo, this is history! But I remember when they did the hundred years for the mill, and that's history.

I thought they shut down the pulp operation as soon as Zellerbach sold.

It was still running during the strikes, because I can remember they got loggers in to work up there during the strike. I've got pictures of some of those, the loggers that came in. And thank heavens I did write some names down, which is kind of cool. But I'd have to look through other stuff.

We got a lot of stories about fish in the mill.

I'll bet. Did you get stories about the marijuana? They found, in one of the tanks—after they quit making the sulfite they just had these tanks empty there—and this was when I was still working there—they found a marijuana mill going on. They were using the mill's water; they were using the mill's electricity, and they were growing marijuana. You know, they had the lights and all this stuff in there. I don't remember how they found it, but I remember shortly after that they had a big thing where everybody had to be tested for drugs. I remember when I had to go over, and the nurse was telling us that, oh, there are some guys that are so scared they'd stop and have a beer on their way over to have their drug test. She'd say, I think you'd better come back another time; this isn't going to work.

So they must have worked there.

Oh, they did. I still in my mind can kind of think who was in on it, but—it's neither here nor there. Yeah, they had a drug mill going, marijuana, at one time. It had to have been in the '80s when that happened.

Somebody said they shut the mill down for an hour at noon every day for a while when the [I-205] blasting was happening, because dynamiting would have wrecked the paper—

You shut a machine down, and in those days you were talking at least $25,000 a whack for a machine to go down. So I would be surprised; I'd be real surprised if they did that. I don't really remember the bridge having any impact at all. Probably the only thing I can think of is the—"new bridge," we call it—the new bridge going in was, maybe Wymore was—it was easier to get his trucks over to the mill that way instead of over the old bridge, because those trucks are pretty good sized. Steve's got the business now, and there was Don and Gene. Gene was the owner and the brother Don, but the dad is the one that started it. Somewhere I have a picture—and it's in the archives at the museum—pictures of their first trucks coming in, taking out one roll at a time over to the "Enterprise Courier" in Oregon City, the newspaper.

People talked about when they first did the parking down below, only supervisors.

Oh, yes, because I was given a very prime place to park. Somehow, the head of the mill figured a way to get that. His car was too big, and mine was small enough that I could park someplace else, or something or other to do with that. Anyway, he got my parking spot. But then I do remember when the union had a real fit: why are your people more special than our people, to park there? And we grumbled—when we'd come in in the morning, the day shift was still there, so there was never a place for us to park, ever. And I started work at seven in the morning, generally, or earlier. And there was never a place for me to park. And I kind of resented that in a way, because a lot of times I was there at dark and had to walk out to my car, clear up, all by myself in the dark up there, and then I thought, well gosh, so do the others. So buck up and get over it. Yeah, I remember that.

Actually, driving down and up that ramp was kind of a—whoo—you had to stay right in the straight and narrow to get down. Oh, and how many times a big truck would go down those, and they're in to here, and no place to turn around. Here are these big semis. Which was really hard for the cars that were parked down there if they got stuck. What they had to do was call their office to send another piece to hook onto the other end, and then they would take them apart, and they could take them up that way. And then I've seen truckers that tried to turn around in there too, and if our car was parked in the way, sometimes we'd have to go and move them, and move them clear down to the other end so that they could jockey them around to get back up there.

How many years at the mill, between your husband, you and your kids?

Oh, my word! Years are 28 years. Bob was there 39 years. Susan worked there three years, long hours. Tim is still working there, and I think he got his 20-year this year. I think it's his 20th year. Jeff worked there off and on a couple of years, something like—well, maybe he was there longer than that. Maybe he even got his five-year.

You're pushing 80, 90 years.

So lots of hours. I mean, it's been a great place to work. Tim still really enjoys working there, and he does good work. And the noise. Oh, my word, the noise in some of those areas used to be just tremendous. That's good, you know, that they do make you wear ear protection now, because—well, all the guys that you've talked to are real hard of hearing.

The first time I talked to George Droz on the phone he said, "Speak up! I worked at the mill, you know!"

He's a sweet man too. There's so many of them that are such nice people, and we've lost a lot. A lot of them are gone already.

That's why we're trying to get some stories before they're gone.

I'll have to see if I can just think about—I haven't thought about it in years—where they took that one paper machine. But they took it intact.

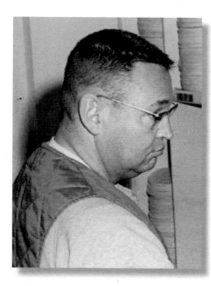

Harold King

Interview July 29, 2006

Died June 20, 2010

I'm Harold E. King, and I'm 84 years old, to be 85 on September 11th—9/11. Those terrorists, they disrespected my birthday.

I guess they did. So what year were you hired?

Nineteen forty-eight at Crown Zellerbach.

How many years did you work there?

Thirty. I worked 30 years at the company.

So if you went in '48, plus 30, you went out—

Seventy-eight; 1978 I took early retirement; bad idea. Bad idea.

Why?

Well, because I had had some compelling personal reasons when I did it; I thought it was a good idea. I had other income, and I was into an enterprise with my son. I was going to help out, and so I took early retirement. Well, things happen, you know, over the years, and I could use the money now, the extra money. But it's too late to look back on that. But that was what happened at the time.

Did you have any jobs before you went to the mill to work?

I got out of the Marine Corps in '45, and I'd had about over a period of a couple of years—three years—I'd had about a dozen, something like a dozen, jobs. I was a guy that would get browned off at the management and quit. And I did that several times, and I sat down and had a talk with myself one day, and I said, you've got to get a job and stay with it if you ever want to build any seniority or any security. And besides that I was a father, so I had responsibilities, you know. You have to have medical care and one thing and—so I decided to go to work— you know, there was plenty of jobs in those days. I got hired at Crown, and I decided, I'm going to make this stick, because I don't care: Come hell or high water, I'm going to work here, one way or the other, because I've got to have the job. So that's the mentality that I had when I went to work at Crown, and I made it stick.

Do you remember the day that you went in and applied for the job?

Sure. I went from the job I was on--it got off in the afternoon--and I went down to the company, and they hired me that day. And I went to work the following day. It was a day job, the yard gang, and it was pretty good, you know. And I didn't have to start right in on shift work, which will kill you. Bad idea, you know, that shift work.

Do you remember what the pay rate was?

At the mill—$1.42 and a half was the base pay at that time, union scale base pay. And damn glad to have it.

So they were union in 1948.

Well, that was one reason I was there, because I knew that. And we had an NLRB election after I'd been there for a few months. It was to initiate the union shop. They'd had a union contract which did not have union shop. And that's the way I remember it; it was about making the union shop part of the agreement, the labor agreement.

What was that first day like?

We were laborers. We were materials handlers, and that's what we did. We hand-trucked, we handled sacked material, we handled barrels, all of that by hand, and the rock pile. We handled rocks, lime rock, by hand. Everything—almost everything—was done by labor, which later on was transformed to use machinery. A personnel man always took you out to introduce you to the boss; I remember that. I'm trying to think of his name, but it slips my mind. He was very nice; he was a gentleman. So anyhow, the foreman—you know, back in those days the foremen were loudmouths mostly; they were loud. And sometimes obscene in their characterizations with giving you orders and so on. But it was a kind of a rough and ready place to work actually. I mean, a lot of tough guys, you know, doing that kind of work; a lot of us just out of the service, too.

You want to know what I did. I don't know what I did the first day; it was labor. And so on. And the first couple of years was pretty tough, especially when you're trying to keep up a weightlifting program along with it. It's really tough. And I asked for split days off so I could lift weights on my days off at that time, you see. One thing about the yard

Harold King, 225 lbs. in 1947, before hire at the mill. He later won the Mr. Oregon lifting competition.

gang: you never did the same thing for more than a couple hours at a time, and then you'd have a different kind of a job, another assignment. So it wasn't dull. I kind of liked it.

Didn't you have to eat enormous amounts to keep your energy up with all the work-- both the weightlifting and the regular job?

Yes. The answer is yes.

How did you handle your meals?

Well, big dinner. I never eat big lunches. Some of those guys would eat half a dozen sandwiches at lunchtime. I never did; my stomach wouldn't handle it with all that serious activity in the afternoon. It wouldn't have worked. So just a very modest lunch, and I was known for going without breakfast as well. That wasn't a good idea, but that's what I did. But I ate big dinners. Yeah, I was a real eater.

I stayed in the yard gang because it was day work. Let me give you some other reasons that have to bear on that, because in 1950 I became the shop steward in the department because nobody would do it any more. There was a lot of strife going on between the management and the men, and a lot of favoritism and a lot of dissatisfaction, and seniority was not being observed in that department at the time. And so I'm one of the rabble-rousers that didn't like that. It was seniority in the contract, but it was not being observed. What we needed was to enforce the contract in our department. I got with Chairman Bill Perrin, president of the union and chairman of the standing committee that administers the labor agreement in the plant for the union. And I conversed with him, and he was very happy to have my interest, because you've got a big department, 65, 70 men, and no shop steward. Nobody wanted it. I stood up and I said, okay, here I go. And I put my name on the board, and that's how I got to be shop steward of the yard department.

You were working your way into a position of some power.

The union made it possible for me to work in that paper mill, and I've told people that literally hundreds of times. I tell them that that's the only way I could have stayed there because I got in the union and I understood it, what it was for, and I understood how to use it. And that saved me, you know, because I was a kind of a volatile son of a gun that didn't like to take a bunch of crap from anybody. And as quick as I became a union official, they treated me with respect from then on; they had to. Anyway, I'm bragging, but it's true. It happened.

And from then on all I did was to calm things down in the department, and I followed the contract. I studied the union contract and the grievance procedure, and I followed the grievance procedure by the letter, and it works. Calmed things down. We had a lot of crybabies that just wanted to complain. And when you're going to take them in the office so you can discuss the problem, then they back out. I had a lot of that. But it worked. And I'm just tickled to death. I was the shop steward for seven years.

So 65 people worked the yard?

It varied, you see, because of the workload they would—in summertime they'd have a bunch of extras. Sometimes 70, depending on how much work was needed, you know.

How long did it stay manual?

Oh, they were beginning to mechanize when I went to work there, and they had three tow motors, lift trucks, and they had two of them working and the third one part-time, and they just gradually began to devise methods to handle all of that materiel on pallet boards, and in what form, and make it work, and how to stack it and all. It was just kind of a learning thing that was going on. There was still a lot of labor going on when I left in 1978, but I understand—they're probably completely mechanized now. I don't know—I'd like to take a tour of the mill sometime, you know. But I don't think they let you in there. Crown Zellerbach used to conduct mill tours any time. Anybody could go in there and go to the personnel—those were the good old days, you know, when everything was just laid back, relaxed, you know. It was a nice time. It was a nice time to be in industry, I can tell you that.

Two people that we've talked to led tours. Paul Miken said his first job was tour guide.

Gosh, I'd forgot about—Paul was a fighter. He was a boxer. My goodness. He married Arnold King's daughter, Paul Miken did. Where's he at today?

He lives on Miken Lane in West Linn, with his wife Rose. They're master gardeners now.

I think he went into the company before he left there, didn't he? Became management? I remember that, because, you know, over the years some of us used to go down there and have mill tours and stuff, and see some of these guys that were working for the company, you know.

Paul was a tour guide, and so was Mary Gifford.

They were getting some women in at that time, and they were bucking rolls, and it's hard, heavy work, and they were having bad backs and one thing and another.

Did your dad or any other members of your family ever work at the mill?

No.

No kids went there, no brothers or cousins?

Well, I had a son who was a temporary employee in the grinder room, and he lasted two weeks and quit. That's it. I mean, the rest of them—my descendants weren't tough enough to work in a paper mill; bunch of wimps.

We've heard how hard it was, especially down in the grinding room and sawmill areas.

Well, yard crew was pretty tough, you know. It was rocks handling and barrel handling, and we had 1,100-pound barrels of glue that we had to hand-truck. We had a ramp that went up—two guys, one pulling and one shoving, and I just pulled mine on up. I didn't stop for any—I was kind of a showoff back in those days, actually, and I used to do all kinds of stunts. I used to lift a 600-pound barrel of zinc dust. Zinc dust was used in the sulfite process, the burners. The burners—they had to melt the rocks and made the sulfite liquor. And anyhow, they had those 600-pound barrels—and they were not big; there were just small; it was such heavy stuff—and I figured out a way to lift one of those just as a stunt, you know, with a couple of hooks.

I've never heard about the rock before.

Lime rock and sulfur were the two ingredients that they made the sulfite liquor from. The rocks were dumped in the tower, and in the bottom of the tower—there was a water shower coming down from the top, and the burners shot a gas from—from burning the sulfur burners, the gas went in the bottom and it went up and it condensed and melted the lime rock to form the liquor, the cooking liquor, that produced pulp from wood chips. They were in…the digesters, right? So lime rock—they had to have it because of the lime content. That was a sulfuric acid—that was a sulfite process, not the sulfate process, which was different, which was black liquor.

How did the lime rock get there? Was that barged or—?

It came by rail from Georgia. We understood it was from Georgia, but it could have come from several other locations in gondolas, and then it was offloaded. We offloaded it by hand at Pulp Siding—that's up the river from the paper mill on this [Oregon City] side—onto the barge, the rock barge. And the rock barge went into the mill, and there again it had to be handled again, by hand and with carts, and—

Did it get offloaded in the locks?

We loaded the rock with crane. We loaded them into a skip, an iron container that was hung with chains on a steam—it was a steam donkey. And we would fill that with rock by hand in the gondola, and it would lift that and take it over and dump it on the barge. The barge would go to the mill, be tied up to the dock, the dock ramp, and that's where we would unload it. And it would be hand-trucked, that is, in a handcart, into the elevator that took it to the top of the tower, and it was dumped in up there. It's all handwork.

Wow. There was an elevator that went up a silo?

A silo, right. Yes.

You did part of the work up at Pulp Siding?

Oh, I worked up there a lot, yeah. Two days a week there'd be a yard crew that would get on the barge, and a tugboat would push us up there. We'd unload those rocks, and it would take us about two or three hours, and then we'd sit on our tailbones for the rest of the day because it was hard work and we weren't required to do any more work the rest of the day. And—on the rock barge—we would come into the mill in time, and if we were a little early getting in, why, we'd go and hide someplace so that we didn't have to take a job. That was by understanding, you understand; I mean, that was the way it was done, see? Go and hide out. But you had done a day's work; I guarantee it; it was a day's work.

So you got to know the tug people pretty well.

Oh, yeah. Yes, oh, yes; we were good friends.

Do you want to talk about the tugs and how they interacted with what you did?

Those tugs, those Crown Zellerbach tugs, were the old style. You ever hear of a marine engine that runs both ways, turns over forward and then turns over backwards, to back up? That was a marine engine. One of the tugboats—can't remember the name of it—had one of those engines, and when you wanted to reverse the propeller, you reversed the motor; it run backwards. It was a very slow-turning engine with a big propeller, and I think its RPMs are like 100 per minute or something; it was very slow. We all knew those tugboat operators, and it was just a kind of a leisurely kind of a job they had moving those barges, because you didn't go very fast, and tie and untie them. They had to pull the logs into the sawmill. But we were kind of friends because we always had a lot of communication when we'd ride up the river, and then back in when we'd be in the tugboat. We'd talk union stuff, you know, grievances and so on.

We've talked with, I think, at least three [tug] people—well, Paul Miken, Milt Wiese and also Art Dorrance.

Oh, yes, yes, yes. Art Dorrance.

You got any stories about those guys?

Well, Milt, he was a typical nice guy. Art Dorrance was a hell of union guy, and he was chairman of the Health and Welfare Committee for years, along with Hank Shaston. Hank Shaston, he died when he was 61. He was alcoholic, an ex-Marine. But Art came along, and Hank showed him the ropes, and Art done a lot of good—I wonder whatever—is he still alive? Is he still around?

Yeah, we interviewed him. He said he used to have pool parties.

Oh, yes. Oh, my! And they'd have a spit with a hog, turning on a spit. A baby hog would be in there, and I couldn't believe it. Dorrance was a smart guy. I mean, he was a smart guy.

I guess he worked on the PACC board for you guys...

Oh, yes, he was our chairman—I mean, he's just one of those kind of guys that he was the guy, you know. He knew it backwards and forwards. And I had something to do with some of that stuff myself, because I was union president for a while.

He was pretty interesting to talk to. He sounded like he had a good time on that job.

Oh, I'm sure he enjoyed that job. He was a digester cook—no, he was an acid maker down in the Mill B that I was telling you about—the sulfite process. He was an acid maker and he got that—he bid on that job up the river, and of course he just—it really liberated him.

Somebody described him as a plumber. He used to work on pipes and stuff too, didn't he?

Guess he was in the pipe shop. I forgot about that. I guess he was, yeah. He was in the pipe shop. But originally he was an acid maker.

A lot of the guys have done a lot more than one job

Well, they bid on those day jobs so you'll have a life. You don't have a life when you're in shift work. Boy, I worked four years in it, and I know it's a tough way to live.

How did you get into the four years of shift work?

Well, that job—we got a shift mule in the department because they had a slush maker back there, and we had to have a mule to feed the slush maker.

Okay, I don't know what a mule is.

A mule is a lift truck, and we always called them mules. I don't know whether anybody else—but it's a fork truck. I was damned instrumental in getting us that job and keeping in the yard department, because it was going to go to—they was going to put it in the beater department, and I fought them, and we got our way, and so when it come time to man it, we had to have four guys that would go on shift work, right? And we didn't have them, so I said I'll step in and I'll do my share, because I had a hand in acquiring this job for us here in the yard department, and so therefore, if I have to do it, I'll do it. So that's how I—you know, and it was a good job. As far as your duties were concerned it was fine. But it was hell for shift work, you know.

It was a mule driver? Is that what they called it?

Yes.

You're a mule driver.

Right. I suppose that's an improper term, but that's the way we regarded the mule drivers. All crazy; all the mule drivers were, you know, a little bit—not playing with a full deck, really.

You're talking about yourself.

Yeah. But that was their—other people's—appellation. I didn't really regard myself as being crazy, but anyway, it was okay.

So you learned how to back up really well?

Oh. You pick up your load, and you mostly back up, yes. You had to turn around and then—I'm stiff in the shoulders—but you know, being a mule driver—and then I got back on days, and look, I was all over the mill driving that mule, and I would do work for all the foremen in the mill that had something for—lift a load here or there. You did what you could, and you helped everybody out, and it made for a good impression that people had of you. All I was doing was my job, but I tried to be cooperative in doing my job, see?

And besides, then I was a member of the standing committee, starting in 1958, I guess it was, and it didn't hurt to be known, because you have to be reelected every year. And besides, I was on Perrin's team, and he was—you couldn't beat him. I mean, he was a working man's intellectual. He had brains. He had a powerful mind. I only wish that I had half of his capability and his understanding and his durability and his determination—all characteristics of William R. Perrin, who was a graduate of Oregon Agricultural College, better known nowadays as Oregon State University. He had a degree in business administration and a degree in journalism from Oregon Agricultural College. He graduated in the '20s.

So what job did he do at the mill?

Journeyman electrician. Union president. And a career labor leader if there ever was one. I sometimes mistakenly tried to follow in his steps, so to speak. Sometimes you're better off not to do that, not to try it, because you can't do it. You can't follow in anybody's footsteps; you have to follow in your own footsteps. But I certainly derived great benefit from my association with him, because we had good relations, and he showed me the ropes. Showed me the ropes.

It meant a lot to me because I was learning; I was learning all the time, and I hope I'm not over, done with that yet. I hope my education is not yet complete.

That's a great attitude. So—you said you knew Rosie Schultze.

He was a shift supervisor, foreman, on the paper machines, and in that capacity I would see him almost every hour. And I did jobs for him as well, on shift work;

just when I was on shift work. I lost track of Rosie after I went back on days, but he was a shift supervisor, and he was a hell of a guy. He's a big man; big, overweight guy.

He was 17 when he went to work at the mill in 1927, right after Crown Z became Crown Z.

Crown Willamette went from Crown Willamette to Crown Zellerbach in '28?

I believe so.

I didn't know that. They had all these mills in the northwest, a whole bunch of them. And what happened—the Zellerbachs all died off, and the people that owned, that run the company—of course, the stockholders—the people that run the company sold the timber to Sir James Goldsmith from Great Britain, and then they gradually began to sell the paper mills because they were not productive; they didn't have timber any more. The timber was a goldmine. The paper mills were not a goldmine. But that's what happened to a damn nice company, a company that was good to work for, and they—I remember J. D. Zellerbach; he was a humane individual, socially responsible, corporate leader. There ain't many—there's short supply any more, for sure.

People say that he would come to the five-year pin presentations and he'd shake everybody's hand and he'd say, "If you've worked for us for five years you're going to have a job for life."

See, that's a different era completely, when we had that kind of people. And J. D. Zellerbach, among other leaders in the pulp and paper industry on the Pacific Coast, were responsible for the original uniform labor agreement with John P. Burke of the International Brotherhood of Pulp, Sulfite & Paper Mill Workers. Back in '32 was when that was established. And J. D. Zellerbach had a hand in that for sure. And I never personally met him. I never met him personally. I think he was in the mill—in '46—'48, you see—I think in '46 he was still around, going through the mills and so on. But I think in—when I went to work at Crown, he was the Ambassador to Italy in the Eisenhower administration.

Who knew? Really?

He was ambassador to Italy under Eisenhower. I think he retired from that and was never with the company again. There were some Zellerbachs, descendants of his, but none of them had his stature. So anyway—that was a rip-off. They just took the money and run, didn't they—didn't care nothing about the company or what happened to the people.

J.D. Zellerbach portrait

Did you have friends that worked there after you were gone…?

Unh-unh. I was one of those kind that disappeared, and I went into a different life.

After you retired?

Right. I didn't maintain—I was living a different life. I was living over in Beaverton and working with my son in the property business, and—it was a long time before—Perrin got me back into things in 1985. Geez, I'd been—'78 to '85—that's almost 10 years, seven or eight years, in the State Council of Senior Citizens and United Seniors of Oregon. And he insisted, you know; he took me by the scruff of the neck and said, "Come with me; I want to show you this." Figuratively speaking, anyway. And I would do anything he asked me. There was nothing I wouldn't have done for him. But that's how he got me started in senior citizen activities, and that's a whole different story.

After people retire they have a lot of different—

You live a different life. It's like the working world is your real life, and then there's this other life that you go into when you leave the working world, and that's separate.

Do you remember like the first day that you didn't have to go to work there after 30 years?

Well, I don't even like to contemplate, because it was such a mistake that I made at the time. It was a divorce that occurred there directly. And it was kind of a bad time for me. I—that's my blank period right there. I really haven't anything to say about that period in my life. I don't want to talk about it at all.

So how many people worked at the mill when you were working there?

It varied because that paper mill was beginning to curtail and cut back on the workforce. And they cut out the converting department completely because the women's rates—the federal government required you couldn't have a "women's rate"; you had to have a rate. And so they shut down. They shut down the converting and they took it to Wauna. That was a new mill down at Wauna, and it was all automated. And the finishing department was practically shut down. So anyway, 800 probably; 750, 800, somewhere along in there.

This was in the early '70s, when you quit?

I'm talking about in the '70s, yeah.

How many strikes happened when you were there?

Several strikes, because—well, I wrote a dissertation here that sort of skimmed over the history of the disaffiliation, the establishment of a new union in '64 and attendant events. But let me say this: that when we disaffiliated the two international unions and established a new union, we applied for certification with the NLRB, and we were certified by vote, by election. Then they shut us down. We were in a dispute over a new contract, okay? And at the time, I guess the manufacturers didn't know whether they wanted to do business with us or whether they wanted to get rid of us, you know; it was just a small union, right?

And I was at the local level, you see, and I didn't have anything to do with that except when they shut us down, and then I was helping to run a strike here in Oregon City. So the bargaining went to Washington, D.C., where it got settled. It was a two-week strike, is all there was, and the whole community in Oregon City, they all—everybody was helping us out. We had a soup kitchen and everything; it worked real good. That was one strike. Then there was a shutdown in '69 of most—that was the year of separate bargaining. That was the year that the separate bargaining came in, and that really—that caused strikes everywhere. So anyway, there was a shutdown. We were not shut down. It was the Crown Zellerbach system, the way I remember it, that was shut down in '69, and we bargained locally at West Linn, in the hotel, and we were nearing an agreement with the manager. Camas was on strike. So anyway, we made a deal with Camas that we won't sign up with the company until you guys get a deal. So that was our position. So we got back—I think '75, '76, and we were on strike when I left. I think we got a contract in '75, and there was some shutdowns in '76. I'm hazy on this; I'm hazy. But anyhow, we were out for months in '78.

I'm still trying to figure out what years these strikes were, but it's hard to remember.

Well, there were some strikes along in the '70s that were lengthy, and largely mistaken on the part of the union, as far as I'm concerned. But that's beside the point—where was I at?

Well, we were talking about the long strikes. What happened to you during the time that you had to be off?

If you're a small-time labor leader, you're busy running the strike. You have to organize the strike, and there's picket—and you have to answer the phone. I spent two weeks—that '64 strike—I was in the union hall 10, 12 hours a day answering the phone. Those are not happy times—strikes—I'm telling you. You have been a failure in bargaining: The minute you're out on strike, you've already failed. You know, there's a lot of union leaders, they get drunk with power. And they think, oh, boy, I'm going to get these people on strike; we'll show 'em. Well, that's bad; that's bad philosophy. Anyway, some of those strikes were not—shouldn't have been.

So you got paid for doing that?

No. Oh, no.

How did you live?

Oh, well—how did I live? I guess I had enough money to live on. Actually, when these strikes occurred, I didn't have any children to worry about. A lot of them did, though. Some guys went out and got jobs; we told people to go out and get a job. We didn't have any strike fund, not in those days. I guess they have now, but we didn't have no strike fund. We were just a fledgling union, you know. We're lucky we didn't fail completely. We had help from AFL-CIO unions while we were in our burgeoning time, and we had the four-union coordinating committee. Bill Perrin was a great diplomat; he was a great ambassador for us, as our leader. There was the four-union coordinating committee; that was the IWA, the lumber and sawmill workers, the printing specialties union, with Ken Young, director, and AWPPW that was in coordination during those tough years where we were just getting going. And not only that, but they provided us with money as well as cooperation. And I love to tell that because it was all under the table.

Oh, nobody knew?

Because we were an outlaw union then, in those periods. That was union warfare.

So when you say "we," was the mill divided into different unions?

No. This was an industrial union setup, and you have the bargaining rights in that mill, in that plant. And so you represent the people in the plant, and provisions are made in contract provisions for the different craft—the crafts, the trades, or whatever you want to talk about, the mechanics, the millwrights. We're all in the same—it's an industrial union setup. No craft unions, and no separation at all. So you've got one bargaining unit the company does business with. You can have mill-by-mill strikes. And that's a bad thing. If you've got everybody signed up under a contract, a uniform agreement—like 21 mills or something like that on the Pacific Coast—24 mills, I guess it was, 21 companies.

Later on in '69 is when the union wrote a letter to the manufacturers' association asking for separate bargaining, see? And that's after Perrin was out of office. Bill Perrin was voted out of office. But when you ask about these craft unions and so on, we didn't have craft unions. We had an industrial union. Even after, when we had the separate bargaining, it was still the same union. They had to negotiate all these separate contracts. Well, you can't do as well with that as you can with a big unit.

Right. You don't have as much leverage.

Because that's power. Well, that's the way it needs to be. But we need those umbrella contracts that we had, like the uniform labor agreement on the Pacific coast that we lost because of our own leadership, in '69. That's when the new president of the Association of Western Pulp & Paper Workers had written a letter to the manufacturers' association requesting separate bargaining instead of uniform bargaining. And that's when it happened. That literally weakened the association at that point, because you don't have strength. You have 19,000, 20,000 people in one unit, for gosh sakes. I mean, that's terrific. That's a terrific tool to have. And labor unions need that. And there's a lot of separate bargaining going on nowadays. You know, well, the UAW—this is not a good time for organized labor, is it? The public is so brainwashed and publicized against the idea of organized labor. Everybody's organized except the workers, you know that? Everybody's organized except us. And we're too goddamn stupid to know that. And my relatives here, they live in a different world.

You said the community in West Linn, in '64 I think, gathered around the strike and had a soup kitchen. Can you talk about that?

That community of Oregon City/West Linn, the whole community—the county, I'd say. It was the first shutdown in this area of the pulp and paper manufacturing since, I think, the 1917 strike, when they broke the union. Now, I don't know—I know that I had guys, the old timers in the mill, telling me about the 1917 strike, where they busted the union. They had to go to work nonunion

Photo courtesy of Harold King: West Coast 1967 labor negotiations in Portland. Labor on the left, management on the right. Harold King at center of the labor table, seventh from the foreground, in white.

and take what the company would give them. And that was when the company that owned the plant at that time built the West Linn Hotel for the purpose of housing strikebreakers. That's how it got there. But the community and the businesses, with the soup kitchen it didn't cost us a dime. I mean, supplies just kept coming in, and the men's wives were down there cooking, and the strike shifts would come off shift, and they'd eat. And the pickets, on their way out, they'd stop and they'd eat. And actually, people were having a riot; they were having a lot of fun, you know. As long as you don't go too long, you know. It's no fun when you go too long. But it was two weeks, and we had an agreement in Washington, D.C. with the—I would think we got help from James R. Hoffa, Teamster President. He went on the evening news, assuring all and sundry that we support the Association of Western Pulp & Paper Workers in the their attempt to sign a new labor agreement with the Pacific Coast Pulp & Paper Manufacturers. And that didn't hurt us a bit, you know—I mean that kind of support.

Wow.

Publicity, you know, and here in Portland we had the cooperation of the ILWU, the Teamsters union with—damn, I can't remember his name; he was Mr. Teamster in the Portland area for years and years and years. And him and Perrin were just like this, you know. If we hadn't had the right kind of leader, that would never have happened.

So where was the [strike] kitchen, actually?

In the basement of the union hall. We had the union hall, 708 Jefferson Street [Oregon City]. That building, it was an old church house, and we had bought the church in '59 and remodeled it with our funds. So there was a basement, and there was a main floor, and there was even a balcony upstairs, and we had contract conference reports, and I participated. I was one of them. God damn it, I can't believe it; I was thinking back on it. And that church, that auditorium and the balcony, was full. Everybody was there. I mean, they wanted to hear the contract report. God damn, I was one of them. It's nice, you know, to think about.

I think I have a picture of people sitting in the audience at that meeting.

Exactly. There is a picture, and I should have a picture. I don't have.

There's women in the audience, too.

Yes, oh absolutely! And by the way, my album has got a picture of the 1965 executive board of Local 68. I believe Perrin is in the picture, and I'm president; I'm the guy; I'm standing in front. I can't believe that I did it! I can't believe it!

William Perrin, left end. Del Herndon, second from left, front row.
Harold King, front row, center.

It seems like this is a bigger part of your life than just the work.

Oh, without a doubt. I mean, the job was incidental. The damn job was
incidental. No doubt about that. But it paid my bills, provided me with a private
pension plan, and those are all things that were due to the union; I mean, the
activities of the union. You don't get that kind of stuff just free of charge without
going after it.

Okay. What you remember about the big flood?

You're not going to believe this, but during the flood cleanup, I was off sick.
I was off on sick leave, and I missed—but I know—the sub-basements were
flooded clear to the ceiling, and—it was just a long process to get it all cleaned up
and replace the motors that were ruined. It was quite a—jeez, I was treasurer of
the union. There was a period of weeks there that I was off work, and I was—the
flood was on the news, and I wasn't down there. And I was saying to myself, my
God, they'll think I did this on purpose. But it wasn't on purpose, believe me. It
just happened.

Yeah, pretty good timing, though.

That type of an event puts a big work load on the yard department, and those guys were working overtime, making money hand over fist, but I was just glad I wasn't there because it's no fun.

So, almost everybody is pretty deaf because of the sound levels and the machinery and stuff.

They didn't introduce earplugs and earmuffs and all that kind of thing until almost to the—several years before I was out of there. And I want to confess that I'm hard of hearing and that several years that I worked there I could have had earplugs in, and I didn't do it because I was stubborn, too stubborn to do it, you know. I thought it was ridiculous. So I've got hearing loss. But it was not advocated by the company as a desirable thing for the workers to do. Some—it was voluntary. Now, the paper machine people, my God, that hearing loss must be terrific, because that's a loud, howling whine, loud noise for eight hours a day in there, and it never diminishes; whereas where I was outdoors and driving a piece of machinery up and down the dock and in and out of the paper machine rooms, it was intermittent that you had an attack on your hearing mechanism.

But it's cumulative--I think about the people in the sawmill. Can you imagine how loud it must have been?

Terrific, yeah. That sawmill was a noisy place. Boy! Gosh. You know, industry was—it was tough on human beings, you know that? But it had to be. It had to be done, and at least people had money working in industry. Now industry is damn near gone; disappeared. You know, people go through life, and that's the only kind of work they can do. There's no jobs. There's no decent employment any more for somebody like me.

Well, when I talked to John Hanthorn, who is a very nice man--

John! Where's John? Hanthorn, he was quite a—you know, he worked for AWPPW for years as a job analysis expert. [referring to John Thorne]

Really? Well, there's two Hanthorns, I think. There's John and there's Ted.

I'm talking about John Hanthorn, and I just barely could recollect him, and now I do recollect him. And he was a very, very well-spoken gentleman. Doesn't he do guns? Repair guns? Gunsmith! I'm sure he does, because I remember in the '80s that he was retired, but he was still doing work for the job analysis committee for the AWPPW. And the information we had at that time was he moved to Enterprise, Oregon and he was a gunsmith; he took it up in retirement.

[clear that it's not the same John] Millwright was a great job--

Something I could have done and made a lot better money, you know. I had plenty of opportunity, and I know my education was good enough; I could have passed the tests, I'm sure. But I didn't do it because I kept thinking about, well, I'm doing all right with this, and I've got the union to think about. Of course, that wouldn't interfere with my union; it might have, I don't know, but it doesn't matter. It's all water under bridge now, that's for sure.

I took this picture of Del Herndon.

Delbert Herndon was a boss.

Here's a picture I took to the picnic. Del is in the middle in the back there. I can tell you who those people are: in the back row…Willis McDaniels.

Oh, my! On the end?

Yeah.

He died young. He did not live long.

Aaron Winegar.

Winegar!

The front row, from the left, is Walter Peterson.

I processed a grievance for Walter Peterson, it was a plus rate. Him and another guy, and they were being passed over for a journeyman plus. You're a journeyman and then you go to journeyman plus, and then you have certain more responsibilities and so on. But it only paid maybe 10 cents or 15 cents an hour; it's up to the company how much they give you, but it's a plus rate, see? So they had this grievance, and so I was going to take it to arbitration. This is a nice experience for me because Ray Dupuis, the mill manager, he's a guy who came up through the ranks. And he was the manager. And it was a third-step meeting; that is to say you got the grievance through the standing committee and you're at the mill manager level. I was called in for an informal meeting. And there I'm sitting across the table, and there's Ray Dupuis and the plant engineer, whose name I can't recall at the moment. Ray Dupuis, he was a friendly guy, and he would like to talk about making paper. So we sit there for a half an hour talking about making paper, right?

And this is going on, and I'm thinking, hmm, what's this about anyway? But anyhow, he finally got down to it and says—I had already appealed; I guess this was after the third step; I don't know—but I had an application in for arbitration on the case. And he said, "If I give these men their plus rates",—it was Peterson and another guy – "will you call off the arbitration?" And I said yes. And that ended the meeting. They about fell out of their chairs because they didn't think that I would go along with that, you know. Why wouldn't I? He gave me what I asked for. What the hell. I mean—so anyway, that's a nice experience. You like to call something like that up. It worked out good. And there was no question about the decision.

It looks like you go to your own retirement group here, inner circle of labor people?

That is the Northwest Labor Retiree Council, an arm of the Northwest Labor Council, established in 1993 by Jess Dranahan and—I can't call up his name—the executive secretary/treasurer of the Northwest Labor Council at that time. But it was established at that time under the auspices of the national AFL-CIO, John Sweeney the president.

Ah. Famous name.

And since then there's a lot of water under the bridge, and I was a part of the people that met there for getting it started and so on. And at the time the idea of the AFL-CIO was to establish one of these chapters at every labor council in the country, which hardly happened at all; it just was not carried out. So since then there's been some different endeavors by the national AFL-CIO. As you know, the national AFL-CIO is busted up now into two factions. And before the break came, an effort was made by the national AFL-CIO to establish an organization for senior citizens named the Alliance for Retired Americans, which was to succeed the National Council of Senior Citizens, which was going to be terminated. I don't understand why it was done that way, but anyway, that's what was done. So subsequently -- and that was some years ago—ARAs have been established in many of the states, including Oregon and Washington.

So I'm on the board of directors of that organization as well, because they asked me. And I stayed away from it for quite a long while, but I finally said okay, I'll become a part of that. It meets once a month, and the board meets as well, and so I'm going to be participating in that. But you know, you never have money enough for these things. The funding is the problem, because you don't have enough dues-paying members. I mean, if it's voluntary, people are not going to join. You know what I mean? These things, it takes a few people to kind of keep it alive and do some lobbying and so on, but—I'm secretary/treasurer of this thing here, and have been since 1994. And all I do is, I convene the meeting, I provide a record of the meeting, which I do. I provide a record of the meeting, and I usually write something besides that; a composite of what I'm reading. All I know is what I read. I mean, I don't know anything other than what I read. So that's what I write, is what I know.

What kind of things could happen to you working on the yard?

Well, the mule driver don't want to run over anybody. You could seriously injure somebody if you're not—I'll tell you, a big part of your job was safety. You know, I run into a foreman one day and knocked him down. Al Steichen was his name, and he was a master mechanic. He was behind me, and Al was hard of hearing, and he'd walk with his head down like this, in deep thought, and wasn't paying any attention to where the hell he was at. So anyway, I knocked him down, but I never got into any trouble. I guess they figured it was his own fault, but—we had mirrors, rearview mirrors right up in front of us here, so that we'd look up and see when we were backing up—you might be fiddling with your load and look up there and see if there was anybody behind you. A lot of guys pulled their backs—we had a lot of heavy lifting. And what they did, if you got injured on the job, they'd bring you to work so there wouldn't be a lost-time accident.

I heard that. People say there were all kinds of ways of getting around it.

And that had to do with being insured, your insurance rates with the Workmen's Compensation Board in Salem; that's what it was in those days. I smashed a finger one time, but I was just moving the forks around on the—it was my own fault. I was never on a safety committee; I just sort of avoided that. I was more of a grievance type.

George Droz said they first started having the safety committee meetings at four?

In standing committee meetings with the company there was a time when quitting time was quarter to five, and we would convene up in the office at quarter to four. That would give us an hour to discuss grievances or just have a meeting. And then Barney Malaney was installed, and he came down from someplace, as the personnel manager, and Ray Dupuis, he was Ray Dupuis' man. Ray Dupuis liked to be the good guy, and Barney Malaney was the bad guy, so the workforce were going to like the mill manager but not like the personnel manager. That's the way that worked. But Barney decided to convene standing committee meetings at 4:30, so then we'd go up there at 4:30 and we'd be there 'til 5:30 on our own time. But that was just a trick the company pulled. That was all right; what the hell.

So somebody was talking about the windstorm. Were you—

Columbus Day!

Yeah--were you there for the wind storm?

I lived on Ninth Street and was protected there by the church that's across the street. The wind didn't bother us very much. I was working. We were working. I can't remember what effect, if anything, it had on the paper mill.

Sometimes people had to work up on the roof…

Oh, yes, absolutely. You had to keep the roof in condition up there—the tarpaper. They had tarpaper up there. Well, I think they had roofers come in and contract putting it down. But then, of course, the yard department, they had to send somebody up there to pick up paper and clean up and so on.

How did you get back out to the yard for the last part of your career?

That general duty, day duty for a millwright, was general materials handling. That is to say, hauling of all kinds of—everything—anything on a pallet board, or machinery, paper machine machinery. We handled all machinery with our lift trucks and devices we had—I didn't do it—but ideas that had been implemented with dollies and fifth wheels and stuff to where we could handle these huge paper machine, 20-ton-type rolls that we did. But the four years that I was on the slush maker, supplying the slush maker, that was just a period of time in there. So before that I'd been a day mule driver, and after that I was a day mule— you've got seniority rights, you know; you can't let them isolate you. You have to keep your seniority rights so you can back and forth and around from one job to the other and still maintain your position. So then after that, I spent the rest of my time there with just general duty, and spent a lot of time on union business. Boy, I tell you. I mean, after the new union got in, sometimes we'd be on delegated duty for months at a time; it was terrible. Terrible waste of money. God, I didn't agree with it; I thought we should have been more businesslike about the way we approached the bargaining of contract negotiations. But I did—I spent a lot of time in that; it was more than I wanted to. I'd rather have been at work.

I've heard a lot of stories about poaching salmon in the mill.

Catching the salmon and taking them home in the lunch bucket and so on?

Did you experience any of that?

Not me. Not me, but those guys in the grinder room, they were getting fish. I don't know how—I'm not an expert on that particular thing, but I was told that this was what was being done, and who cares?

People that worked down in the pocket grinders, in the sawmill, needed the food for their family—and so they divided it up.

Yes. Did anybody ever mention to you that they stole toilet paper and took it home in their lunch buckets? Nobody mentioned that?

No. Why? What was that about?

It's free. Well, you know, there were times when there were lunch bucket inspections too, at the gate. And we opposed that, of course. Guys stole stuff out of the welding shop. I'm telling you, those guys stole everything. Machines, welding machines, went out. This was a terrible thing—there were millwrights; I ain't going to name any names—there were millwrights that went up, say, to Pulp Siding on the east side of the river and to Warehouse 3 on the west side of the river to do work. And the guy would drive his truck. They had a pickup truck, and there'd be two or three millwrights that were loaded in there in the front end. And the back end was loaded down. Guess what? With stuff that was required and needed on the job, but how much of it ever got back in the paper mill? That's the question, you see. It got to be kind of a known activity. And especially the mule drivers that worked in both of those locations up the river, we could see what's going on. We'd just zip our lip, you know. Stealing toilet paper, yes. That was quite a routine for some of them.

John Hanthorn loved having that big shower room in the mill, where they could shower before they went home.

Change clothes and shower and put on clean clothes.

Yeah. That refreshed him and washed all that mill away.

I now know my mistake: Hanthorn was a papermaker on the supercalenders. Who I was thinking of was John Thorne. You ever hear of John Thorne? John Thorne was who I had in my head, and so he's the guy with the diabetes; he's the guy that worked for AWPPW as a job analysis expert.

But now you know who John Hanthorn is.

Yeah, I know who he is. He's a tall papermaker, yes. Kind of a sandy complexion, I think, as I remember.

Yeah. The best thing he could think about working at the mill, was the showers.

It was a real service, you know. I bet they don't do things like that any more, huh?

I think it's still there. So for you, what was the best thing about working at the mill?

I have to tell you that I'm looking back with rose-colored glasses on a lot of that. I mean, at the time I didn't think it was so great, to tell you the truth. But now, looking back on it, I say I think it was great, because how could I have done any better, really? All I can say to you is that I was a union man more than I was a company man. And that was my forte, and that was my endeavor, was—and I wanted to be a labor leader. That's what I wanted to be. I don't think I was, but that's what I wanted to be. I've got some credits since I'm retired that I don't think I deserve, but I've got them. I'm a member of Oregon's Labor Hall of Fame, for example. I was one of the first put into that category. That's an honor roll that is fostered and made possible by this organization here--one of the things that we decided to do when we started out years ago, when we first formed, and got some of the really famous labor leaders that we thought deserved that kind of recognition. Well, I think we've cheapened it in recent years. I mean, you know, I mean you put everybody into something, why then it's not as exclusive any more.

Maybe it's that there aren't as many opportunities to be a great labor leader any more.

That's a good point. That's probably true, because organized labor hasn't got a leg to stand on any more with this cheap labor and with shipping, exporting— the exporting of America.

What was the worst thing about working at the mill or being a mule driver? What was the worst thing about spending 30 years at Crown Z?

I don't know what to say, because the only bad thing that ever happened to me, I had some ill health and some sickness.

[Break]

How did the [yard] operations that you were doing relate to the locks?

I spent many a day parked beside the locks out there, watching the pleasure craft go through on Sundays and that sort of thing. If we're hauling, you always had to wait, and if the bridge is up then you get a break, right? And it was kind of nice in sunny weather. Those guys had good jobs there on the locks. It was a nice thing. I don't know if they've still got—I guess there still must be—government locks in those days; I suppose they still are. Or are they company owned? I don't know.

They're government.

And it was just—it was scenic, a placid, pastoral scene.

Do you remember when they were closed to be worked on?

Yes, yes, yes. They'd have to repair—major job—and the locks would be closed, yes.

They'd have put quite a crimp into the barge and tug thing.

Well, at a time like that, if the locks were down any length of time, they'd have to truck. Load out there at the warehouse and truck—is that warehouse still there? I'd just like to go down there sometime. Is the hotel gone? I guess it is.

The West Linn Inn? Yeah. I ask people how they related to the hotel, too.

Oh, those shift workers, they would park and they would go in there and spend an hour or so. And they had slot machines, you know, and they'd play—some of those guys would put their paycheck in the slot machine. Terrible thing. It was awful.

I didn't know they had gambling back then.

They had slot machines that paid off, but they wasn't supposed to pay off, and they did. That was going on. The sheriff finally raided the place and took them all—Joe Shobe destroyed all the—took them out of there because we had some guys, poor guys, that worked down in the grinder room, and they were somewhat limited in their capabilities, and would put their money in those machines and be broke. I remember one guy—I can't remember his name any more—he was a very limited individual, lucky to have a job, really. But he could work in the—he could feed a grinder. So that's what he did.

Nobody else has talked about that.

The only time I spent time in there was service award dinners. I was to many a service award dinner as vice president and as president of the union. I would meet these people that I never met before in my life. But we'd have those service awards up there. I got my own service awards, but I'm talking about the union officer would be there—just a courtesy to ask you—you're invited, you know. So Perrin lots of times would be out of town or gone or something, and I'd stand in in his place. The same thing happened with the mill tours; I wanted to mention that. Always ready for a mill tour. We'd have the dignitaries from San Francisco—One Bush Street, San Francisco—would come, and they're going to tour the mill. So there'd be a retinue of people touring the mill, and the leader would be in front, you know, and the guide and so on from the office, and they frequently would ask, if a union standing committee member was handy, you'd be invited to meet the dignitary and go on the tour and everything. And just a grand thing, you know—public relations, right? Never see anything like that now, would you?

I've been on tours, under different circumstances.

Did you stop by Mill B? I was trying to explain to you about the towers and the digesters and so on.

I don't think they took us through Mill B. So when they were doing the tours that you would go on as shop steward or whatever, was the old mill open then?

Oh, that was a thriving concern back in those days. Everything was running full-blast. Mill A, Mill B, ten paper machines in operation, and then they told us, oh, we're not making no money; we're losing money. Every day we're losing money. Just propaganda, you know. Losing money. Why were they running if they were losing money? Better shut her down. It didn't make any sense, but that's just the way...

...

A guy I learned to know and like enormously was Gene Wymore, whose trucking agency was there all those years that I was there. And we would talk, you know, and he was just a swell guy, and he was more a union guy—he hated unions, to hear him tell it. He hated unions. But I don't think he really did. But of course, his drivers were all Teamsters. But Gene Wymore was a good guy. I don't know, he died years ago. Then his brother Don Wymore continued to run it, I think. But there's no more Wymore Transfer, I don't think. But I used to spend a lot of time and stop by out there. He'd want to talk, you know, and we'd talk. And tremendous respect for him. The other Teamster business agent that I was trying to think of was Clyde Crosby, Clyde Crosby. And he was Mr. Teamster in the area, and he helped out the new union enormously by money and help and cooperation. I just wanted to get that in, because I have a terrible time remembering names any more, you know. If you get to thinking about certain things, sometimes it'll come to you, you know, and it's like a flash.

Right. And sometimes it's like dominoes: one memory triggers another.

Right. I'm glad I got that John Thorne thing straightened out, because I was wrong on that.

Let's see. What happened when you hurt your finger?

I forgot to tell you about the elevator event.

Okay.

I forgot about it. I got caught in an elevator and was being smashed, and the guy stopped the elevator, and I got out all right. And they took me up to the doctor's office to X-ray me. I was caught crossways, my back and chest crossways in the elevator, and so the machine, the X-ray machine in the doctor's office, wouldn't satisfactorily take a picture of my chest area, chest and back. I was too thick, they said. They took me to Willamette Falls, and I think I might have had a cracked rib; I don't know. I was kind of stove up for several days, and they brought me in and had me sit in the office. I forgot about that, you know. And I was told that— the doctors told me that if I hadn't had such massive—see, I weighed 275 back then—and I was so massive, and there was enough give there that nothing broke.

Where was the elevator?

Well, the Mill C elevator, and it was one of those where you could lean in and touch the buttons and so forth. Somebody had the elevator downstairs. It was my fault. It was my own damn fault. But nobody said anything to me about it.

You weighed 275, and you were in pain and cranky.

I just kept my mouth shut, and everybody else kept their mouth shut. Nobody said anything, and just told me to come to work. They didn't want a lost-time accident, right? So I'd come to work and sit in the office for three days, and then I went back to work.

People would do some sort of little job where –

Okay. Those suction rolls. They'd drill out the holes in the suction rolls. That was the—you always had one of them handy, right, to put somebody to work that had been hurt or sick or something, so they don't get a lost-timer. That went on the whole time; for 30 years that was going on.

I wonder if they still have suction roles. I don't know—

Suction—you have a suction roll on a paper machine, yes. Yes, because the water has to be sucked out of the pulp before it goes onto the felt.

Nowadays they have rooms that they put the operators in that are air-conditioned.

A booth. And they pay them $50,000, $60,000 a year to run the machine. And these machines run—they run by themselves now. They don't make hay, I don't think. I don't know; I haven't been around. When I worked, they didn't have no computers. There wasn't any. Now the computer does the work.

So people talk about making hay.

Oh, that's paper maker terminology.

People tell me that there would be fires from time to time, mostly in the paper machines.

We had a fire department, for sure. I don't recall any serious problem. We had a fire department—Joe was his name; I can't remember—we had a couple of firemen, but there'd be a gang in every department that were the firemen, and they knew what to do, because there were fire protection terminals in all the places in the mill. No, I don't think we ever had any problem with fire that I can recall.

So your job was mostly outside.

Yes, it was dock work outdoors and into the machine rooms and out. Breathed a lot of carbon monoxide, and I have emphysema, I think from the carbon monoxide that gas-powered lift trucks. All those years that I spent in smoke-filled rooms, in conference rooms, and the smoke would be so heavy that there'd be a layer of smoke about eye level, all the way across the room -- just smoking up a storm. And I'm sitting in there breathing that. And I never smoked, but I smoked all right. So I've got emphysema.

So do you remember any of the other people that worked the yard?

No, we didn't socialize. No. We just worked. I'm afraid I can't answer that, because I remember the little guy, German guy, that came over from Germany, and he was about my age. Little short guy, and he had a band. And he used to play at the Mount Angel Oktoberfest. I can't think of his name. Perrin and I, we'd go to meetings together and so forth.

So did you get a lot of people that just came in and out off the extra board?

A lot of that. A lot of extra board. Every summer we had extra board people because Crown Zellerbach wanted to help the kids in school and going to college. And so they would hire them to come in there and take the summer vacation work. And you know, we had a lot of paid vacation. I mean, it cost the company a lot of money, those vacations did. The company used to use that extra board—misuse it. I mean, until you go on regular at 30 days or something, some of your benefits weren't there; seniority rights, for example. And so—but I think my son worked the extra board for a while. You have to be on call. I mean, when they call you, you go in and go to work. The railroads have that, you know, the extra board.

So that was kind of a way to test people, whether they could do hard work or not.

Well, my goodness. That's the way they could sort them out, the ones they didn't want. They just didn't call them. But there was nothing the union could do; I mean, we didn't have them under our—you know, 30 days and then you join the union. Okay, from then on they've got to deal with us. But until then they don't have to deal with us. We didn't have a hiring hall. Nobody does except longshoremen.

Well, can you think of anything I've forgotten?

Covered the waterfront.

So you'd like to go back and see what it's like today?

Absolutely, but I don't anticipate that's going to happen. I'm short-winded.

You do pretty well for a short-winded guy.

I'm long-winded, yeah. I'm short-winded when I have to get up and walk, and it doesn't bother me weightlifting because you just do a few repetitions and you rest, right? I do what I can do. But when I get out—and I used to do a lot of roadwork. In the '80s I did a lot of roadwork, walking and jogging and all that.

But the doctor diagnoses me with—I don't have a pump problem; I have emphysema—dark, black lungs. That's what the X-ray shows: darkened lung area. How did that happen, you know? I didn't smoke, but I was in smoke-filled rooms for years: conference rooms, conventions, conferences; all entailed inhaling a lot of smoke.

🏭

Chuck Calhoun

Interview August 16, 2006

Died June 8, 2008

Okay. Well, I'm Charles Calhoun--Chuck, Charlie, whatever you want to call me. I went to work in Crown in 1948, October 4th. I was 20 years old. I worked there for 36 years, eight months; retired June 1st, 1985. I'm 78 years old; I was born November 20th, 1927, and I worked for a great company.

My gosh, well, I don't have to ask you any questions! Okay, so you started working there in 19--

Forty-eight.

Forty-eight. What made you interested in applying for a job at the mill?

I was a dumb kid without an education, and I needed a good job. I had a wife and I needed a job, and that was as good a place to go as any, you know. Can't beat good industry for work.

Had you had another job before that?

Yeah, but I'd been unemployed. I worked at a door factory in Portland for a while after I got out of the military, and I'd been unemployed for about six months. And jobs were kind of hard to come by there for a couple of years, because all the servicemen was coming home. So I was very happy to get a job down there.

I've had somebody tell me that after you got out of the service, it was like unemployment but not unemployment; it was—

Uh, 52/20. You got $50 a week—no, fifty-two weeks for $20 a week or something like that. I forget.

Which branch were you in?

Well, I was in the signal corps for a while, went to radio school. And then I volunteered and went in the airborne, and I was in there about a year. And I was only in the military for a year and a half. I went in 1946 and got out in '47, August of '47. Yeah, I was in the signal corps for a while. When I went in on February 1st of '46, it was right after the end of the war, and they were starting to cut back a lot. And they put me in the signal corps, and I started a radio course that was a five-month course, and I was only going to be in for 18 months. They kind of decided after three months that was a bad deal, so they took me out of it.

Were you disappointed?

No, no. I had a good experience. Learned how to jump out of airplanes. It's pretty exciting.

Had you worked before you went in the service?

Yeah, farm labor, pretty much, before, because we were farmers. And I went in the service right after I turned 18, so—I turned 18 in November of '45.

Olaf talked about his service and he said that he's in the Round the World Club.

He is. He had probably a very interesting tour of duty. And he's old enough, see—I think Ole's 83 or something, so he's probably—he seen probably most of the war. I didn't turn 18 until the war was over, so.

So 36 years…

I didn't have an education, you know, and when you don't have an education, any work is good.

Did you have any relatives that worked there?

Nope. I did have a cousin which worked there after I did—John Calhoun—but he's deceased.

So you just heard about it from your friends?

I don't remember that part. I just went down there looking for a job, and that's where I ended up.

Do you remember your first day?

Yeah, absolutely. I went to work on swing shift, wrapping plug rolls.

What is a plug roll?

It's like wrapping paper for meat or groceries or stuff, like they used in them days, and they were in rolls about that big around and probably about that long. And then they come off the machine onto a deal, and you rolled them in paper and crimped the ends to ship them.

What machine made them?

I think one and two, maybe, was making them at that time. I worked that two or three nights, not a lot of nights.

So were you extra-board at first? How did that work out for you?

Good. Yeah, I didn't have any trouble.

They gave you enough hours?

Yeah, mm-hmm.

What happened after that?

Well, after that I went to work in the chippers, where they chipped wood to make the pulp. And I worked in there for eight years. That was pulling wood off a conveyor into a chipper, and that was pretty hard work, but, you know, when you're young and you've got a wife and a kid, you don't think too much about that. We worked days and swing, but we worked a lot of overtime, too, which— you liked at that time, you know.

But no graveyard?

No graveyard where I worked.

Is that the same as the pocket grinders?

No, that's different. Grinders is up in the other end of the mill, and we was in the mid-part of the mill, in Mill B, they called it, where they chipped the wood. They chipped it in three-quarter-inch squares, pretty much the way it came out. And it went up, and they put it in a digester and cooked it and made pulp out of it. There's two different things: the grinders made pulp, and we made stuff that had fiber in it. They used the fiber to make the sheet, the pulp to fill it in. They'd get the strength from the fiber of the chips. So the digesters is like a big pressure cooker. It held anywhere from seven to 11 cords of wood. They put that stuff in there with an acid and cooked it for six, seven hours.

So where does the stuff that goes in the pocket grinders go?

It goes to a completely different place. It's ground, and they have a mixing room, and they have a screening room that all this stuff goes through, and it goes to the mixing room. And the digesters cooked it in sulfurous acid with steam for six, seven hours—whatever it took—and they pumped that out of there, or blew it out of there, into a chamber and flushed it to get the stuff out of it. Then they washed it out of there, and it went through screens to clean it up, and then it went to the mixing room. And the mixing room had a crew that—they mixed all kinds of stuff in it to make the paper, and, you know, they put a lot of different stuff in it, you know, starch and different stuff like that, depending on what kind of paper they're making.

What does a chipper look like?

Well, a chipper has a big long chute on it, and you pull the wood off the conveyor end of that deal, and there's a big huge wheel in there that's got knives on it that set at about that far out. And it just slices the ends of that wood off as it comes in there. They've got, you know, two, 300-horse motors on them.

There's an old picture of the mill, this guy in a T-shirt with a cigarette hanging out of his mouth, throwing things into the door of something up high—

Well, that's probably the pocket grinder you're talking about. A grinder has a got a big wheel in it that's made out of stone or whatever. And they've got three pockets on—I think one on each side and one at the top. I never worked up there, so I don't really know. It's just from what I've seen. They'd throw that wood in there, and there's a deal that pushes down on that wood, and it lays in there flat like this, and it's cutting it off the side of the wood, so it's not taking it off the end like we did. I'm not positive about this. Anyway, it made a pulp. It made just a fine pulp. And the pulp's a filler, you know. You've got to have something to hold that paper together. So you've got a fiber, and then you've got a pulp that fills it in. And then they put all kinds of starches and different types of stuff in it.

Photo courtesy of West Linn Paper Company archive. Circa 1950s.

Did you lose your finger at the mill?

No, I did not. I lost that when I was a kid and was doing a dumb thing. I lost that when I was two years old, and I lost that when I was about 13 or 14.

So you worked three years—

Eight years in the chippers.

Eight years in the chippers. And that was a union job.

Yeah, I worked at all union jobs.

And you bid on something else after that?

Well, yeah, I bid and worked part-time up in what they call the blow pits, where they blew that stock out—just on and off. That was just kind of part-time. I never did get a permanent job in there. And then I worked up on the river rafting logs one summer about eight months. And that was pretty interesting.

Did you work with Paul and Milt and Art?

I was up there before they were up there. This was in—this would have been in about 1956 or '57, somewhere in there; probably '56.

How did you like that?

It was fun. I mean, all I wore was a pair of shorts and a pair of cork boots all summer. But I worked down there until December, and that was the year we had about 10 inches of snow in November, and that kind of took the fun out of it. Actually, they had a log dump up past Canemah up there at the pulp mill – 'Pulp Siding' they called it—and they shipped logs in there and dumped them in the river. And at that time they'd take them over to different stations, and we'd sort them and rewrap them. And some of the logs they sold, and the rest of them went to the mill. And they'd wrap them up, and they stored them alongside the river, and then they'd take them down to the mill and use them. But like some of the better logs, the big peelers and stuff, they sold, or sent to the mills for plywood.

What kind of log was it, then? Was it spruce, cottonwood?

We got all of it, you know. We never got any cottonwood. We got all fir or hemlock. And spruce; we got some spruce. Most of it was fir and larch, mountain larch. A lot of it was pretty good-sized timber in them days. Better than they have now, I guess.

Did you go in those little shacks along the bank where the people that worked the rafts could eat their lunch and stuff?

Yeah, yeah.

Did you ever fish?

Never had time for that. Just fall off a log and go swimming, when you wasn't supposed to do that, though.

We've heard stories about going in with all your clothes on, too, when it's really hot.

Yeah.

Winter ended that particular—

Well, that's about the time bundles started coming in. They started bundling everything, and that done it, kind of put an end to sorting logs on the river. They done everything a little bit different. They don't do any of that any more. Everything was done in bundles, and they'd put them in big bundles with big metal wraps around them, and that's the way they floated them down the river, and it's the way they done everything.

So somebody upriver just dumped their log truck bundle right into the river?

Well, I don't remember how they bundled them, whether they done it before they took them off their log trucks or what, but that done away with the work that we were doing on the river.

Did you get to choose what you did after that?

No, I went back to the chippers, and then I bid on a millwright job.

What kind of qualifications did you have to show them?

You had to fill out a form, and I guess they used it from seniority and from your qualifications--maybe you had in the past or what. But I bid on jobs for six, seven years before I got one, so—and to give you an example, like in the millwright gang when I was in there and I'd been in there for 25 years, there probably wasn't a guy in there who had less than 20 years, 25 years of seniority, so you know, there wasn't much turnover.

Those were really good jobs.

They were a good job. The mill was a good job. They had great benefits, they treated you good, and I never had no complaints about the mill.

Ole talked about how—he was there for I want to say 36 or 38 years—and he said it was just the last seven when he was really happy.

Yeah, he had a good job. He kind of was in charge of all the winders and, you know, any problems they had or anything. But he kind of was his own boss, so to speak. Yeah, that was great. It was a good job, yeah.

It sounds like—the bosses were pretty nice in general.

I think back in the '40s and early '50s that things were a lot different. The bosses were more hard-nosed back in them days from the old times, when everybody kind of ruled with an iron hand. And as the years went on they kind of got away from that.

Do you remember some of the older-timers that were there, or did they use to ever share stories about the really early days at the mill? I've talked to Rosie…

Yeah, Rosie probably would have been the best one to talk to, yeah. He was pretty much an old-timer around there, yeah. You know, there's so much automation went on. Everything was done pretty much by hand when I went to work there, like the rolls coming off the machines. But I never worked on any of the machines like that 'til I got to be a millwright, anyway.

Yeah, he started in '27. He was 17.

The year after I was born. He was probably one of the best supervisors they had down there, one of the best. They had several, but a lot of them old paper mill bosses were pretty ornery.

What makes a good supervisor? I mean, because he worked his way up through the jobs?

Somebody like him, who can communicate with the guys and don't down-beat them or cuss them out when they're working on something. My theory always was, a boss was just as good as his crew made him. You know, if you don't have a crew that works with you, you've got a problem to start with.

So after you were there, then you had cousin that decided to work there?

Yeah, I had a cousin who was in the Navy. Him and I grew up together and was always pretty close. He'd been in the Navy for seven, eight, nine years, and when he got out he didn't know what to do. So I told him to come on over and live with us for a while, and he went down and got a job in the mill. He went there a couple years after I did, I think, maybe two or three years after I went to work there. When I went to work there, there was probably about 1,500-plus people working in the mill, when they had 10 machines running.

Do you remember when they started phasing out machines and shrinking? I mean, you lived through that.

Yeah. Probably in the late '60s—you know, I think No. 10 was probably the first one they shut down. They phased that out. I don't remember just what years— probably it might have been in the early '70s that they phased out the others, like one and two—seven and eight, I think, were really, really old machines, and they sit off kind of by themselves. They was really antiques; they was from back in the 1800s. And they phased them all out, and I think a lot of those old—one, two and three—they phased down, and that would have been in the '70s, too. And No. 4 they kept; and I think four is even gone now, but I'm not sure.

I think they have three that are operating now. They've renumbered them.

Yeah, nine, five and six. But they don't call them than any more. They've got different names for them.

Yeah, they're one, two and three. I've been really curious about the one that came around the Horn. I think it was No. 8, but I'm not sure.

Yeah, I think you're probably right, yeah. Pretty crude old machine.

It was slow.

Yeah, it was. It only run a couple, three hundred feet a minute, compared to nine running a couple thousand feet a minute.

What happened to these machines when they went away? Where did they go?

A lot of them went to South America.

Did they?

Mm-hmm. They tore them down and catalogued them, and they sent—I don't know how many of them went, but like one, two and three, I'm sure they all three went to South America. And they even sent—not from our mill—but they hired people from up here to go down there and help set them up and get themgoing.

Was it to Crown Z mills down there or different—?

No, just a company bought them, took them down there. Because they're a little behind the times down there yet in places, I guess. I don't know.

Paul Miken had said that Crown Z, when it was a really big international, corporation, actually had a floating sawmill on the Amazon.

I'll be darned. Never heard that one.

I'd never heard of such a thing either.

Well, Paul worked in the sawmill—he worked in the yard gang for a lot of years, and he worked in the sawmill afterwards, so he probably would know that.

Mary said she thought No. 8 was one of the first two machines, from an old mill in 1869.

Seven and eight were probably the first two machines. They were both from the—I don't know—1870s, 1880s. No. 8 was the oldest one, yeah.

She said that she thinks it was given or sold to a museum, and we want to figure out—

That very well could be. I'm sure they didn't go to South America; they was too obsolete. And No. 10 I think probably was just junked out, but I don't know that either. But one, two and three—at least I know a couple of those machines did go to South America.

There's a picture of a little machine—and it's maybe like this wide or something—and I think, that's got to be seven or eight.

They weren't very wide. I don't remember how wide they was—
60 inches, maybe.

Did you have special assignment as a millwright?

Oh, no, I worked in what they called general maintenance for probably a year or something, And every morning you had a job to go to, or if there was a machine down, you was assigned to go down to a machine, and the machine boss or millwright of that department would assign you to a job, whatever he's going to do, yeah. So they had schedules laid out for you every day; that was the bosses' job.

After the first year or two, did you manage to get promoted up?

They had a process. You had to wait to get a base rate, a D rate, I guess they called it. They changed it a little bit after that. But I think I worked in there seven years before I got to be a journeyman, because you had to wait till somebody retired and there was an opening to move up from a helper to a C rate. And then you had to go through a four-year process, D, C and B, to an A rate. So then by the time I got to that, you had to take a test in between each one of those to move on to the next level.

They were qualified people.

Oh, most of them were, yeah. Everybody's different. I mean, you've got different people that are good at better things, you know. But after about that first year, I got to go into what they called a Flying Squad. There was eight or 10 or 12 people, and George Droz was in that, and that's when I started to work with George. But it was kind of like an emergency crew. If they had something that went wrong, right away they had a crew; they'd pull us off of -- set us on a breakdown or whatever, you know. And the other things we did then was like— they was doing a lot of improvements at that time, putting in new machinery and stuff, and we was doing a lot of that, doing a lot of layout work and putting in new machinery. I really liked it; that was a lot of fun.

That sounds like fun.

Yeah, it was, because, you know, it's interesting. You're doing something different all the time, you know.

I don't know why they called it the Flying Squad, because you're supposed to be able to go any time, you know. You had a portable toolbox, and if they had a machine or had something wrong, you'd drop what you was doing, and two or three or however many guys they needed was sent there to work on it, to do whatever they needed.

Now, would that be for things like a blown bearing, or—

Absolutely. Yeah, anything whatever. We was strictly paper machines. We didn't work on anything on the other end of the mill at all. Each department had their own assigned millwrights, and like general maintenance had a crew of I don't know how many—20, 30, 40 guys, whatever it was—and if the sawmill was going to need guys, they'd draw out of that general maintenance pool. And when I was in general maintenance I did work up there some. I never did work much in the grinder room. The Flying Squad, that strictly worked on the paper machines.

So, when did the floor collapse?

That was before my time. That was probably '46, somewhere there. Yeah, I wasn't there. It might have been '47. But that was up at the grinder room, at the other, upper end of the mill. Mill was divided up in three parts. Mill A was the grinder room and a sawmill. Mill B was where the digesters and the mixing room and that stuff was. And Mill C was machines one, two and three, and nine. And then there was Mill D, where five and six was. So it was divided up like into four categories.

And then on top of that there were silos and storage areas for materials, weren't there?

Oh, Mill B, like where I worked. Yeah, like in the chippers—this stuff was sent up an elevator, on a conveyor, and they had great big bins up above these digesters that would hold like 10 or 15 cords of wood. I mean, they were huge, yeah. And then the acid room, where they made acid, that's where they used the lime rock and all that kind of stuff, yeah. See, a lot of that stuff back then was all done by hand. I mean, you know, it's hard work.

Somebody told me that the Flying Squad didn't last very long.

I worked in there probably six, seven years. Yeah, times change. George went down to five and six, and then a few months later there was an opening down there, and he asked for me. And so I went down there and worked with him, and that's where we spent the rest of our days. But the Flying Squad probably was 10 years, lasted 10 years. I would say at least that long, maybe.

Did they do fires, too, or was that the fire department?

Well, that was kind of a volunteer deal. I belonged to the fire department, yeah. But if they had a fire, you dropped whatever you was doing if you belonged to the fire department,—and about half the guys did—and you went to where that fire was, because you couldn't get outside help in there.

You had hoses, big hoses.

Well, yeah, they had fire hydrants. They had our own fire system in the mill, sprinklers and everything, you know, on the machines. If something happened, they had sprinkler systems on them.

When you were millwright, you got to work days except for overtime?

The fact is, I kind of forgot about that. I got stuck with shift millwright for a few years too, and you worked around the clock on that. So that's kind of a troubleshooter, like when you're on a swing shift at night, you had an electrician and a millwright on duty 24 hours a day. And like on nights, if we had a breakdown, between you and the electrician you figured out pretty damn quick what's going on, because they didn't want them machines down. If you had to have help, you called your supervisor, and they got help for you. But I think back at that time, like five and six was probably running about $82 a minute or something like that. So they don't want that machine down very long, you know. It runs 24 hours a day, seven days a week.

Do you remember what your first pay rate was?

I think it was about $1.23 or $1.25 an hour, which was pretty good money in forty-eight. And then when I went in the chippers, I got a nickel raise. I think I went to $1.25 or $1.27. I don't remember what it was, you know, but it was pretty good.

So when did you have to join the union?

When I went in the mill, you had 30 days to get in the union. It was a closed shop. Yeah, you had to be in the union unless you was salaried. And salaried, of course, you didn't go in the union.

Who was your steward for millwrights?

Boy, I don't remember back then.

Doesn't matter.

Yeah, it does, because this guy was a great guy, and he worked as a maintenance guy and went to become salaried later. God, I can't think of his—I'm sorry, I can't think; I can't remember some of them guys' names, you know.

You must have been there for at least three significant strikes...

Nasty things. Nasty.

Do you want to talk about the first one that you remember?

The first one I don't think was that bad. It was the last two that was bad because they lasted six, seven months. And it never devastated us that bad, but it ain't good for nobody. But some of the other guys, you know, that don't have anything to fall back on, you know, it just about ruined them. It wasn't good for the mill either, because after the first strike you had guys that had scabbed, and nobody ever forgive them. It's just one of those things where—it was bad. I mean, you had people that broke up friendships. It was not good. And I'm sure it wasn't good for the mill either. They went in there with extra help and tried to run the mills. But the problem was, when we went back to work we had bearing trouble for two or three years after that because it wasn't maintained while they was trying to run it, and we had a lot of down time from bearings that wasn't lubed, or whatever the situation was, or from other things that probably hadn't been paid attention to and should have been.

I can't remember who the shop stewards were back in there. Harold Bancke was a shop steward for a lot of years later on, and he done a really good job. I think Olaf Anderson worked in a lot of that, too. I'm not sure. Irvin Lavier was a—he was a strong union man, and I think he was the president of the union down there for a while. And he was just a great guy. I mean, he was good to me. He moved on up and became a supervisor, and he was just a good guy, you know.

Did he have to go through the foreman job before he—

Yeah, he did, he did. You know, he worked his way up through the ranks like a lot of the guys did. George did the same thing. People that wanted to do it—I had no desire to be a salaried person. It didn't appeal to me. But, you know, they kind of pushed me that way a few times, but I—

John Hanthorn—

Oh yeah. I know John. He worked in the supercalenders, didn't he?

He said that he, you know, chose to stay hourly.

Yeah. He's a good guy too. He's smart, you know…you know what? I think it was how you appealed to the other supervisors and the bigger shots in the mill.

Some people, it just seemed like they had really good people skills.

We did. We had a lot of them. George Droz is one. Leonard Makey was another one. And as far as maintenance was concerned, them guys were just—you know, nothing was a problem to them. And I'm not like that. I could do things, but I wasn't that good a natural mechanic like some of those guys were. Olaf's another one. He's a good guy. But, you know, he—Olaf's—he liked to kind of sit in the back row. And I'm that way, too. Do you want to talk about them strikes any more?

I do. So we've talked to supervisors, and we've talked to the guys who were off. We've heard from supervisors that, "Oh God, we had people from Alabama all over, and they were working 12-hour shifts, and they didn't know what they were doing."…

They was doing what they had to do to try to make some paper, and I don't think they done a good job at it either, from what I hear. You know, there's another fascinating thing about these supervisors. Most of these guys like Paradis, Bob Gifford, Irvin, all them guys come up through the ranks. They weren't educated supervisors that was hired from the outside, so these guys knew what the working man was going through, you know. I mean, they'd been there and done it. So them guys that become supervisors, most of them, they come up the hard way through the ranks. You know, so however they describe the strikes, I don't know, but they were terrible, for everybody. It was probably harder on those guys than it was us, you know.

They worked ungodly hours. Of course, everybody worked terrible hours in the old days. Del said the longest he ever worked was 20, 26 hours or something.

Yeah, I think 20 hours was about the longest I ever worked. But then they changed it so they couldn't work you more than 16 hours without a relief, or they could work you a couple more hours if you had a relief coming in. But yeah.

We haven't heard much talk about the fact that some people did break the picket lines.

Oh yeah. I don't think there was a lot of guys, but there was some guys. And you've got to understand: maybe they've got a family; it's hardship; they needed to work, you know. It was really hard. People just shunned them when they went back to work. It was pretty sad.

Some people went up to the woods and cut wood and sold wood so they could live.

Yeah. Yeah, I don't know; when I was growing up, at the end of the pay period we didn't have any money left. You know, you've got three kids and a house payment, a car payment, a washer and dryer, davenport, you know. You ain't got no money left over. So when you go on strike, there ain't nobody there giving you a handout. Well, we did have a union fund that one time, and we got so much money a week, which, you know, kept you eating, but—

Was that the last one?

Probably. And that was about—that was in '78? And I'll tell you something else. You know, most of the guys, including myself—you know, a lot of us guys then were 18, 20, 25, 30 years—you know, you didn't realize what it was like, because we was making good money, had good benefits, and when the strike come and we went out looking for jobs, they was seven, eight dollars an hour. And you didn't realize because you'd never had that experience before that—you know, you didn't know how to look for a job, and a lot of them didn't have the skills to do them anyway, you know. That was a big eye opener for me, and I thought man, if something ever happens and they do away with the union or they shut the mill down, what are you going to do, you know? This is the only place you've ever been.

As a millwright you had to figure out what they let go to hell while they were running it?

Well, by this time I was working on five and six. George and I was working in the department, and John Pickle also worked down there with us. And that was our responsibility—five and six—to take care of them. And then on repair days you brought a crew in for whatever work you was going to do, anywheres from 10 to 30 guys, or whatever, you know. But they had—Walt Riggs would be a good guy for you to interview, if you haven't interviewed him yet. He was kind of like an inspector. He went around checking bearings and stuff. That's all he did was walk around, and he had a deal he could listen to the bearings and listen to the machines and stuff. And we done an inspection every day. One of us would make an inspection through all of our pumps and listen to everything, but we had a lot of night down time from deals with bearing failures and stuff. They may not admit that, but—

Everybody says that, you know, strikes are mostly lose-lose…

Yeah, there wasn't nothing good come out of those strikes.

Okay. When you were working the river, did you ever work in the lock area?

No. Some of the older guys—the guys from the upper river, where I worked that one summer, they did. They took the barges and stuff down through the locks, but I never was involved in any of that, because I was just a gofer one year when I was up there.

I always ask people about the West Linn Inn, because you know, the story goes that the West Linn Inn was actually built by the company in like '17 or '18 because of a strike.

Yeah. I don't know why they built it, but it was owned by Crown, and a lot of people, employees from different times, did live there. But they had a nice restaurant and lounge in there, and we used to furnish meals for guys, you know. Like if we worked overtime, if we worked over 10 or 11 hours, they actually brought catered meals, or brought a meal in to you.

Just in regular overtime?

Oh yeah, yeah. If we had a machine down that was going to be down 12 to 18 hours, they fed you, and they took good care of you.

We were told that they built the Inn because they wanted to have a place inside the picket lines for replacement workers.

That had to have been way back when. But the Inn was still open when I went to work in the mill, and I don't remember what year they shut it down. But they did shut it down, and like if there was just three or four of us working late or something, they had restaurants around town that would fix meals and send over to us, taxis.

After they shut it down?

Oh yeah, after they shut the Inn down, yeah. Art's Café, Shari's downtown, and Sid's Restaurant, they had a deal set up. They'd make meals, good meals. For the most part it was pretty good. And if they had a big major shutdown, they'd bring in a catered service and set it up for you. You know, if they had like 100 guys working overtime or something, a big shutdown, yeah, they took good care of you.

We think that they probably took it down in '82—

I would have thought it went down before then. Well, you know, the hotel part probably went down before then, and the restaurant part stayed open a lot longer. Yeah, a lot of the guys would meet there and have coffee. They had a nice lounge in there. It was a beautiful place in there, actually. They had a nice fireplace in there. Well, you know, it was built way back when, and it was so obsolete, that's why they closed the hotel part of it down, because, you know, I guess the rooms were too old and weren't adequate, or whatever the situation was. I think they probably had community bathrooms that the older hotels had.

And it had a bowling alley in the basement. Did you ever bowl down there?
No. I wasn't a bowler.

Most mill workers probably didn't get into that.
I never had time for it.

What do you think about safety issues? Was there protective gear when you started?
Yeah, I went to work there in '48. And I'm wearing hearing aids. And the chippers was a super, super noisy place. I mean, you know, probably 100-plus decibels, maybe 120. I don't know; it was pretty noisy. And if you wanted ear protection, like just simple little earplugs, they had to order them for you. So consequently, the safety programs didn't really get into it until probably 15 years later, probably the last 15 or 20 years I was in the mill. We had a full-time nurse's station in the mill and a full-time nurse, first-aid station. And probably the last 15 years or so I was in the mill, they would give you a hearing test every year. They had a chamber there where they could test you. So they had records of what you—and you couldn't go on the job without safety equipment— safety shoes, hat or ear protection or whatever. And then they got real into the safety programs, because if somebody got hurt it cost them money. Then you had to replace them or whatever, you know. They actually had a pretty good safety program.

We've heard a lot about claims that say "99 days without an injury" -- well, without a—
Recordable injury.

So did you ever see any bad accidents or anybody get killed down there?
No, I never seen anybody get in a bad accident. I got my foot broke.

On the job?
Yeah, I was working with a couple guys, and one guy made a little mistake and dumped a pump on my foot. No, it's okay. I was off work for about six weeks

Oh-oh. That would have been a workers' comp thing.
Well, you know, they kind of got away from that towards the end, because I don't know if they didn't consider it was cost-effective. I don't know why. Boy, you can't ask why management done anything—

So you didn't have any close calls. What was your experience in 1964 with the flood?

That was a terrible experience. We worked a lot of long hours without hardly any sleep, and we went in there and covered the motors up, and went back the next day and pulled it all—pulled motors up, raised them up. Went back the next day and raised them up right almost against the ceiling and wrapped them in plastic, and they all still got wet. So then when the flood went down, we went in there, and they didn't have no electricity. The mill was like a morgue—that's a bad word maybe. But there was mud, dirt like this deep everyplace, and it was cold, it was wet, it was nasty.

So they started cleaning it up, and we started pulling motors and shipping them out. They'd ship them out, we'd get them back, we'd put them in, and before we could get them hooked up and get them warmed up, they'd go ground, and we had to pull them out and ship them out again. So they finally got big space heaters in there, so when we put a motor in, we could put a space heater on it to keep it from accumulating moisture to short it out, because we'd pull some of them motors two or three times before we'd get them in there working, and you're talking about motors as big as this table, you know, electric motors.

The dampness shorted them out?

Oh yeah, there was just so much moisture and humidity in there that—and then they had big hoses in there, trying to wash everything down and get it cleaned up, you know, and it just devastated everything. All the big electrical panels were shot because that silt is so fine, it just filtered into everything, and there was no way they could clean it up. You can't get it out of there so—Crown said if they knew later what the problem was they'd have shut her down right then.

People have talked about great big logs down in the sub-basement, and how they got there nobody knows.

Oh yeah, it was just a devastated mess. I mean, it was terrible, terrible.

So everybody became general maintenance.

They had everybody working, you know, it was a big job. It took a long time to get everything going again.

Were you watching the water go up that night, or—

We was working so many hours you didn't have time to think about it. And then the other thing is, we had a big flood on the Clackamas, and my father-in-law had a farm down there, and he had 39 head of cattle down here along the river on the bottom level. And they got trapped, and he was wanting me to come down. So I come over and tried to help him, and I wasn't getting no sleep anyway. Yeah, it was a bummer.

I'm sure you were glad that they decided to start up again.

Oh yeah. Very grateful.

There have been other high-water times…

You know, I don't remember any others so bad like the '64 flood, but yeah, we had some—we had some times we shut the mill down, a couple other times.

Okay. Did the union guys have their own Christmas parties and stuff that compared with the management parties?

Oh, we weren't invited to management parties. No, we didn't have any— unless it was just something that somebody got together or something. I don't remember any significant Christmas parties or anything. They maybe did, but I don't remember.

Mary's [Gifford] is the only one that's mentioned that—she's the most positive person.

Well, she married Bob Gifford, and Gifford went into management. So maybe she's relating to the Christmas parties and stuff that they throwed. But I don't— I don't know; maybe they did. I can't remember.

So did you ever hear stories about the Horseshoe Club?

No, but there was some horseshow players in the mill, but I don't know about management. I don't know anything about the Horseshoe Club.

Did you ever get over to Sullivan, the electric plant, or have any interactions over there?

No, no.

What happened if the shift changed and the bridge was up? How did people—

You'd go down and walk across the locks. You know, everybody didn't leave at the same time. They kind of strung out like a bunch of cattle, you know. People walked across the locks. It was two feet wide and had railings on it. The bridge wasn't usually up all that long anyway, most of the time.

Lots of people talked about fish in the mill. Do you have any stories about that?

Oh yeah. I don't know if I should talk about them or not.

Well, sure. Everybody else has.

Oh yeah, we fished the locks. When I was working in the chippers, yeah, we had time, yeah. I was trying to think of an old ship millwright -- and this would have been back in the '40s or early '50s and I was still working in the chippers. And I was down fishing, and we'd make snag hooks, great big ones like this, you know. And you'd go down and get on the bank and get you a rope and throw them out and snag a fish. And there was thousands of fish in there. I was trying to think of what this old-timer's name was. But anyway, he come down there—he was a ship millwright—and he come down there, and I was fishing, and he started fishing, and he got hung up. Something happened, anyway. He said, oh crap! So he jerked his knife out and he cut my line off, cut the wrong one. And I had a fish on the hook, the problem was.

And I know a guy—I'm not going to mention names or anything about this—but I know of a guy that was fishing off of the bridge—they used it to bring the bark and stuff over to the steam plant for cooking, you know; it had a conveyor across there. And he was fishing off of there one night, and he hooked a big, like a nine-foot, sturgeon. And he said the only thing that save his butt, he had a rope tied under the bridge. And he said that thing jerked so hard it swung that bridge.

My God. I never thought about sturgeon in there—what do they call that area up by the turbines from Mill A?

Power plant. I don't know. That's where we fished, down behind the power plant, pretty much.

Salmon fishing at Willamette Falls, circa 1950s. Photo courtesy of West Linn Paper Company archive.

Somebody said if fishing had been good that after night shift it would smell like a cannery.

Yeah, that's possible. I don't know when I was working—but I was over to Mill A, and they had like six, eight fish laying out there on the deal, you know. They'd divide them up between everybody. Oh gosh, I can remember walking out on that conveyor behind the PGE generating plant, and you could look down off of that into the water. And they would be in there just like that, just like maggots in there, thousands and thousands of them. And that was before they put the fish ladders in. And I don't know if it's still like that after they put the fish ladders in or not, but –

I heard they had to shut down some of the turbines for the grinding room because of the fish.

Oh, that was after I left. I don't remember them doing that when I was there.

They were still running the pulp, making pulp, while you were there...

Yeah, right.

Any other animal stories? We've heard about beaver, about ducks swimming in the locks.

Oh yeah, I've seen ducks with little babies going down the locks, yeah. Female ducks, certainly.

One lockmaster who just died, Arnold Nodurft, worked 40 years, I guess.

Yeah, I knew him.

And he said he would let this mama and her baby ducks go down.

That's right. Yeah, I've seen that happen, right. I was trying to think—I'm thinking that flying goose that's out of McMinnville—

The Spruce Goose?

I think they took that up the river on a barge. And I remember that going up through the locks. Very huge. I think that's what it was. It might have been something else that went up. That might have went up after my time, I'm thinking now, because I retired in '85, but—

They used to take the Canby ferries up.

This was something else. I can't remember what the heck it was now.

I was going to ask some questions about if you remember when affirmative action and the hiring of women started?

I tell you something I remember more significant than that to me, is when they said you couldn't discriminate with hiring, and they hired women and everything, of course, but the biggest thing was they hired minorities, blacks and Mexicans or whatever. And after 20 years of all those people they hired, I think we had like three of them left in the mill. And the women—I remember the women being there, and most of them are good workers like everybody else. I mean, they didn't have to do any super heavy work; they didn't have to pick anything up over 60 pounds. And most of them were on the machines anyway, and they wouldn't have to because most all that stuff was automated.

Mary Lefebvre was bucking rolls, but she said, you know, people always helped each other.

I don't remember anybody having any problem with the women. I mean, women, they went to work there. I remember that, but you know, I always remember the minority thing more than anything.

Do you remember long hair coming in?

Yeah, but I didn't wear it. Oh yeah.

And Mary said they'd uncovered a big pot-growing operation down there.

Yeah, that was after my time. My understanding is—the guys I talked to—over in the mixing area, and then you had to go down and outside and around and underneath a bunch of stuff, and it was pretty concealed. And there was a big—I call them tanks or whatever you want to call them, but there was like a big room, probably four times as big as this is, for a stock deal that had been obsolete. And somebody, most likely an electrician who had access to the electrical stuff to get the lights and stuff down in there, and they had a major growing problem down there. But they had left; they never caught anybody that I know of. But they had all kinds of expensive lights down there and everything, the stories I got from the guys.

She said they were using mill water, mill power…

Major growing lights down there, yeah.

Nobody got arrested.

I heard about some guys that were bringing in pot and selling the stuff to some guys that got arrested, but I don't know if it had anything to do with that or not. To my knowledge, they didn't ever find out who was doing that. That was after I was out of the mill.

One of the supervisors said that when he got there in the morning, he'd go around and say hi to every man on the shift just so, you know, he could check their breath.

He could smell the pot on them?

No, the breath from alcohol.

Oh, from alcohol, to see if they were drinking or not.

Yeah, because that would be real dangerous.

Yeah. Well, Henry Herwig, he had the clay plant, and there was a lot of different machinery in there, guys working with, that probably it wasn't a good idea to be—well, even on our job, you know, it wouldn't be a good idea to be drinking on the job. There was guys down there that did, though, but not me. I know guys that did it.

Okay, what was the worst thing about working in the mill?

Shift work. Shift work.

When you worked it, how did your family cope with it?

Well, it was the idea that I'm the guy bringing home the money. I mean we had to deal with it, you know. I was very grateful for working at Crown. I mean I wasn't like a lot of guys; I never complained about it; I liked my job. Good job. I only had a night school, one-year, ninth-grade education, but I went to night school for five winters to get through my school for the blueprint reading and stuff. And at the same time I got my GED, but I didn't have an education. I was pretty happy to have the job I had, pretty lucky. I didn't mind swing shift; graveyard was terrible because I never got enough sleep. I never got more than four or five hours' sleep a day. You know, from three to five in the morning was just terrible, terrible, you know.

Did you get sleepy at work?

Oh yeah, you're up there—like I was working in the blow pits up there, and you've got a big water, hydraulic gun, you know, and you're washing that stuff down, and I'd lay there and go to sleep and wake up and—yeah, there would be times when you couldn't keep your eyes open. It was really hard. I hated that graveyard.

Somebody told a story about a shift sleep deprivation problem. And the guy lay down somewhere to take a nap because he was really tired.

Well, you know, you had connections, like shift millwrights and electricians, you know. There was times when you had really hard days, and you could sneak off in a corner and catch a 15-minute nap. You know, sometimes it didn't take much to—but you know, we kind of looked out for one another. If somebody needed a nap, why—we didn't neglect our job, you know.

Somebody said that one thing they really hated was efficiency experts—

Well, that just about ruined the mill, I think. I don't think they gained anything from that. And the sad part about that was, I think I had about 19 or 20 years in the mill at that time when the brought them efficiency experts in, and they started laying off percentages of everybody. And I was the next guy on the list to go, and I'm sitting there with 20 years' seniority. So I mean it was pretty, pretty hard. A lot of the guys that laid off came back to work later on, but I mean, you know, that was nasty.

It happened several times, didn't it?

It happened more than once, but I just remember that one time specifically. It was bad. Well, you know, my theory is, they hire these college-educated efficiency people to come in, and it don't matter right or wrong; they're going to find a way to—they have to justify their job. And they're going to find a way that they're going to need to lay off so many people, whether it's right or wrong or indifferent. In my opinion, I think they should have done their own homework.

So it was about workforce reduction, not just greater production?

Yeah, basically it was too many people. Yeah, that's—they weren't dealing with how to increase the production or efficiency. I mean, you can't do that laying off people.

[Stretch break]

They have tours, and seeing the drum debarkers, I can't imagine how noisy --

That was up in Mill A, in the sawmill. I'll tell you where it was noisy, was in the basement of five and six, or even the other machines. Five and six and nine had the big vacuum pumps. They were huge, you know. I mean, they was—and they create a vacuum on the machine that sucks the water out of the paper on them deals. They would go across up there, and they're probably 130 decibel or even more. And you know, we worked down in the basement around them things a lot. You're working—even with hearing protection, you're working down there with just tremendous high-pitched noise, very high-pitched noise.

George talked about when they had to do the firebrick inside these cookers--being in there with a chipping gun, and how he coughed blood afterwards.

Yeah. I think about the worst—we had to go down where they had these big stock pits—they pumped it into for the machines, and the machines pumped it out up onto the machine. It was probably say this wide and maybe this long or longer, but it had a partition in the center open here, and it had a big impeller on the other end that circulated that pulp to keep it moving, because the pulp is about 90 percent water to start with, and you have to have that much water to float it, to make it move. And we went down and worked in them, and that pulp is hot. And you get down there and it's probably 110 or 130 degrees, and the humidity's outrageous.

I had a guy pass out on me down there one day with me. And you try crawling up a ladder with a guy on your shoulders—it's not a fun deal. You can't stay down there for more than 15, 20 minutes. And the same way working up on top of the paper machines, where the vent hoods were. If you had a motor you had to change up there, it was 130, 140 degrees up there, and you couldn't stay up there more than 15, 20 minutes.

In the wintertime it would be bad, but in the summertime, it had to be awful.

We worked around a lot of hot machinery on them machines, yeah.

So what was the best thing about working for the mill?

You know, I don't have very many complaints about the mill. I'm one of them guys that's pretty grateful for the job I had. I don't have anything bad to say about any of it, except for the strikes. I think the company—maybe both of us—made a mistake, but—and yeah, I don't get a very good pension. See, that's one thing—that's what we were striking for on them deals, and we never did get it, so I've been retired for 21 years and never had a raise.

No cost of living.

Nothing. So I'd starve to death if I had to depend on my retirement. That's the only bad part about it that I can think of. As far as working in the mill, you know, I needed a job, and that was a good job. And for a dummy kid that wasn't bad.

A lot of people say the best part was the people.

Well, it is, because everybody'd been there a long time. You know, you didn't have any turnover. Like I said, in the millwright gang we probably never had a guy in there that had less than 25 years. We didn't have any turnover. There was one time—and I don't remember what situation it was—they hired some outside millwrights, you know, but that was the only time. Other than that, you know, we didn't—and even the guys that bid in for helpers, most of them had eight, 10, 15 years' service. So you kind of get a bond, like George Droz and I. And these guys that worked together for 30 years, you know, you've got to get along, you know. George and I are both kind of hotheads, so we probably got along better than most guys. We didn't take any crap off each other.

I had no idea!

Oh yeah.

Harold said that the mill saved him because he said he was a hothead.

I had a terrible, terrible temper when I was a young guy. I mean terrible temper. And we was working a lot of overtime and I was tired. And we had a boss—he was a lead man; I guess we'd call him a boss, but he's a lead man. And I accused him of manipulating things to get a couple extra hours of overtime, and I give him a cussing one night, and I told him, now, I said, if you got any—you go run up to the boss and tell him. Well, that's just what he done. So I got called up to the office the next day and got chewed out. He said, "You know, I could fire you for this." And I said, "Don't let nothing but fear stop you." Because I didn't figure to pull wood in that chipper was the best job in the world anyway, so. He said, "Well, I'm just going to give you a little advice." He said, "When you start getting mad," he said, "you back off and you count to 10 and you think about it." And that kind of stuck with me all my life, and I've learned to deal with it. Ed Roake was his name. He was a supervisor in Mill B.

Like Roake's hot dog stand?

I don't know if he was related to them or not, but he worked in the mill and—I'm trying to think of the supervisor there. He was really funny. He spoke with an accent; I don't know where he was from. Ed Roake was his assistant. Anyway, something had happened at the sawmill and we wouldn't be getting any wood, so we'd be down in the smoking area smoking, you know. Everybody smoked back in them days. And he'd come by, and he was always just as serious as anybody was. And he'd get 20 feet past, and he'd stop and listen. And here he'd come back; he'd realize them chippers weren't running. And he'd come back: "Vat are you doing?" Chew us out. We ought to be doing something instead of standing there smoking. I wish I could remember his name. But Ed Roake was a good guy.

Sounds like that was a good rule.

It was good for me. It probably saved my butt a little bit, because I was terrible; I had a pretty bad temper.

People talked about the fact that the pension, you know, even though it's gone through however many hands, just keeps coming. So they're grateful for that.

Yeah, at least we haven't lost it.

Can you think of anything we haven't talked about?

I had something I was just going to say, and I kind of lost it. I can't remember what I was going to say.

Yeah, well, I never had any complaints in the mill. You know, you get fed up like everybody else does, and when they was bringing in them people, those efficiency experts, and I was the next guy on the list to get laid off. I went to Publisher's. I was offered a lead man job at Publisher's in Newberg when they was putting them new machines in over there. And I was just coming to 20 years' service; I was going to get four weeks' vacation, and you got your retirement. And I turned it down, and I thank God I did now, because when I got 25 years I got six weeks' vacation. By the time I retired I was getting six weeks' vacation plus five floating holidays, so I was getting seven weeks a year off. How can you beat that? I never had any trouble. You know, we had enough vacation. I never had trouble getting the time off, I thought it was pretty nice. A lot of hard years.

I mean the years I worked in the chippers. I mean, you know, you're raising a family and trying to survive.

You had to be physically really strong.

We had some bad times in '49 and '50, '51 in there. We had two or three really bad winters. It got down to zero one winter, and we had pipes freezing up in the mill. Yeah, some really major problems. They couldn't keep their stock lines thawed out.

Somebody that used to be a plumber talked about freezing [weather] everywhere.

Yeah, those guys—when it starts thawing out is when they've really got the problem. They'd take welders and hook them up and run that heat through the pipes through the welding leads and try to keep them thawed out, yeah. But those years in there, up to '53, was all three really bad winters that I remember that was nightmares to us.

I've seen some pictures of the falls with ice all over them.

Oh yeah, yeah. Well, the Willamette froze over. I've been here since '47, and I can remember three times that it's been zero. And that year it got down to three below, because I remember. I don't know if it was '49 or '50, '51 in there somewhere—'52, '52, I don't know. But it was nasty.

Were people walking across, driving their cars?

Oh no, I don't think it froze over that good.

Willamette Falls January 1930.
Photo courtesy of West Linn Paper Company archive.

Because in the real old days, I know the Columbia froze--

Well, I worked in that door factory. I've talked to guys down there that said they remembered back in the thirties, '37 or somewhere in there they had a really bad winter. They drove their Model T's across the river.

Paul Miken showed me the greatest book, Modern Pulp & Paper Making, a Treatise…

It's got stuff in it for translating reams into sheets and, oh, different math changes, formulas, yeah. But I have no idea where they're at at this point. Well, you know, if I come across any of this stuff I'll give you a call.

Acknowledgments

This book would not exist without the trust and support of the Kinsman Foundation, whose generous grants allowed us to capture the stories on video and produce a two-part, three-hour documentary saga on West Linn's historic paper mill. Those boxes of tapes could not have moved into print without the trust and support of the Oregon Cultural Trust, the Clackamas County Cultural Coalition and, again, the Kinsman Foundation. We are deeply indebted to those key grantors for believing and investing in the creation of this important archive on Oregon papermaking—a treasure trove of mill worker oral history.

Our deep gratitude also to Melody Ashford and Willamette Falls Television and Community Media Center, for her and her staff's technical assistance and patience through the movie-making stages of the project and then pulling the sound track off the 30 hours of video; to David Donnell, whose flying fingers accomplished the tedious miracle of 30 hours of audio transcribing; to Lauri Burns, a wonderful work study student from a mill family who helped us create the index; to Anita Jones for book and cover design; to Ken Cameron, who donated our striking cover photo; to Bob Smith, for invaluable advice and printing coordination; to West Linn Paper Company, for access to photos from its archive and assistance in fact-checking the name spellings; to the board of directors of the Willamette Falls Heritage Foundation, which turned me loose, believing I could actually pull this off; and to my sweetheart, who heard about little else, conquered every computer challenge and, by my best reckoning, has sacrificed several thousand hours of relationship time to my preservation projects over the last twelve years. He is a saint.

About the Author

Sandy Hickson Carter left the University of Oregon Journalism School in 1982 with her Bachelor of Science degree and that year's "Outstanding Graduate" award. A descendant of 1845 Oregon Trail pioneers, she has been deeply involved in and committed to Willamette Falls-area preservation issues since 1998. Prior to her retirement from a grant-writing career with the Clackamas County Department of Human Services in 2003 she had placed numerous articles in local newspapers and published stories in magazines with national distribution. She worked at Employment Training and Business Services of Clackamas County during the Simpson-to-West Linn Paper Company transition period in 1996–97.

Since 2004, Ms. Carter has been a board member, the executive director and the creative force behind the projects of Willamette Falls Heritage Foundation. She is a contributing scholar to the Oregon Historical Society Quarterly (January 2005) and produces an intermittent history column for the "West Linn Tidings," as well as teaching grant writing for Chemeketa and Clackamas Community Colleges and generating grants and publicity for the projects of the Foundation. She is the primary coordinator for Lock Fest—a non-commercial history festival at Willamette Falls Navigation Canal and Locks. The first Lock Fest was awarded a Crystal Award as the Willamette Valley Development Officers' "Outstanding Event of the Year" in 2004.

In 2007 the West Linn Tidings named Ms. Carter "Citizen of the Year" and in 2010 she received the Les DeJardin Community Service Award from the West Linn Lions Club International.

Ms. Carter lives in a modest 1954 house in West Linn's Bolton neighborhood, surrounded by other small 40s-50s homes (many of which were built by mill families), on a street slanting down to the Willamette River. She is slightly older than her house and has two adult children she loves very much. She regrets she was never able to have this kind of conversation with her own grandparents.

About Willamette Falls Heritage Foundation

Willamette Falls Heritage Foundation became an education and preservation-focused non-profit corporation in 2002. Its board members are partners, stewards and advocates for the culture and heritage at Willamette Falls. Its mission is to preserve and promote the magnificence and history of Willamette Falls so that it may live in the minds and imaginations of people of all ages. Learn more at willamettefalls.org.

Index of Names

*Name spelling could not be verified. Transcribed from audio.

Index of Subjects